Proteomics and Protein-Protein Interactions

Biology, Chemistry, Bioinformatics, and Drug Design

PROTEIN REVIEWS

Recent Volumes in this Series

VIRAL MEMBRANE PROTEINS: STRUCTURE, FUNCTION, AND DRUG
DESIGN
Edited by Wolfgang B. Fischer

THE p53 TUMOR SUPPRESSOR PATHWAY AND CANCER
Edited by Gerard P. Zambetti

PROTEOMICS AND PROTEIN-PROTEIN INTERACTIONS: BIOLOGY,
CHEMISTRY, BIOINFORMATICS, AND DRUG DESIGN
Edited by Gabriel Waksman

Proteomics and Protein-Protein Interactions

Biology, Chemistry, Bioinformatics, and Drug Design

Protein Reviews Volume 3

Edited by

Gabriel Waksman

Institute of Structural Molecular Biology
Birkbeck and University College London
United Kingdom

 Springer

Library of Congress Cataloging-in-Publication Data

Proteomics and protein-protein interactions : biology, chemistry, bioinformatics, and drug
    design / Gabriel Waksman, editor.
        p.   cm. — (Protein reviews)
    Includes bibliographical references and index.
    ISBN 0-387-24531-6
    1. Protein-protein interactions.   2. Proteomics.   I. Waksman, Gabriel.   II. Series.

QP551.5.P765 2005
612'.01575—dc22   2005040224

ISBN-10: 0-387-24531-6   e-ISBN 0-387-24532-4   Printed on acid-free paper.
ISBN-13: 978-0387-24531-7

Printed in Singapore.   (TB/KYO)

9 8 7 6 5 4 3 2 1

springeronline.com

Preface

Gabriel Waksman

Institute of Structural Molecular Biology, Birkbeck and University College London, Malet Street, London WC1E 7HX, United Kingdom

Address for correspondence:

Professor Gabriel Waksman
Institute of Structural Molecular Biology
Birkbeck and University College London
Malet Street
London WC1E 7H
United Kingdom
Email: g.waksman@bbk.ac.uk and g.waksman@ucl.ac.uk
Phone: (+44) (0) 207 631 6833
Fax: (+44) (0) 207 631 6833
URL: http://people.cryst.bbk.ac.uk/~ubcg54a

Gabriel Waksman is Professor of Structural Molecular Biology at the Institute of Structural Molecular Biology at UCL/Birkbeck, of which he is also the director. Before joining the faculty of UCL and Birkbeck, he was the Roy and Diana Vagelos Professor of Biochemistry and Molecular Biophysics at the Washington University School of Medicine in St Louis (USA).

The rapidly evolving field of protein science has now come to realize the ubiquity and importance of protein–protein interactions. It had been known for some time that proteins may interact with each other to form functional complexes, but it was thought to be the property of only a handful of key proteins. However, with the advent of high-throughput proteomics to monitor protein–protein interactions at an organism level, we can now safely state that protein–protein interactions are the norm and not the exception. Thus, protein function must be understood in the larger context of the various binding complexes that each protein may form with interacting partners at a given time in the life cycle of a cell. Proteins are now seen as forming sophisticated interaction networks subject to remarkable regulation. The study of these interaction networks and regulatory mechanism, which I would like to term "systems proteomics," is one of the thriving fields of proteomics.

The birds-eye view that systems proteomics offers should not, however, mask the fact that proteins are each characterized by a unique set of physical and chemical properties. In other words, no protein looks and behaves like another. This complicates enormously the design of high-throughput proteomics methods. Unlike genes, which, by and large, display similar physicochemical behaviors and thus can be easily used in a high-throughput mode, proteins are not easily amenable to the same treatment. It is thus important to remind researchers active in the proteomics field of the fundamental basis of protein chemistry. This book attempts to bridge the two extreme ends of protein science: on one end, systems proteomics, which describes, at a system level, the intricate connection network that proteins form in a cell, and on the other end, protein chemistry and biophysics, which describe the molecular properties of individual proteins and the structural and thermodynamic basis of their interactions within the network.

Bridging the two ends of the spectrum is bioinformatics and computational chemistry. Large datasets created by systems proteomics need to be mined for meaningful information, methods need to be designed and implemented to improve experimental designs, extract signal over noise, and reject artifacts, and predictive methods need to be worked out and put to the test. Computational chemistry faces similar challenges. The prediction of binding thermodynamics of protein–protein interaction is still in its infancy. Proteins are large objects, and simplifying assumptions and shortcuts still need to be applied to make simulations manageable, and this despite exponential progress in computer technology.

Finally, the study of proteins impacts directly on human health. It is an obvious statement to say that, for decades, enzymes, receptors, and key regulator proteins have been targeted for drug discovery. However, a recent and exciting development is the exploitation of our knowledge of protein–protein interaction for the design of new pharmaceuticals. This presents particular challenges because protein–protein interfaces are generally shallow and interactions are weak. However, progress is clearly being made and the book seeks to provide examples of successes in this area.

I would like to thank all contributors for their participation to this book, which, I believe, is timely and provides a good overview of the field. It is their hard work that has made this book what it is, a fascinating foray into the complex world of protein science. Proteomics and Protein-Protein Interactions: Biology. Chemistry. Bioinformatics. And Drug Design.

Gabriel Waksman

Contents

1

Introduction: Proteomics and Protein–Protein Interactions: Biology, Chemistry, Bioinformatics, and Drug Design

Gabriel Waksman and Clare Sansom

ABSTRACT

In this chapter, a general introduction to the book is provided. We explain the organization of the book and how the chapters are interconnected to each other. We also provide some illustrations and highlights that will complement the various chapters.

1. INTRODUCTION

The formidable advances in protein sciences in recent years have highlighted the importance of protein–protein interactions in biology. Before the proteomics revolution, we knew that proteins were capable of interacting with each other and that protein function was regulated by interacting partners. However, the extent and degree of the protein–protein interaction network was not realized. It is now believed that not

GABRIEL WAKSMAN AND CLARE SANSOM   •   Institute of Structural Molecular Biology, Birkbeck and University College London, Malet Street, London WC1E 7HX, UK.

Proteomics and Protein–Protein Interactions: Biology, Chemistry, Bioinformatics, and Drug Design, edited by Waksman. Springer, New York, 2005.

only are a majority of proteins in a eukaryotic cell involved in complex formation at some point in the life of the cell, but also that each protein may have on average six to eight interacting partners (Tong et al., 2004). We start this book with two chapters that describe the successes and limitations of two of the most productive methods used to study protein–protein complexes on a large "proteome" scale: the yeast two-hybrid method (Auerbach and Stagljar, Chapter 2) and mass spectrometry (Wang, Yazdi, and Qin, Chapter 3). A brief description of the principle of the methods is provided in Section 2 of this chapter.

The study of protein–protein interactions predates the proteomics revolution. The pioneering work on antigen–antibody and protease–protease inhibitor complexes has provided insight into protein–protein interfaces and their properties (Ruhlmann et al., 1972; Amit et al., 1986). However, more recently, the structure of larger complexes that function as molecular machines has been determined, shedding light into important cellular functions such as transcription (the structure of the RNA polymerase II core complex [Cramer et al., 2001]), translation (the structure of the ribosome [Ban et al., 2000; Wimberly et al., 2000]), replication (the structure of the γ-complex in bacteria [Jeruzalmi et al., 2001]), or the cytoskeleton (the structure of the Arp2/3 complex [Robinson et al., 2001]), to cite only a few. Section 3 of this chapter provides highlights for some of these structures and describes some of the most striking advances that these structures have contributed. It is not the purpose of this book to make an exhaustive list of all protein–protein complexes, the structure of which has been determined to date, but instead to provide general concepts on the common and distinctive features of protein-protein interfaces and their roles.

Protein–protein interactions can be classified into approximately three subtypes, depending on their stability and the mode of interactions (see Walker-Taylor and Jones, Chapter 5). Although any one protein may be involved in interactions with many others, they may form stable interactions only with a few. These form core complexes, which are stable, can be purified, and are amenable to structural studies. A number of core complex structures have been determined and these structures have been instrumental in understanding how these core complexes carry out their functions (see list above). A second category of interactions are transient, and the proteins involved in transient associations are often regulatory proteins, the role of which may be to confer short-lived, physiologically regulated properties to other proteins or to core complexes. The complexes that proteins form transiently may be unstable and difficult to purify. Successes in determining the structure of transient complexes have been dependent on the affinity of the various constituents participating in complex formation. One historical breakthrough in this regard has been the determination of the structure of the first antigen–antibody complex (Amit et al., 1986), which defined the architecture and chemistry of protein–protein interactions in this versatile structural framework. Sundberg and Mariuzza (Chapter 4) in this book provide an exhaustive review of the antigen–antibody and MHC–TCR complexes. Finally, protein–protein interactions in core complexes or transient ones may be mediated by specialized, small, domains dedicated to protein–protein recognition (Pawson and Scott, 1997).

The proteomics revolution (and before it, the genomics revolution) has in many ways overwhelmed our ability to keep up with the sheer volume of data that it has generated. In this regard, the rapid development of computer science in general and bioinformatics in particular is crucial for us to be able to make sense of the data. In two chapters, by Walker-Taylor and Jones (Chapter 5) and by Marshall and Vasker (Chapter 6), the authors describe the bioinformatics tools that have been deployed to understand the general principles of protein–protein interactions (Walker-Taylor and Jones, Chapter 5) and how this knowledge has been exploited to design predictive docking algorithms to model protein–protein interactions from structure (Marshall and Vasker, Chapter 6).

Unfortunately, prediction of protein–protein interaction is a rather difficult endeavour, not least because proteins may undergo vast conformational changes on association with other proteins. One particularly relevant example of such a case is the interaction of the human immunodeficiency virus (HIV) envelope glycoprotein gp120 with its human receptor CD4. Structural rearrangement in gp120 serves not only as an allosteric trigger for cellular invasion, but also as a mechanism for evading the host immune system. Doyle and Hensley in Chapter 7 show how large conformational changes in proteins can be probed thermodynamically (see also Section 3 of this chapter for the structure of the gp120/CD4/anti-gp120 antibody ternary complex [Kwong et al., 1998]). Another example of the use of thermodynamic investigation of conformational changes, this time in the unbound state, is provided in Section 4 of this chapter.

One outcome of such studies is the observation that protein–protein interfaces possess a high degree of versatility and plasticity (Jones and Thornton, 1996; Lo Conte et al., 1999). This is in part due to the fact that protein–protein interactions encompass a wide range of affinities. However, even within a particular range of affinities, the structural features underpinning binding may vary. For example, protease inhibitors appear to use main-chain–main-chain interactions, whereas antigen-antibody interaction is mediated by side-chain side-chain interactions (Jackson, 1999). Side-chain–side-chain interactions may be more likely to determine specificity. In contrast, serine protease inhibitors must bind tightly to their target proteases. This may be best achieved using constrained "main-chain–main-chain" conformation, and means that the inhibitor will be highly committed to the enzyme. Similar observations corresponding to similar requirements have been observed in the interaction of pilus subunits with bacterial chaperones (see Section 5 of this chapter and [Choudhury et al., 1999; Sauer et al., 1999]). Proteins may be able to use the same template for interactions with different proteins or with different parts of the same protein. For example, in the growth hormone–growth hormone receptor complex (GH–GHR), two receptor molecules bind to different parts of the same ligand (see Section 6 of this chapter and de Vos et al., 1992). This complex structure and the subsequent site-directed mutagenesis studies have defined an important concept in protein–protein interaction, that of "hot spots" (Section 6 and Clackson and Wells, 1995). In the protein–protein interface observed in the GH–GHR complex, although a myriad of hydrogen bonds, van der Waals contacts, and electrostatic interactions is observed, only a limited few of these interactions ("hot spots") have been shown to play an important role in binding. Similar observations have been made in other systems (Dall'Acqua et al., 1996; Bradshaw et al., 1999). One important property of protein–protein interfaces is the shape complementarity between the two regions coming together in the interactions. Remarkably, such complementarity is very often mediated by water molecules, judiciously placed at the interface to fill in holes and increase contacts (Bhat et al., 1994; Lubman and Waksman, 2003). The role of water in both the structural and thermodynamic basis of protein–protein interactions is essential and yet very poorly understood.

One remarkable feature of protein–protein interactions is that they are often mediated by small domains that specifically bind small sequence motifs on proteins. The Src-homology 2 (SH2) domain was the first such domain to be recognized (Sadowski et al., 1986). SH2 domains are involved in the building up of large complexes at and around signaling receptors. SH2 domains bind specifically sequences containing phosphorylated tyrosines and are able to discriminate between tyrosine-phosphorylated sites by exercising some preferences for residues located C-terminally relative to the phosphotyrosine (Bradshaw and Waksman, 2002). Since the discovery of SH2 domains, a large number of protein domains with specialized roles in protein–protein interactions have been found (see http://www.mshri.on.ca/pawson/domains.html for an exhaustive list of such domains). SH3 and WW domains specifically recognize and bind sequence motifs containing prolines (Musacchio, 2002). Proline-rich motifs are among the most common motifs identified, and thus SH3 and WW domains play major roles in protein–protein interactions. PDZ domains are essential for integrity of the postsynaptic density, a large protein complex formed around glutamate and *N*-methyl-D-aspartate (NMDA) receptors in the nervous system (Sattler and Tymianski, 2001). Finally, bromodomains are similar to SH2 domains in that both bind (and thus induce recruitment of proteins that contain them) to sites of protein modifications. Bromodomains, unlike SH2 domains that bind to tyrosine-phosphorylated sites, bind specifically acetylated lysines and thus play important roles in chromatin remodeling during transcription and replication (Dhalluin et al., 1999). Three chapters of this book are dedicated to the review of the protein–protein interaction domains listed above. Ladbury in Chapter 8 provides an account of the structural and thermodynamics work that has enhanced our understanding of SH2 and phosphotyrosine binding (PTB) domain recognition of tyrosine-phosphorylated sites during transduction of cellular signals. Bedford and Sudol (Chapter 9) describe the roles and functions of SH3 and WW domains. Finally, Yan and Zhou (Chapter 10) provide a detailed account of the discovery of bromodomains and also of their struture–function relationship.

There is no unifying theme among the structures of the protein–protein interaction domains listed above. However, as their structures have been characterized, intense efforts have been devoted to designing specific binding inhibitors capable of disrupting protein–protein interaction mediated by these modules. For example, the SH2 domain of the Src kinase has been targeted for molecular design and binding competitors able to inhibit osteoclast function have been found (Sawyer et al., 2002). As the major phenotype in Src knockout mice is a thickening of the bones, it is hoped that a Src SH2 domain binding inhibitor could be used to combat osteoporosis, a devastating disorder in elderly women. Sawyer et al. in Chapter 11 review the field of SH2 domain binding inhibitors and also provide a fascinating account on the remarkable progress made in designing new tyrosine kinase inhibitors.

Using peptides or peptide mimics to disrupt protein–protein interfaces is not a novel idea, but this approach has benefited from structural information. Notably, the molecular design of rigid peptidomimetics is believed to enhance greatly the potential of peptides as therapeutics by not only locking the peptide in a defined binding conformation but also by preventing or slowing degradation (Patani and LaVoie, 1996). However, peptides or peptide-based compounds do not readily cross the cell membrane barriers and thus are not as effective as hoped. Recently, a peptide derived from the NR2 chain of the NMDA receptor known to interact with the second PDZ domain of PSD95, an essential component of the postsynaptic density, was made effective in reducing cerebral infarction in rats subjected to transient focal cerebral ischemia by fusing it to the HIV1-Tat translocator peptide (Aarts et al., 2002). Thus, the use of translocator peptides may be a promising avenue of research for the delivery of therapeutic peptides or proteins (Becker-Hapak et al., 2001). In this book, Aarts and Tymianski (Chapter 12) provide a detailed account of this work and place it in the general context of NMDA receptor signaling and its more general role in the response to the devastating damages to the human nervous system caused by stroke, epilepsy, and head and spinal injury.

As our knowledge of protein–protein interactions increases, such interactions will be more frequently targeted for drug design. Chapter 13 by Freire provides a guide and a general strategy to improve the hit-to-lead route that is so often paved with multiple insurmountable obstacles. However, we should not ignore the vast amount of work that is being achieved in the design of novel forward or reverse genetics methods or target-guided self assembly methods (reviewed in Alaimo et al., 2001) and how these can be exploited to probe the living cell in general and protein–protein interaction networks in particular.

With this book, we have attempted to provide a multifaceted account of the research taking place in protein science. We have attempted to cover the field in a most exhaustive way, choosing highlights from leaders in the field in such a way that our choice should reflect the enormous diversity and complexity of the principles underpinning protein–protein interactions. We hope we have achieved this goal.

2. PRINCIPLES OF MASS SPECTROMETRY (MS) AND OF THE YEAST TWO-HYBRID METHOD

MS and the yeast two-hybrid method are two key proteomics technologies that have been developed in recent years and that can be applied to the study of protein interactions and protein–protein complexes.

The development of MS as a precise technique for the identification of proteins and peptides from their molecular mass and charge has led to an explosion in high-throughput protein analysis methodologies. The basic principle of MS (shown in Fig. 1.1, Ia) involves the production of a stream of ions—in proteomics experiments these will typically be ionized peptide fragments of a protein or complex—from the sample. The ion stream then passes through an analyzer where the ions are sorted by mass/charge ratio, and then enters the detector where the mass/charge spectrum is recorded. The most widely used type of analyzer is the time-of-flight (ToF) analyzer. In this, the ions are accelerated so ions of like charge have the same kinetic energy: it can be seen from basic physical principles that there is an inverse relationship between the time taken for an ion to travel between the source and the detector (measured in microseconds) and its mass/charge ratio.

Currently, the two most commonly used ionization techniques are MALDI (Matrix Assisted Laser Desorption Ionization), shown schematically in Figure 1.1, Ib, and ESI (ElectroSpray Ionization), shown in Figure 1.1, Ic. Of these, the gentleness of the ESI method makes it particularly suitable for the study of protein–protein interactions, as noncovalent interactions may sometimes be retained during this ionization process. In MALDI the sample of peptides is embedded in a matrix. When an ultraviolet laser beam ($\sim$337 nm) is shone onto this matrix the laser energy is transferred into the peptides, releasing a pulsed stream of ions. In contrast, in ESI the peptide sample, dissolved in a suitable solvent, is sprayed out of a needle with a thin tip at a high voltage into an inert, drying gas. This produces charged droplets containing the sample ions, and the droplets then evaporate leaving the ions in the gas phase. These are swept through a sampling cone toward the (usually ToF) analyzer and detector systems.

The yeast two-hybrid method, shown in Figure 1.1, II, is now a well-established methodology for detecting proteins or domains that are capable of interacting. One important advantage of this method is that it results in the cloned genes for the interacting proteins becoming immediately available. Plasmids are constructed containing, respectively, the DNA-binding domain of the yeast transcriptional activator GAL4 fused to the known protein for which interacting partners are sought (top diagram) and the GAL4 activation domain bound to a library of proteins or fragments (middle diagram; n fragments). Interaction between a protein from the library (say protein i) and the known protein (bottom diagram) causes the GAL4 activation domain to bind to DNA, leading to transcriptional activation of a reporter gene that contains a binding site for GAL4. β-Galactosidase is typically used as the reporter gene.

3. EXAMPLES OF LARGE PROTEIN COMPLEX STRUCTURES

Recent determination of the structures of high molecular weight protein complexes has shed light on the mechanism of several important cellular processes, including transcription, replication, cell motion, and viral adhesion. The mechanism of these complexes is often driven by conformational changes both within and between subunits.

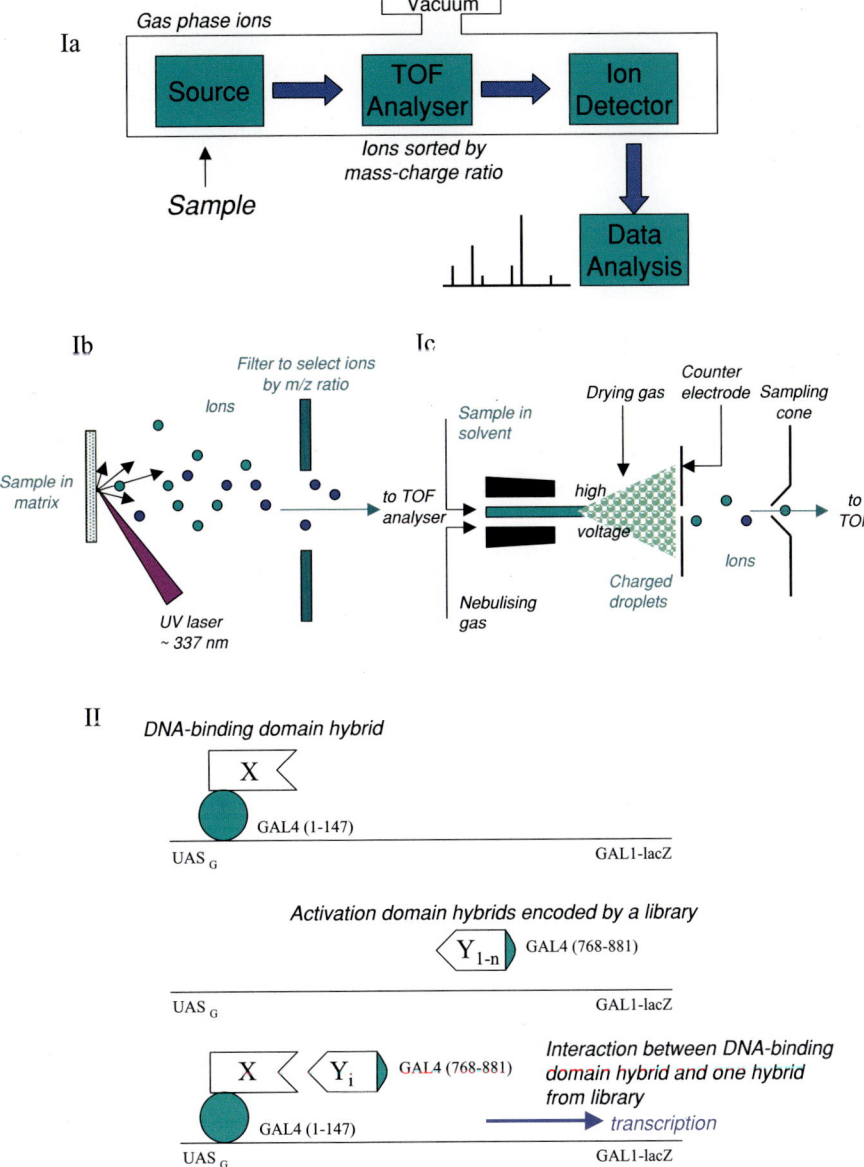

Figure 1.1. Description of the techniques of mass spectrometry and of the yeast two-hybrid method to study macromolecular complexes.

RNA polymerase II is responsible for all mRNA synthesis in eukaryotes. The structure of 10 of the 12 subunits of this enzyme from yeast, at 2.8A resolution, is shown in Figure 1.2, I (Cramer et al., 2001). The authors propose that the DNA double helix could enter the cleft of this "open" structure and be held in place for transcription by a massive protein "clamp" consisting of parts of subunits 1 (shown in *red* in this figure) and 2 (shown in *yellow*). The DNA cleft runs from the bottom right to the top left of the molecule as shown here. A second structure, solved at lower resolution, shows the clamp swung round toward the active centre of the molecule in a "closed" conformation (Gnatt et al., 2001). The models suggest that the most likely route for RNA exit would be via a groove at the base of the clamp.

In bacteria, an ATPase known as the γ complex plays an important part in DNA replication. It is the part of the larger DNA polymerase III complex that loads the pol III β-subunit (the sliding β-clamp) onto the DNA; it is therefore a homolog of eukaryotic replication factor C. Once attached to the loaded clamp, the catalytic α-subunit will move along the DNA and catalyze replication. The *Escherichia coli* γ complex, shown in Fig. 1.2 II, is a pentamer of five subunits, each with the same fold (an N-terminal recA-like domain followed by two helical domains) but with different interdomain orientations in each subunit (Jeruzalmi et al., 2001). The nucleotide binding sites of the subunits are arranged to face the inner surface of the complex. The three "middle" subunits, which are the most similar, are termed γ and the two outer ones δ (the wrench) and δ' (the stator); it is the wrench that binds to the β-clamp. The diagram shows an "open" form of the enzyme in which the wrench is free to bind to the clamp; this structure suggests a mechanism for replication in which the complex switches between this and a closed form where the wrench is occluded.

The Arp2/3 complex is an assembly of seven proteins that initiates actin poly-merisation in eukaryotic cells. This process generates the network of branched actin filaments that pushes forward the leading edge of motile cells. This assembly, shown in Figure 1.2, III, consists of two central subunits with the same fold as actin, Arp2 (cyan) and Arp3 (*blue*), and five peripheral subunits (p40 to p16) with different folds. Part of the Arp2 subunit was not observed in the crystal structure. The complex is inactive until it is activated by nucleation-promoting factors such as members of the WASp/Scar protein family (WASp, the first member of this family to be discovered, is the Wiskott–Aldrich syndrome protein). The structure as shown suggests that ac-tivation could take place by motion of the ARPC subunits inducing a conformation change bringing the Arp2 and Arp3 subunits in contact with each other to form a nucleation site.

One key target for anti-HIV drug design is the binding of the HIV virus envelope glycoprotein (gp120) to the CD4 receptor of the host T cells, prior to viral entry into the cells. The structure of the gp120–CD4 complex has been solved, in complex with a neutralizing human antibody (Fig.1.2, IV). The presence of a hydrophobic cavity at the CD4/gp120 interface in this structure is a clear indication that the formation of the complex must induce a conformational change in gp120. The structure also shows a conserved chemokine receptor-binding site, and illustrates possible points for intervention by potential anti-HIV entry drugs. This section of Figure 1.2 also serves as a figure for Chapter 7 by Doyle and Hensley.

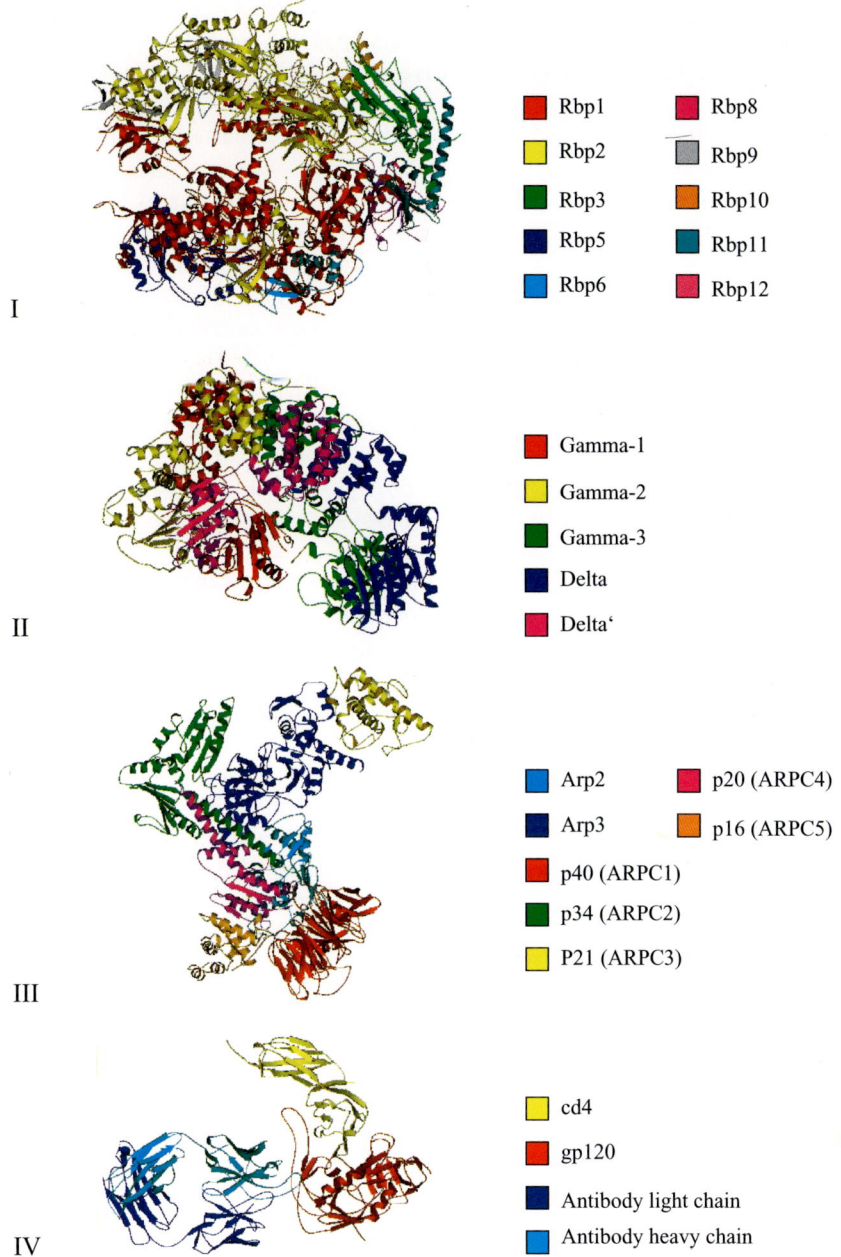

Figure 1.2. Macromolecules complexes formed during transcription, replication, actin polymerization, and HIV host recognition.

4. PROBING CONFORMATIONAL CHANGES IN THE UNBOUND STATE: BINDING THERMODYNAMICS OF THE TANDEM SH2 DOMAIN OF THE SYK KINASE

As illustrated in Chapter 7 by Doyle and Hensley, "induced fit" conformational changes occurring on binding can be probed using calorimetry. Indeed, such conformational changes have distinct thermodynamic signatures, that is, they are generally characterized by a large heat capacity change. However, the heat capacity change can also be used to characterize other types of conformational transitions, and notably as demonstrated recently by Kumaran, Grucza, and Waksman, to characterize conformational transitions occurring in the unbound state (Kumaran et al., 2003).

The macromolecular system studied by Kumaran et al. (2003) is the tandem SH2 domains of the Syk kinase. The Syk kinase is involved in signal transduction pathways mediated by immune receptor. It contains two SH2 domain located in tandem (termed Syk-tSH2; I in Fig. 1.3). Syk-tSH2 allows recruitment of Syk to tyrosine-phosphorylated sites termed ITAMs (*I*mmuno-receptor *T*yrosine-based *A*ctivation *M*otifs) on the immune receptor. The sequence of three such receptor ITAMs is shown in VII (Y* indicates a phosphotyrosine). Syk-tSH2 is remarkable because it can bind with high affinity to doubly phosphorylated ITAMs (dpITAMs) that have widely different sequences and also very different lengths of sequence (indicated as "spacer region" in VII) between the two phosphotyrosines (compare dpITAM binding of FcR-γ and FcRIIA in VIII for wild-type Syk-tSH2). How can Syk-tSH2 do this?

The answer to this question was provided by a calorimetric experiment in which the binding enthalpy was measured as a function of temperature (results in Fig. 1.3, II, open diamonds.) Note that these results were obtained using the CD3-ε peptide shown in Fig.1.3, VII). As can be seen, the binding ΔH has a nonlinear dependence on temperature. This is odd, as most binding reactions involving proteins display a linear dependence. These results were interpreted in light of a two-conformer model, which is shown in Figure 1.3, III. In this model, the unbound form of Syk-tSH2 exists in a temperature-dependent conformational equilibrium involving two forms, A and B. The A form predominates at low temperature, and is the preferential binding form for the CD3-ε dpITAM peptide. At low temperature, the observed binding enthalpy is that of binding alone. As the temperature is raised, the B state becomes populated, and the binding enthalpy is now composed of two terms, the enthalpy arising from the B to A transition and the intrinsic binding enthalpy (A to AX). This model fits the data very well (see line through experimental data points). But is it right?

The model was proven to be correct by a second series of experiments. It was hypothesized that the A and B forms of Syk-tSH2 corresponded to a closed and an open form of the protein, respectively. The closed form corresponds to the two SH2 domains being close together. The open form corresponds to the two SH2 domains

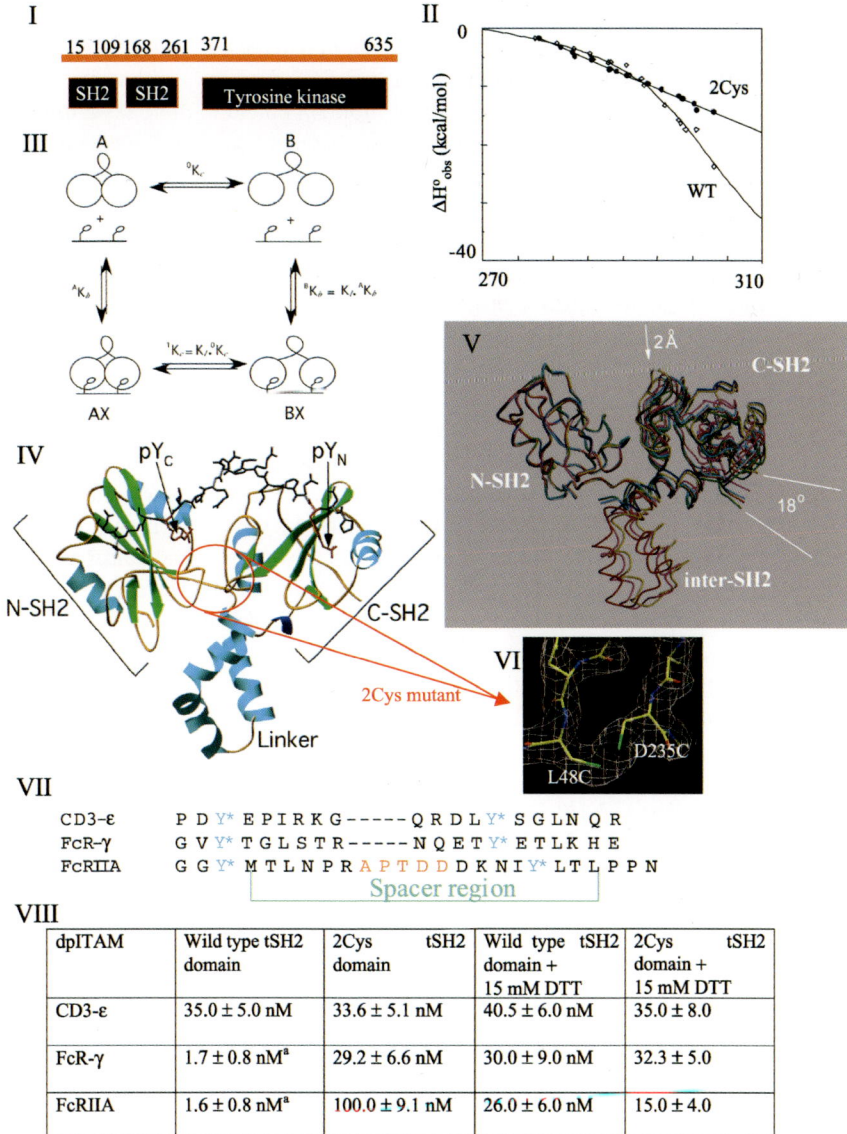

Figure 1.3. Thermodynamics of conformational changes: the case of the tandem SH2 domain of the Syk kinase.

being far apart. This hypothesis was suggested by experimental X-ray crystallographic evidence that shows that in the bound state (i.e., bound to the CD3-ε dpITAM peptide of Fig.1.3, VII), the relative orientation of the two SH2 domains is subject to significant variations (see structure of the Syk-tSH2 domain bound to the CD3-ε dpITAM peptide in Fig.1.3, IV; this structure contained six independent views of the complex; when superimposed (see Fig.1.3, V), there is evidence of variability in the relative orientation of the SH2 domains with the two extreme conformations differing by an angle of 18° and a translation of about 2 Å). Thus, it was hypothesized that in the unbound state, the relative orientation of the two SH2 domains could vary even more, perhaps explaining how Syk-tSH2 can bind with equal affinity to dpITAMs with widely different spacer length between the two phosphotyrosines. To test these hypotheses, a mutant (termed 2Cys) containing two judiciously-located Cys residues was engineered (see *red circle* in Fig.1.3, IV). These Cys residues, when oxidized, form a disulfide bond (see Fig.1.3, VI showing the electron density for the bond), which locks the two SH2 domains in the "closed" conformation, that is, when the two SH2 domains are close to each other. If the above hypotheses are correct, such a mutant, being unable to transition to the open form, should no longer display a nonlinear dependence of the binding enthalpy on temperature (it should be linear), and moreover, this mutant should see its affinity for "long" dpITAMs (such as FcRIIA in VII) considerably reduced. This is indeed the case. In II, closed circles, the 2Cys mutant is shown to display a linear dependence of its binding enthalpy on temperature, and as shown in VIII, its binding affinity is considerably reduced for the FcRIIA dpITAM peptide but not for the CD3-ε dpITAM peptide.

5. PROTEIN–PROTEIN INTERACTIONS IN THE CHAPERONE-USHER PATHWAY OF PILUS BIOGENESIS

Bacterial pili (shown in Fig.1.4, I) are important organelles used by pathogenic bacteria to sense their environment and attach to host tissues in a specific manner. Thus, pili play important roles in initiating infections. Of all pili, the P pilus has been the most thoroughly investigated. P pili are found on the surface of uropathogenic *Esherichia coli* (UPEC) and are known to initiate attachment of UPECs to the kidney epithelium. This specific recognition event is due to the presence of an adhesion protein called PapG at the tip of the pilus (see Fig.1.4, I), which binds to a kidney receptor called globoside (GbO4). The P pilus is composed of a thick part at its base that is formed by the subunit PapA and of a thin and flexible part that is formed by the subunit PapE (see Fig.1.4, I). Two subunits act as adaptor subunits: PapK inserts between the thick and thin part of the pilus (i.e., between PapA and PapE), and PapF inserts between the terminal PapG subunit (or adhesin) and PapE. All subunits are assembled into a pilus by a conserved mechanism involving a chaperone, PapD, and an outer-membrane protein, the PapC usher (see Fig.1.4, I). After translation, each subunit is translocated to the periplasm by the Sec pathway where each is taken up by the chaperone PapD, forming a binary chaperone–subunit complex. Each complex is then targeted to the usher PapC where it is assembled within the pilus. The first subunit to be processed is PapG, hence its location at the tip of the pilus.

The molecular basis of pilus assembly has been elucidated (Choudhury et al., 1999; Sauer et al., 1999; Barnhart et al., 2000; Dodson et al., 2001). Each subunit has an Ig fold where the last strand (strand G) is missing. The result of this missing secondary structure is a large groove on the subunit surface (see Fig.1.4, II, where the

I

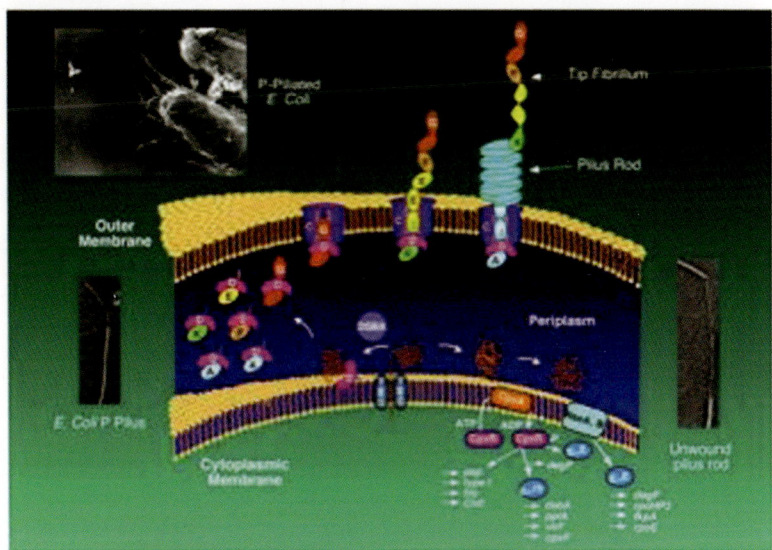

II

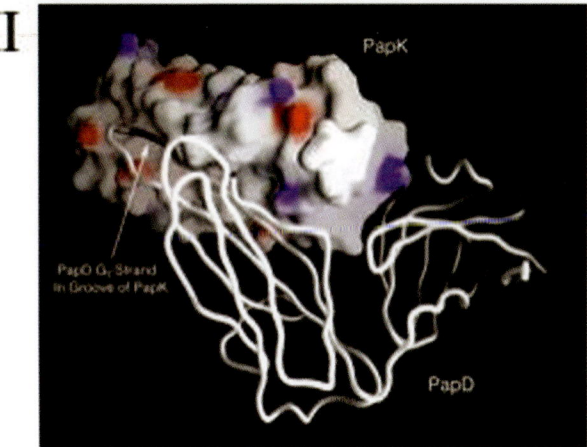

Figure 1.4. Protein-protein interactions in the chaperone-usher pathway of pilus biogenesis.

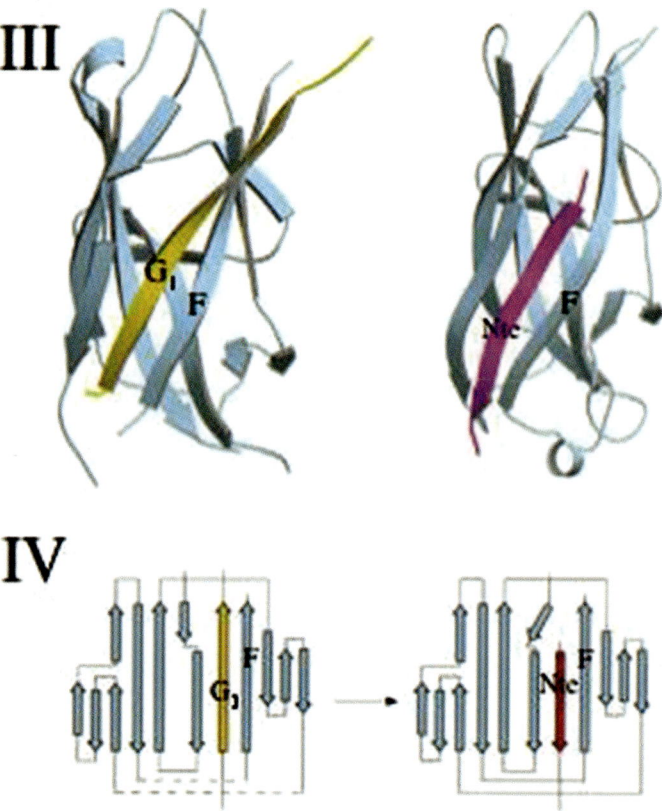

Figure 1.4. (Continued)

subunit is represented in a surface diagram while the chaperone is shown in a ribbon diagram). The missing secondary structure is provided in *trans* by the chaperone, which "donates" its G1 strand (indicated in Fig.1.4, II). The mechanism by which the chaperone provides the missing secondary structure is called "donor-strand complementation." During pilus assembly, the chaperone–subunit complex is targeted to the usher where the complex dissociates. The chaperone is released in the periplasm while the subunit binds to the subunit that was assembled in the previous round of assembly. This process occurs through a mechanism termed "donor-strand exchange" which is explained in Fig.1.4, III and IV. In Fig.1.4, III at left, the complex of the PapE subunit (in cyan) with the G1 strand of the chaperone (in yellow) is shown. In Fig.1.4, IV at left, the same complex is shown but in a topological representation. Both show that the G1 strand of the chaperone is complementing the fold of the subunit PapE. However, note that the G1 strand is running parallel to strand F and thus the reconstituted Ig fold is atypical. During donor-strand exchange, the G1 strand dissociates and is replaced by the N-terminal sequence of the subunit which comes next in the assembly line. Indeed, all subunits contain in their N-termini a sequence containing alternating hydrophobic residues that can insert in the groove of the subunit assembled previously. This is shown in Fig.1.4, III, right, in a ribbon representation and in Fig.1.4, IV, right, in topological representation. Note that, this time, the N-terminal sequence peptide (termed "N-terminal extension" and labelled Nte (red) in Fig. 1.4, III and IV at right) runs anti-parallel to strand F and thus a typical Ig fold is reconstituted. It is this transition from an atypical fold to a typical fold that provides the energy for assembly.

6. STRUCTURE OF THE COMPLEX BETWEEN GROWTH HORMONE AND ITS RECEPTOR

The structure of the complex between human growth hormone (hGH) and the extracellular domain of its receptor (hGHR) is a clear illustration of the principle that proteins are able to use the same template for different interactions. The complex has a 1:2 stoichiometry, with one molecule of the hormone bound to two molecules of the hormone, as shown in Figure 1.5, I (de Vos et al., 1992). The hormone (*red*) folds into a four-helix bundle with an unusual topology; the receptor molecules (*blue* and *green*) are all-beta structures, each containing two separate immunoglobulin-like domains. Although the interactions the receptor molecules make with the hormone are completely different—and the surface area buried by interactions with the "right-hand" (as shown here; *green*) receptor molecule is much greater than that buried by interactions with the "left-hand" receptor molecule, the residues that each receptor contributes to hormone binding are equivalent. The complex is also stabilized by interactions between the C-terminal domains of the receptor molecules.

Mutation of a single residue of the hormone, glycine 120, to arginine, is sufficient to turn the hormone into an antagonist that can only bind one receptor molecule. This 1:1 complex binds its single receptor molecule in a conformation that is extremely similar to that of the tight-binding receptor molecule in the active 1:2 complex. Detailed analysis of the crystal structure of this 1:1 complex has, with alanine scanning, revealed the molecular basis for hormone-receptor affinity (Clackson and Wells, 1995).

The hormone-binding site of the receptor is centered on a hydrophobic patch at the junction of the two immunoglobulin-like domains, and which interacts with hydrophobic residues on the receptor surface. This hydrophobic patch, shown as *blue/green* in the 1:2 complex in Fig. 1.5. II, is centered on two tryptophan residues, Trp 104 and Trp 169, and surrounded by more hydrophilic residues. Alanine scanning showed that the two central tryptophans contribute most of the binding energy to the hormone–receptor complex. The receptor surface area associated with these two residues is shown in *blue* in Fig. 1.5. II; surface area associated with other residues in the "hot spot," including the crucially important Arginine 43, is shown in *green*. Kinetic analysis has shown that electrostatic interactions of this positively charged residue with the receptor play a peripheral role in binding: this residue's importance

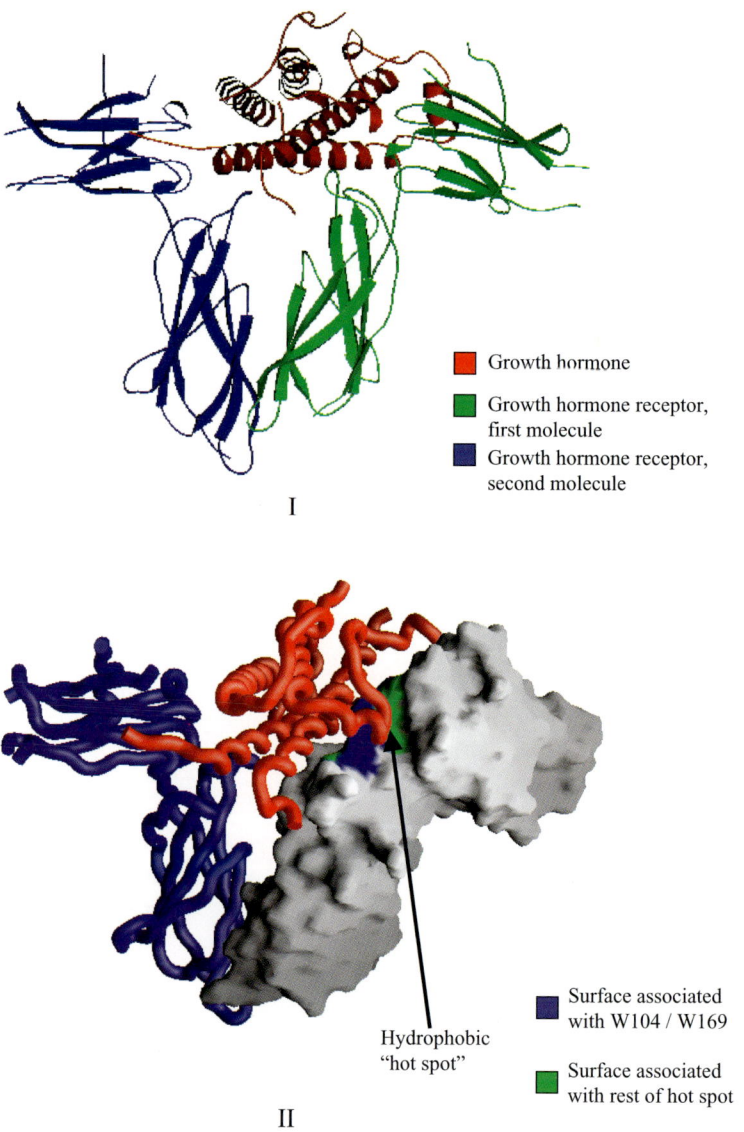

Figure 1.5. Structure of the complex between growth hormone and its receptor.

derives from the fact that its aliphatic side chain packs against Trp 169, wedging it into its required binding conformation. Several other hydrophobic residues in the patch, including proline 106, also serve to "lock" the crucial tryptophan residues into position.

Analyses of many protein–protein interaction surfaces have shown that, although an interaction surface typically involves many residues and buries many hundreds, if not thousands, of Å^2 of surface area, "hot spots" of a few residues that contribute most of the binding energy to the interaction may be the rule, rather than the exception. This study of the growth hormone–receptor complex highlights the important role of peripheral residues in orienting the residues of the hot spot in an optimum position for binding.

REFERENCES

Aarts, M., Liu, L., Besshoh, S., Arundine, M., Gurd, J.W., Want, Y.T., Salter, M.W., and Tymianski, M. (2002). Treatment of ischemic brain damage by perturbing NMDA receptor-PSD-95 protein interactions. *Science* 298:846–850.

Alaimo, P.J., Shogren-Knaak, M.A., and Shokat, K.M. (2001). Chemical genetic approaches for the elucidation of signaling pathways. *Curr. Opin. Chem. Biol.* 5:360–367.

Amit, A.G., Mariuzza, R.A., Phillips, S.E., and Poljak, R.J. (1986). Three-dimensional structure of an antigen-antibody complex at 2.8 A resolution. *Science* 233:747–753.

Ban, N., Nissen, P., Hansen, J., Moore, P.B., and Steitz, T.A. (2000). The complete atomic structure of the large ribosomal subunit at 2.4 A resolution. *Science* 289:905–920.

Barnhart, M.M., Pinkner, J.S., Soto, G.E., Sauer, F.G., Langermann, S., Waksman, G., Frieden, C., and Hultgren, S.J. (2000). PapD-like chaperones provide the missing information for folding of pilin proteins. *Proc. Natl. Acad. Sci. USA* 97:7709–7714.

Becker-Hapak, M., McAllister, S.S., and Dowdy, S.F. (2001). TAT-mediated protein transduction into mammalian cells. *Methods* 24:247–256.

Bhat, T.N., Bentley, G.A., Boulot, G., Greene, M.I., Tello, D., Dall'Acqua, W., Souchon, H., Schwarz, F.P., Mariuzza, R.A., and Poljak, R.J. (1994). Bound water molecules and conformational stabilization help mediate an antigen-antibody association. *Proc. Natl. Acad. Sci. USA* 91:1089–1093.

Bradshaw, J.M., and Waksman, G. (2002). Molecular recognition by SH2 domains. *Adv. Protein Chem.* 61:161–210.

Bradshaw, J.M., Mitaxov, V., and Waksman, G. (1999). Investigation of phosphotyrosine recognition by the SH2 domain of the Src kinase. *J. Mol. Biol.* 293:971–985.

Choudhury, D., Thompson, A., Stojanoff, V., Langermann, S., Pinkner, J., Hultgren, S.J., and Knight, S.D. (1999). X-ray structure of the FimC-FimH chaperone-adhesin complex from uropathogenic *Escherichia coli*. *Science* 285:1061–1066.

Clackson, T., and Wells, J.A. (1995). A hot spot of binding energy in a hormone-receptor interface. *Science* 267:383–386.

Cramer, P., Bushnell, D.A., and Kornberg, R.D. (2001). Structural basis of transcription: RNA polymerase II at 2.8 angstrom resolution. *Science* 292:1863–1876. Epub 2001 Apr 1819.

Dall'Acqua, W., Goldman, E.R., Eisenstein, E., and Mariuzza, R.A. (1996). A mutational analysis of the binding of two different proteins to the same antibody. *Biochemistry* 35:9667–9676.

de Vos, A.M., Ultsch, M., and Kossiakoff, A.A. (1992). Human growth hormone and extracellular domain of its receptor: crystal structure of the complex. *Science* 255:306–312.

Dhalluin, C., Carlson, J.E., Zeng, L., He, C., Aggarwal, A.K., and Zhou, M.-M. (1999). Structure and ligand of a histone acetyltransferase bromodomain. *Nature* 399:491–496.

Dodson, K.W., Pinkner, J.S., Rose, T., Magnusson, G., Hultgren, S.J., and Waksman, G. (2001). Structural basis of the interaction of the pyelonephritic *E.coli* adhesin to its human kidney receptor. *Cell* 105: 73–743.

Gnatt, A.L., Cramer, P., Fu, J., Bushnell, D.A., and Kornberg, R.D. (2001). Structural basis of transcription: an RNA polymerase II elongation complex at 3.3 Å resolution. *Science* 292:1876–1882.

Jackson, R.M. (1999). Comparison of protein–protein interactions in serine-protease and antibody-antigen complexes: implications for the protein docking problem. *Protein Sci.* 8: 603–613.

Jeruzalmi, D., O'Donnell, M., and Kuriyan, J. (2001). Crystal structure of the processivity clamp loader gamma (gamma) complex of *E. coli* DNA polymerase III. *Cell* 106:429–441.

Jones, S., and Thornton, J. M. (1996). Principles of protein-protein interactions. *Proc Natl. Acad. Sci. USA* 93:13–20.

Kumaran, S., Grucza, R.A., and Waksman, G. (2003). The tandem Src homology 2 domain of the Sky kinase: a molecular device that adapts to interphosphotyrosine distances. *Proc. Natl. Acad. Sci. USA* 100:14828–14833.

Kwong, P.D., Wyatt, R., Robinson, J., Sweet, R.W., Sodroski, J., and Hendrickson, W. A. (1998). Structure of an HIV gp120 envelope glycoprotein in complex with the CD4 receptor and a neutralizing human antibody. *Nature* 393:648–659.

Lo Conte, L., Chothia, C., and Janin, J. (1999). The atomic structure of protein-protein recognition sites. *J. Mol. Biol.* 285:2177–2198.

Lubman, O.Y., and Waksman, G. (2003). Structural and thermodynamic basis for the interaction of the Src SH2 domain with the activated form of the PDGF beta-receptor. *J. Mol. Biol.* 328:655–668.

Musacchio, A. (2002). How SH3 domains recognize proline. *Adv. Protein. Chem.* 61:211–268.

Patani, G.A., and LaVoie, E.J. (1996). Bioisosterism: A rational approach in drug design. *Chem. Rev.* 96:3147–3176.

Pawson, T., and Scott, J.D. (1997). Signaling through scaffold, anchoring, and adaptor proteins. *Science* 278:2075–2080.

Robinson, R.C., Turbedsky, K., Kaiser, D.A., Marchand, J.B., Higgs, H.N., Choe, S., and Pollard, T.D. (2001). Crystal structure of Arp2/3 complex. *Science* 294:1679–1684.

Ruhlmann, A., Schramm, H.J., Kukla, D., and Huber, R. (1972). Pancreatic trypsin inhibitor (Kunitz). II. Complexes with proteinases. *Cold Spring Harb. Symp. Quant. Biol.* 36:148–150.

Sadowski, I., Stone, J.C., and Pawson, T. (1986). A noncatalytic domain conserved among cytoplasmic protein-tyrosine kinases modifies the kinase function and transforming activity of fujinami sarcoma virus p130$^{gag-fps}$. *Mol. Cell. Biol.* 6:4396–4408.

Sattler, R., and Tymianski, M. (2001). Molecular mechanisms of glutamate receptor-mediated excitotoxic neuronal cell death. *Mol. Neurobiol.* 24:107–129.

Sauer, F.G., Fütterer, K., Pinkner, J.S., Dobson, K.W., Hultgren, S.J., and Waksman, G. (1999). Structural basis of chaperone function and pilus biogenesis. *Science* 285:1058–1061.

Sawyer, T., Bohacek, R.S., Dalgarno, D., Eyermann, C.J., Kawahata, N., Metcalf C.A., III, Shakespeare, W., Sundaramoorthi, R., Wang, Y., and Yang, M.G. (2002). Src Homology 2 inhibitors: peptidomimetic and nonpeptide. *Mini Rev. Med. Chem.* 2:475–488.

Tong, A.H.Y., Lesage, G., Bader, G.D., Ding, H., Xu, H., Xin, X., Young, J., Berriz, G.F., Brost, R.L., Chang, M., et al. (2004). Global mapping of the yeast genetic interaction network. *Science* 303:808–813.

Wimberly, B.T., Brodersen, D.E., Clemons, W.M., Jr., Morgan-Warren, R.J., Carter, A.P., Vonrhein, C., Hartsch, T., and Ramakrishnan, V. (2000). Structure of the 30S ribosomal subunit. *Nature* 407:327–339.

2

Yeast Two-Hybrid Protein–Protein Interaction Networks

Daniel Auerbach and Igor Stagljar

ABSTRACT

The availability of complete genome sequences of numerous model organisms has initiated the development of new approaches in biological research to complement conventional biochemistry and genetics. Consequently, high-throughput methodologies also need to be applied in the emerging field of proteomics. Here, we discuss several methods that have been developed in the past years in order to characterize proteins and their functions on a large scale. We focus on the yeast two-hybrid system, which is the most widely used method to study protein–protein interactions and which has been used several times now to sucessfully map entire interaction networks on a large scale. We discuss small-scale pilot projects and how they have been upscaled to genome-wide screens, such as for the budding yeast *Saccharomyces cerevisiae*. We then compare the yeast two-hybrid system with several other screening methods that have been developed to investigate interactions between proteins in a high-throughput format, such as affinity purification methods coupled to mass spectrometry. Efficient adaptation of such methods to a high-throughput format, coupled with the increasing use of databases to compare interaction maps generated with different methods, will help in elucidating protein-protein interactions on a scale that would have been unthinkable just a few years ago.

DANIEL AUERBACH • Dualsystems Biotech Inc., Winterthurerstrasse 190, CH-8057 Zurich, Switzerland.

IGOR STAGLJAR • Institute of Veterinary Biochemistry and Molecular Biology, University of Zurich, Winterthurerstrasse 190, CH-8057 Zurich, Switzerland.

Proteomics and Protein–Protein Interactions: Biology, Chemistry, Bioinformatics, and Drug Design, edited by Waksman. Springer, New York, 2005.

1. INTRODUCTION

The availability of complete genome sequences of numerous model organisms has initiated the development of new approaches in biological research to complement conventional biochemistry and genetics. For example, only one third of all 6200 predicted yeast genes had been functionally characterized when the complete sequence of the yeast genome first became available (Goffeau et al., 1996). At present, only 3800 yeast genes have been characterized by genetic or biochemical means and there still remain approximately 1800 genes encoding proteins of unknown function (Kumar and Snyder, 2001). The same observation holds true for the human genome: approximately 80% of all predicted human genes have not been characterized to date (Aach et al., 2001). To answer this challenge, researchers have developed different high-throughput strategies to characterize unknown genes on a large scale.

To date, most interaction maps have been created by genetic screening in yeast, namely by using the yeast two-hybrid system (Fields and Song, 1989). The reasons for the success of the yeast two-hybrid system in large-scale screening projects are manifold: as an in vivo genetic screening system, it is easily scalable, no purification steps or optimizations with regard to binding or washing conditions are involved, and automatization using robotic platforms is very easy. On the other hand, false positives and false negatives remain a problem of the yeast two-hybrid system; consequently, large-scale interaction maps derived by such methods require stringent selection criteria to yield useful information. Below, we first discuss the protein–protein interaction maps from various organisms that have been created using the yeast two-hybrid system and then discuss the advantages and disadvantages of this method. Finally, we briefly describe what has been done in analyzing those interaction maps to date.

2. THE YEAST TWO-HYBRID SYSTEM

The yeast two-hybrid system originally created by Fields and Song is a genetic system wherein the interaction between two proteins of interest is detected via the reconstitution of a transcription factor and the subsequent activation of reporter genes under the control of this transcription factor (Fields and Song, 1989). As depicted in Figure 2.1A, a protein X is expressed as a fusion to a DNA binding domain (DBD). The DBD–X fusion is commonly termed the "bait." Because of the affinity of the DBD for its operator sequences the bait is bound to a promoter element upstream of a reporter gene but does not activate it because it lacks an activation domain. A second protein Y is expressed as a fusion to an activation domain (AD) and is commonly termed the "prey." They prey is capable of activating transcription but usually does not do so because it has no affinity for the promoter elements upstream of the reporter gene (Fig. 2.1B). If bait and prey are coexpressed and the two proteins X and Y interact, then a functional transcription factor is reconstituted at the promoter site

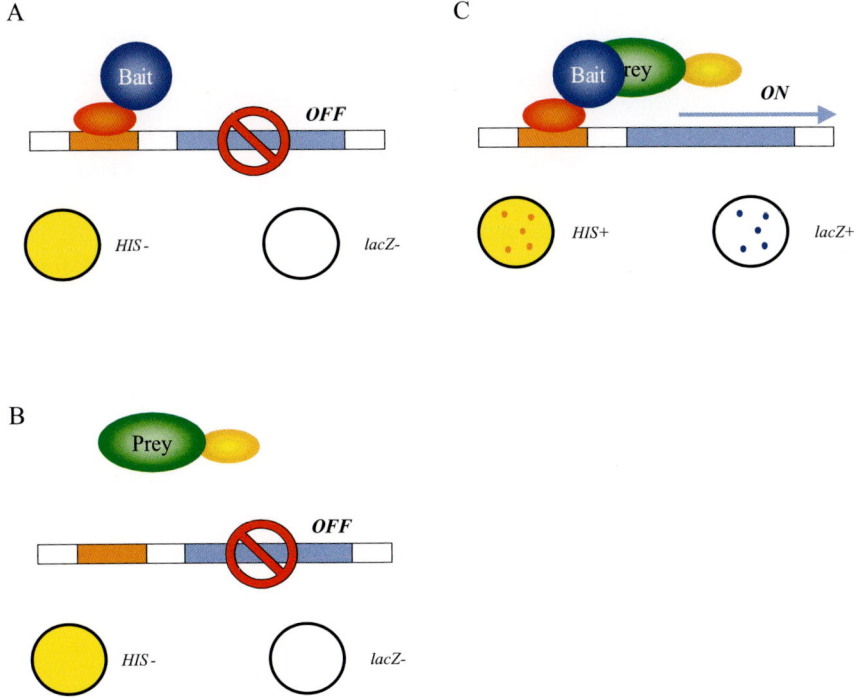

Figure 2.1. The yeast two hybrid system. (**A**) A bait is expressed as a fusion to a DNA binding domain (DBD), for example, the *Saccharomyces cerevisiae* GAL4 transcription factor or the *Escherichia coli* LexA protein. The DBD–bait binds to the operator sequences present in the promoter region upstream of the reporter gene but does not activate its transcription since the DBD–bait does not contain an activation domain. (**B**) A prey is expressed as a fusion to an activation domain (AD), for example, from the GAL4 transcription factor or from the Herpes simplex virus protein VP16. The AD–prey fusion has the capability to activate transcription in yeast but because it is not actively targeted to the promoter it does not activate transcription of the reporter gene. (**C**) The interaction between bait and prey targets the AD–prey fusion protein to the promoter, thereby reconstituting an active transcription factor. The hybrid transcription factor is bound to the promoter upstream of the reporter gene and therefore activates transcription. The readout of the activated reporter gene is measured either as growth on selective medium (auxotrophic selection markers, such as *HIS3*, *URA3*, or *ADE2*) or in a color reaction *(lacZ)*. Yeast expressing only the DBD–bait or the AD–prey on its own do not grow on selective medium *(HIS−)* and do not display blue staining in a color assay *(lacZ-)*, whereas yeast harboring an interacting DBD–bait and AD–prey display growth *(HIS+)* and blue color *(lacZ+)*.

upstream of the reporter gene. Consequently, transcription of the reporter gene is activated. Thus, in a yeast two-hybrid assay a protein–protein interaction is measured through the activation of one or several reporter genes in response to the assembly of a transcription factor by the said protein–protein interaction (Fig. 2.1C). In common yeast two-hybrid screening schemes the prey is usually replaced by a collection of unknown preys expressed from a cDNA or genomic library. Screening of entire libraries against a defined bait may then lead to the discovery of novel interaction partners. For large-scale screenings, two approaches are commonly used: the library screening approach, in which multiple baits are screened against a library, and the matrix approach, in which an array of defined preys is substituted for the library.

3. LARGE-SCALE SCREENS USING THE LIBRARY APPROACH

The library approach is schematically shown in Figure 2.2A. A particular bait is expressed in a yeast reporter strain of the mating type a, whereas a collection of preys (the library) is transformed into a yeast reporter strain of the mating type α. The bait-bearing strain is then mated with the mixture of library strains, and clones expressing an interaction pair are isolated on selective media. To determine the identity of the interacting prey, the library plasmid encoding it has to be isolated from the yeast strain and amplified in *Escherichia coli*. The region encoding the prey is then sequenced.

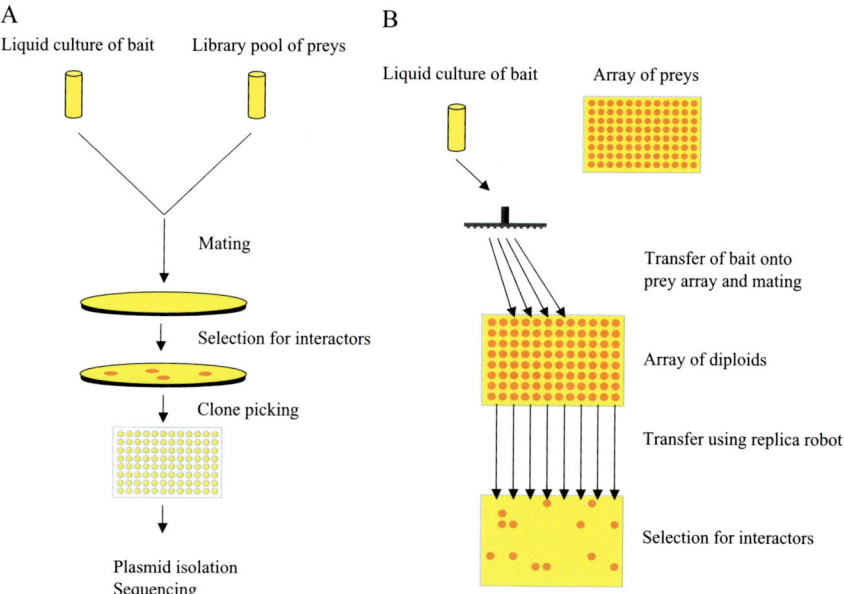

Figure 2.2. High-throughput approaches utilizing the yeast two-hybrid system. (**A**) The library screening approach. A yeast strain expressing a bait under investigation is mixed with a collection of yeast strains each expressing a random prey from a library. Incubation in rich medium allows the two strains to mate and diploids expressing bait and prey are selected. The diploids are then transferred to selective medium to isolate those clones containing interacting baits and preys (selection for interactors). Yeast clones that display growth on selective medium are picked up, transferred into multiwell plates, and processed for plasmid isolation and insert sequencing to identify the interacting prey. (**B**) The matrix or array approach. An array of preys is prepared by spotting yeast clones each expressing a known prey onto plates. The colonies on the array are then picked up by a robot and mated with a yeast strain expressing the bait under investigation. An exact replica of the array is transferred to a fresh plate to select for diploids expressing bait and prey and then to selective medium to select for interacting baits and preys. The identity of the prey in colonies that grow under selection is determined by its position within the array.

Although very powerful when used to investigate a single bait, the library approach is technically challenging when screening a large collection of baits: every yeast clone from the primary selection that may contain a valid protein–protein interaction has to be picked up, the library plasmid isolated, and the identity of the interacting protein established by sequencing of the cDNA insert. Even when using robotic equipment the screening of several thousands of baits against a library still presents a formidable challenge. Furthermore, false-positive interactors, for example, preys that activate the reporter genes without actually binding to the bait are common in library screens and are difficult to identify and eliminate. Several recent improvements in vectors and screening strains have led to a considerably lowered rate of false positives (James et al., 1996; Walhout et al., 2000b; Ito et al., 2001). For instance, when using only one auxotrophic selection marker, 10% to 70% of all clones in a screen may represent false positives. When using two auxotrophic selection markers, either in a simultaneous or in a sequential selection scheme, false positives are virtually eliminated from the screen (Fang and Macool, 2002).

4. LARGE-SCALE SCREENING USING THE MATRIX APPROACH

In the matrix approach a collection of defined preys is used instead of a random collection of open reading frames (ORFs) or ORF fragments. Each prey is separately introduced into yeast and the transformants are arrayed on plates using a robot (Figure 2.2B). A bait-bearing strain of the opposite mating type is then mated with every prey-bearing strain and the resulting diploid strains are replicated onto selective medium. If a particular diploid within the array grows on selection medium, its prey must interact with the bait under investigation. As opposed to the library screen, no plasmid isolation or sequencing is necessary since the position of the growing diploid on the array identifies the prey it expresses. In essence, a matrix screen consists of a series of defined interactions between a bait and a number of preys, rather than a screen of a bait against a collection of unknown preys. An advantage of the matrix screen is that repeated screenings will identify false-positive interactors present in the array. If a particular prey interacts with every bait being tested, chances are high that the interaction is either nonspecific or that the prey activates the reporter genes in the absence of a real protein–protein interactions. Such preys become apparent after several screens and can then be discarded.

5. ADVANTAGES AND DISADVANTAGES OF THE TWO APPROACHES

Both the library screen and the matrix screen have advantages and disadvantages, and ultimately they complement each other. This is underscored by the fact that most large-scale screening projects have been carried out using both approaches (Flajolet et al., 2000; Uetz et al., 2000; Walhout et al., 2000b; Ito et al., 2001). A screen using a cDNA or genomic library will find not only full-length interactors but also interacting subdomains of a given protein. This may be an advantage where a domain is capable of interacting with a bait but the entire protein is not, for instance, because other parts of the protein inhibit the interaction. It has been argued that library screens are more sensitive than matrix approaches (Ito et al., 2001). On the other hand, they often yield small protein fragments that interact nonspecifically with many unrelated baits, so-called false positives. In conventional yeast two-hybrid screens, in which only one protein is used as a bait, false positives are commonly eliminated by testing each prey against a set of bait proteins. Preys that interact nonspecifically with all baits tested are labeled as false positives and are discarded from the screen. However, when testing thousands of different baits in the context of a large-scale screening project, no such test can be carried out because the actual number of interactions to be assayed would be too high. For this reason, large-scale library screens probably yield a high number of interactions that do not occur in a physiological context and that must therefore be labeled as false positives. Mrowka et al. (2001) have compared results from the two genome-wide screens of *Saccharomyces cerevisiae* with published interaction data from the research community and have concluded that unless there is a severe bias in published interactions, the genome-wide datasets contain an estimated rate of false positives between 44% and 91%.

Many proteins are modular in nature and contain domains with specific functions, such as protein–protein interaction motifs. Often it is important to know which particular part of a prey interacts with the protein of interest. Since libraries contain random fragments of proteins, a library screen usually gives information on the minimal protein fragment necessary for interaction, provided a sufficiently large number of preys are rescued from the screen. A recent study has undertaken a genome-wide screen of the prokaryote *Helicobacter pylori*. A total of 261 baits were screened against a complex *H. pylori* library of small genomic fragments. More than 1200 interactions were identified, resulting in a protein interaction network that connected close to 50% of the entire proteome (Rain et al., 2001). The use of a small fragment library allowed the alignment of preys that were identified in each screen and the consequent identification of minimal interacting domains from these alignments. Such strategies may prove useful for the identification of as yet unknown interaction motifs in genome-wide screens. Matrix screens do not yield this type of information because they commonly use full-length proteins. Using computer algorithms in combination with high-throughput cloning strategies to create arrays containing mixtures of full-length proteins and protein fragments may be envisaged, however.

When compared to library screens the matrix approaches have an important advantage: false positives can be easily identified because every prey has a defined position within the array. Preys that interact nonspecifically with multiple baits can be identified easily and removed from the dataset. Furthermore, the matrix approach works well with sequenced genomes, in which every ORF has been predicted by computational methods and can be cloned easily using high-throughput methods (Uetz et al., 2000; Ito et al., 2001). For genomes of higher eukaryotes such as human or mouse, however, the situation is more difficult: genome annotation is less reliable, and because of the presence of genomic introns, ORFs cannot simply be cloned using high-throughput genomic polymerase chain reaction (PCR). For this reason, library screens will probably remain the method of choice for constructing interaction maps of higher eukaryotes for several years to come.

Interestingly, screens using the library approach tend to yield a higher number of interactions than those carried out with the matrix approach. The reasons for this are probably manifold. One explanation is that in matrix screens many ORFs have to be discarded for reasons inherent to the proteins under investigation. For instance, interactions in the yeast two-hybrid system have to take place in the nucleus. Consequently, proteins that possess hydrophobic transmembrane domains will be unable to reach the nucleus. This problem is partially circumvented in library screens: libraries that express protein fragments have a greater chance of containing prey fragments that lack "difficult" protein regions, such as transmembrane domains. Consequently, they may detect interactions that are lost when using arrays of full-length proteins. On the other hand, library screens have the aforementioned problem of reproducibility. Whereas any interaction in a matrix screen can be repeated multiple times, library screens are essentially irreproducible. If the resulting interactions are not verified, for example, using multiple control baits, a dataset derived from a library screen may contain a very high number of false-positive interactors that would not occur in a physiological setting (Mrowka et al., 2001).

In the following, the large-scale studies that have been carried out to date are discussed in detail. They range from small-scale interaction maps, which draw protein interaction networks of protein complexes such as the proteasome, to truly genome-wide screens that manage to interconnect a large fraction of the genome of the organism under investigation. The results of each study are summarized in Table 2.1.

6. SMALL-SCALE PROTEIN INTERACTION MAPS

Initially, large-scale yeast two-hybrid screens were used to generate small interactions networks that focused on protein complexes or components of a defined signaling pathway. In 1994, Finley and Brent generated a protein interaction map of cyclin-dependent kinase interacting proteins in *Drosophila melanogaster*. They identified 19 interactions and demonstrated that each cyclin-dependent kinase interacting protein associates with a specific spectrum of Cdks (Finley and Brent, 1994). In another example, a large-scale screen was used to identify interactions between several yeast proteins involved in mRNA splicing. The investigators carried out library screens with 15 defined baits and identified 170 interactions (Fromont-Racine et al., 1997). The screen identified nine preys that encoded known pre-mRNA splicing factors, as well as five preys that were homologous to human splicing factors.

In 2000, Walhout et al. created the first partial protein interaction map for a multicellular organism, the nematode *Caenorhabditis elegans*. Focusing on proteins involved in vulval development, they screened a cDNA library with a total of 27 baits and identified 148 interactions, including 15 previously known interactions and 109

Table 2.1. Large-scale screening projects carried out using the yeast two-hybrid system

Organism	Predicted ORFS	Methods	AD hybrids	DBD hybrids	Interactions	References
Bacteriophage T7	55	Matrix screen	11 ORFs	34 ORFs	3	Bartel et al., 1996
		Library screen	Library	34 ORFs	22	
Vaccinia virus	266	Matrix screen	266 ORFs	266 ORFs	37	McCraith et al., 2000
Hepatitis C virus	~10	Matrix screen	11 ORFs	10 ORFs	0	Flajolet, et al., 2000
		Library screen	Library	200 ORFs	15	
Helicobacter pylori	1590	Library screen	Library	261 ORFs	1280	Rain et al., 2001
		Library screen	Library	27 ORFs	148	Walhout et al., 2000
Caenorhabditis elegans	19,099	Matrix screen	29 ORFs	29 ORFs	11	Davy et al., 2001
		Matrix screen	30 ORFs	30 ORFs	17	
		Library screen	Library	30 ORFs	138	
Drosophila melanogaster	13,600	Matrix screen	5 ORFs	9 ORFs	19	Finley et al., 1994
		Library screen	Library	15 ORFs	170	Fromont-Racine et al., 1997
Saccharomyces cerevisiae	6200	Library screen	159 ORFs	159 ORFs	183	Ito et al., 2000
		Matrix screen	192 ORFs	5345 ORFs	281	Uetz et al., 2000
		Library screen	5345 ORFs	5345 ORFs	692	

interactions that had been predicted based on the *C. elegans* genome sequence (Walhout et al., 2000b). Another group has focused on proteins of the 26S proteasome in *C. elegans*. A matrix screen using 30 baits and preys yielded 17 interactions, whereas library screens with the 30 baits resulted in 138 interactions (Davy et al., 2001).

7. COMPREHENSIVE PROTEIN INTERACTION MAPS

Not surprisingly, the first genome-wide interaction maps were created using bacteriophages and viruses as model systems. Their small, well-characterized genomes make a comprehensive screening study relatively straightforward. The first genome-wide yeast two-hybrid study was carried using bacteriophage T7 as a model system, which has a total of 55 predicted ORFs (Dunn and Studier, 1983). Using a combination of matrix and library screening approaches, random bait and prey fragments were screened against each other and 25 interactions were identified in this way (Bartel et al., 1996).

In the subsequent study, McCraith et al. applied the matrix approach to examine interactions between 266 ORFs of the vaccinia virus as bait and prey fusions (McCraith et al., 2000), resulting in a total of 70,000 combinations of DBD and AD fusions. This study identified 37 protein–protein interactions, including 28 that had previously been identified. The same matrix approach has also been applied to build a protein interaction map of 200 ORFs of the hepatitis C virus (Flajolet et al., 2000). Interestingly, no interactions were identified using this approach, most probably because of incorrect folding or mistargeting of full-length DBD and AD fusions. To circumvent this problem, the authors applied the exhaustive library screening approach in which the above-mentioned 200 DBD–ORF fusions were screened against a random genomic library. This approach yielded 15 interactions that included both previously known and novel interacting pairs.

Similarly to bacteriophages and viruses, bacterial genomes, because of their small number of protein coding genes, also represent ideal model organisms for generation of protein interaction maps. However, only one systematic yeast two-hybrid approach has so far been undertaken in bacteria to analyze protein interactions at a global level—the one of the human gastric bacterial pathogen *H. pylori*, whose genome encodes 1590 predicted ORFs. Using the matrix approach and exhaustive genomic library screening on 261 ORFs fused to DBD, Rain et al. identified a total of 1280 interactions, resulting in a protein interaction map covering much of the *H. pylori* proteome (Rain et al., 2001). This study indicated that building a protein interaction map of a pathogenic bacterium may represent a powerful new tool for understanding the molecular mechanisms of infection and drug resistance and for developing novel innovative therapies.

The most comprehensive yeast two-hybrid screenings to date focus on the yeast *S. cerevisiae*. In 2000, two groups completed the comprehensive yeast two-hybrid mappings on all 6000 yeast ORFs as baits using both matrix and library screening approaches (Ito et al., 2000, 2001; Uetz et al., 2000). In an effort led by Ito et al., large-scale matrix screen using 159 ORFs cloned as DBD and AD fusions were performed and resulted in identification of 175 interactions, of which 163 had not been reported previously (Ito et al., 2000). Recently, the same group completed their exhaustive yeast two-hybrid screenings on yeast ORFs and identified 841 interactions in total (Ito et al., 2001). The second comprehensive study performed by the Fields group utilized both array and library screenings (Uetz et al., 2000). Using the array method, 192 ORFs were created as DBD fusions and then mated with the 6000 ORFs of yeast fused to the AD. Only 20% of all interactions were found in both screens, resulting in 281 protein pairs. For the exhaustive library screen, a library was made by pooling the 5345 AD-fused ORFs. These were then mated separately to the same 5345 ORFs fused to the DBD, yielding a total of 692 protein–protein interactions.

8. COMPARISON OF INTERACTION MAPS CREATED BY DIFFERENT SCREENING METHODS

When comparing the datasets of Ito et al. and Uetz et al. it is interesting to note that despite the fact that both groups used the same 6000 ORFs in their experiments, only 20% of all interactions in the two datasets actually overlap (Ito et al., 2001). The reasons for this small overlap are difficult to explain. A significant factor may have been the use of different experimental systems: bait and prey plasmids used in the two studies differ with regard to copy number within the cell, selection markers, and promoters driving expression of bait and prey proteins. The use of PCR to amplify the yeast ORFs may have introduced mutations that abolish interactions and, most importantly, the stringency of selection may have been different, eliminating interactions seen by one group from the other group's dataset (see discussion in Ito et al., 2001). The small overlap can be taken to mean that even when using exhaustive library screens that potentially cover all interactions in a genome, the subset of protein–protein interactions that can be identified using the yeast two-hybrid system is far from representative.

How do large-scale interaction maps therefore compare with the interaction data gathered by the research community in the past decades? A recent publication has compared the data from several large-scale studies on yeast with data available from public protein interaction databases (von Mering et al., 2002). They also compared yeast two-hybrid screenings with data from several other high-throughput methods, such as affinity purification coupled to mass spectrometry (Gavin et al., 2002; Ho et al., 2002). Their findings highlight again the problem of small overlap between different interaction datasets: despite the fact that high-throughput methods in yeast have generated some 80,000 interactions to date, only a small fraction (2400 interactions) are supported by more than one method. Thus, every method probably has a bias toward certain protein–protein interactions and may fail to detect others. As already mentioned, any screen based on the yeast two-hybrid system will have difficulties in detecting interactions between integral membrane proteins or membrane-associated proteins since these proteins are unable to reach the nucleus. Consequently, membrane proteins are underrepresented in yeast two-hybrid datasets (von Mering et al., 2002). On the other hand, methods based on copurification of protein complexes often miss signaling pathways and transport pathways. In this respect it is also important to note that the yeast two-hybrid system is a method for detecting binary interactions, because under normal circumstances only two proteins are assayed against each other. In contrast, complex purification schemes aim at isolating native protein complexes; for example, they will identify several interaction partners of a given bait. As the two large-scale yeast two-hybrid studies were carried out on yeast proteins, this problem may have been partially circumvented because endogenous yeast proteins may have formed complexes with bait and prey proteins and may thus have acted as bridging partners. Therefore, the Uetz and Ito datasets may also contain annotated interactions between two proteins that may in fact have been mediated by a third bridging partner.

9. THE CHALLENGES AHEAD

The future of proteomics is the definition of the exact function of every protein in a cell, and how this function may change in different cellular conditions, with different modification states of a protein, and with different interacting partners. For more than 10 years, the yeast two-hybrid system and its variations have played an important role in the study of physical protein–protein interactions. The recent application of the yeast two-hybrid system to large-scale screenings has culminated in the construction of several protein interaction maps that manage to connect the majority of proteins encoded by the organism under investigation. These screens facilitate the understanding of gene function in several ways. First, they provide insight into the possible functional roles of previously unknown genes by linking them to already characterized proteins. Second, they help to assign additional, novel functions to many previously characterized proteins. Third, they identify novel interactions between proteins that have previously been assigned to common biological processes based on circumstantial evidence such as transcriptional coregulation or subcellular colocalization.

Recently, alternative approaches for the identification of protein–protein interactions on a genome-wide scale have been developed, which are based on the characterization of protein complexes using mass spectrometry (Gavin et al., 2002; Ho et al., 2002). As opposed to the yeast two-hybrid system, which largely detects binary interactions, this method relies on the selective purification of entire protein complexes from the cell, followed by separation of its subunits and their identification by mass spectrometry. The methodology has been applied to characterize multiprotein complexes systematically on a large-scale in yeast and has identified hundreds of novel protein–protein interactions and protein complexes (Gavin et al., 2002; Ho et al., 2002). The future progress of interactive proteomics will involve refinement of such approaches, as well as the integration of data sets derived from as many different methods as possible. Ultimately, protein interaction networks that have been constructed from different datasets will hopefully show little bias with regard to protein classes or functions, will represent the entire proteome under investigation, and will contain only interactions that have been proven by several methods, for example, they are likely to represent actual interactions that occur under physiological conditions within a cell. Once such representative maps have been constructed it will also become feasible to address the issue of regulation: by comparing representative protein interaction maps from the same cell type or organism under different growth conditions it may become possible to dissect cellular reactions in response to changing environmental conditions at the level of protein–protein interactions.

REFERENCES

Aach, J., Bulyk, M.L., Church, G.M., Comander, J., Derti, A., Shendure, J. (2001). Computational comparison of two draft sequences of the human genome. *Nature* 409:856–859.

Bartel, P.L., Roecklein, J.A., SenGupta, D., Fields, S. (1996). A protein linkage map of *Escherichia coli* bacteriophage T7. *Nat. Genet.* 12:72–77.

Davy, A., Bello, P., Thierry-Mieg, N., Vaglio, P., Hitti, J., Doucette-Stamm, L., Thierry-Mieg, D., Reboul, J., Boulton, S., Walhout, A.J., Coux, O., Vidal, M. (2001). A protein–protein interaction map of the *Caenorhabditis elegans* 26S proteasome. *EMBO Rep.* 2:821–828.

Fang, Y. and Macool, D.J. (2002). Development of a high-throughput yeast two-hybrid screening system to study protein–protein interactions in plants. *Mol. Genet. Genomics* 267:142–153.

Dunn, J.J. and Studier, F.W. (1983). Complete nucleotide sequence of bacteriophage T7 DNA and the locations of T7 genetic elements. *J. Mol. Biol.* 166:477–535.

Fields, S. and Song, O. (1989). A novel genetic system to detect protein–protein interactions. *Nature* 340:245–246.

Finley, R.L., Jr. and Brent, R. (1994). Interaction mating reveals binary and ternary connections between *Drosophila* cell cycle regulators. *Proc. Natl. Acad. Sci. USA* 91:12980–12984.

Flajolet, M., Rotondo, G., Daviet, L., Bergametti, F., Inchauspe, G., Tiollais, P., Transy, C., Legrain, P. (2000). A genomic approach of the hepatitis C virus generates a protein interaction map. *Gene* 242:369–379.

Fromont-Racine, M., Rain, J.C., Legrain, P. (1997). Toward a functional analysis of the yeast genome through exhaustive two- hybrid screens. *Nat. Genet.* 16:277–282.

Gavin, A.C., Bosche, M., Krause, R., Grandi, P., Marzioch, M., Bauer, A., Schultz, J., Rick, J.M., Michon, A.M., Cruciat, C.M., Remor, M., Hofert, C., Schelder, M., Brajenovic, M., Ruffner, H., Merino, A., Klein, K., Hudak, M., Dickson, D., Rudi, T., Gnau, V., Bauch, A., Bastuck, S., Huhse, B., Leutwein, C., Heurtier, M.A., Copley, R.R., Edelmann, A., Querfurth, E., Rybin, V., Drewes, G., Raida, M., Bouwmeester, T., Bork, P., Seraphin, B., Kuster, B., Neubauer, G., and Superti-Furga, G. (2002). Functional organization of the yeast proteome by systematic analysis of protein complexes. *Nature* 415:141–147.

Goffeau, A., Barrell, B., Bussey, H., Davis, R.W., Dujon, B., Feldmann, H., Galibert, F., Hoheisel, J.D., Jacq, C., Johnston, M., Louis, E.J., Mewes, H.W., Murakami, Y., Philippsen, P., Tettelin, H., Oliver, S.G. (1996). Life with 6000 genes. *Science* 274:563–567.

Ho, Y., Gruhler, A., Heilbut, A., Bader, G.D., Moore, L., Adams, S.L., Millar, A., Taylor, P., Bennett, K., Boutilier, K., Yang, L., Wolting, C., Donaldson, I., Schandorff, S., Shewnarane, J., Vo, M., Taggart, J., Goudreault, M., Muskat, B., Alfarano, C., Dewar, D., Lin, Z., Michalickova, K., Willems, A.R., Sassi, H., Nielsen, P.A., Rasmussen, K.J., Andersen, J.R., Johansen, L.E., Hansen, L.H., Jespersen, H., Podtelejnikov, A., Nielsen, E., Crawford, J., Poulsen, V., Sorensen, B.D., Matthiesen, J., Hendrickson, R.C., Gleeson, F., Pawson, T., Moran, M.F., Durocher, D., Mann, M., Hogue, C.W., Figeys, D., and Tyers, M. (2002). Systematic identification of protein complexes in *Saccharomyces cerevisiae* by mass spectrometry. *Nature* 415:180–183.

Ito, T., Tashiro, K., Muta, S., Ozawa, R., Chiba, T., Nishizawa, M., Yamamoto, K. Kuhara, S., Sakaki, Y. (2000). Toward a protein–protein interaction map of the budding yeast: A comprehensive system to examine two-hybrid interactions in all possible combinations between the yeast proteins. *Proc. Natl. Acad. Sci. USA* 97:1143–1147.

Ito, T., Chiba, T., Ozawa, R., Yoshida, M., Hattori, M., Sakaki, Y. (2001). A comprehensive two-hybrid analysis to explore the yeast protein interactome. *Proc. Natl. Acad. Sci. USA* 98:4569–4574.

James, P., Halladay, J., Craig, E. A. (1996). Genomic libraries and a host strain designed for highly efficient two-hybrid selection in yeast. *Genetics* 144:1425–1436.

Kumar, A. and Snyder, M. (2001) Emerging technologies in yeast genomics. *Nat. Rev. Genet.* 2:302–312.

McCraith, S., Holtzman, T., Moss, B., Fields, S. (2000). Genome-wide analysis of vaccinia virus protein–protein interactions. *Proc. Natl. Acad. Sci. USA* 97:4879–4884.

Mrowka, R., Patzak, A. (2001). Is there a bias in proteome research? Genome Res. 11:1971–1973.

Rain, J.C., Selig, L., De Reuse, H., Battaglia, V., Reverdy, C., Simon, S., Lenzen, G., Petel, F., Wojcik, J., Schachter, V., Chemama, Y., Labigne, A., Legrain, P. (2001). The protein–protein interaction map of *Helicobacter pylori*. *Nature* 409:211–215.

Uetz, P., Giot, L., Cagney, G., Mansfield, T.A., Judson, R.S., Knight, J.R., Lockshon, D., Narayan, V., Srinivasan, M., Pochart, P., Qureshi-Emili, A., Li, Y., Godwin, B., Conover, D., Kalbfleisch, T., Vijayadamodar, G., Yang, M., Johnston, M., Fields, S., Rothberg, J.M. (2000). A comprehensive analysis of protein–protein interactions in Saccharomyces cerevisiae. *Nature* 403:623–627.

von Mering, C., Krause, R., et al. (2002). Comparative assessment of large-scale data sets of protein–protein interactions. *Nature* 417:399–403.

Walhout, A.J., Boulton, S.J., Vidal, M. (2000a). Yeast two-hybrid systems and protein interaction mapping projects for yeast and worm. *Yeast* 17:88–94.

Walhout, A.J., Sordella, R., Lu, X., Hartley, J.L., Temple, G.F., Brasch, M.A., Thierry-Mieg, N., Vidal, M. (2000b). Protein interaction mapping in *C. elegans* using proteins involved in vulval development. *Science* 287:116–122.

3

The Use of Mass Spectrometry in Studying Protein–Protein Interaction

Yi Wang, Parvin Yazdi, and Jun Qin

ABSTRACT

Mass spectrometry has now become a mainstream technique in biology research. We first give a brief account highlighting the most important developments in mass spectrometry that may be useful for the study of protein–protein interaction; next, we discuss some interesting issues that are starting to emerge as we learn more about protein complexes; finally, we discuss in some detail one example of using mass spectrometry to study protein–protein interaction in the area of human genome maintenance, and propose a new concept that we termed "network analysis proteomics" that aims to identify modular protein interaction networks.

1. EVOLUTION OF MASS SPECTROMETRY AS A POWERFUL TOOL

Mass spectrometry (MS) has become a mainstream technique used in biology research as a result of technology developments over the last decade. The concurrent development of the human genome sequencing project accelerated the acceptance

YI WANG, PARVIN YAZDI, AND JUN QIN • Verna and Marrs McLean Department of Biochemistry and Molecular Biology and Department of Molecular and Cellular Biology, Baylor College of Medicine, Houston, TX 77030, USA.

Proteomics and Protein–Protein Interactions: Biology, Chemistry, Bioinformatics, and Drug Design, edited by Waksman. Springer, New York, 2005.

of MS in the biological community and provided the foundations for the birth of the proteomic era. MS coupled with protein database searching has dramatically improved both the sensitivity and the speed with which proteins can be identified. This technique now has replaced Edman sequencing for identifying proteins from species whose genomes have been sequenced. With a complete set of human proteins in the database, we can identify the composition of large, endogenous protein complexes in their native states with proper assembly and modification, if they can be biochemically purified. This technology now can be used for high-throughput analysis, allowing tens to hundreds of proteins to be analyzed and identified in a single day. There are excellent reviews on the technology development of MS and technical details about using MS data to identify proteins with database search (Mann et al., 2001; Aebersold and Mann, 2003). Database searching using MS data has been developed to a point such that the end user does not need to know how the searching program works, and the search is largely automated (Fenyo et al., 1998; Yates, III, 1998; Krutchinsky et al., 2001; Wolters et al., 2001). False positives (the program identifies the wrong protein) and false negatives (the program fails to identify the protein) are rare as the associated statistical analysis of the search program can almost make certain that the identification is the correct one.

Such sensitivity, speed, and ease of use have made MS a powerful tool in biochemistry and cell biology. Single protein complex identification is no longer a technical challenge, and is routine in many research institutes. Protein complex purification followed by MS identification on a genomic scale has been demonstrated in yeast with a good success rate (Gavin et al., 2002; Ho et al., 2002). Organelles such as the nuclear pore (Rout et al., 2000) and the nucleoli (Andersen et al., 2002) have already been analyzed. In special cases, the whole proteome of an organism can be analyzed and a large percentage of the proteins identified (Corbin et al., 2003). This ability is unprecedented and is revolutionizing biology research, as it is now possible to carry out discovery driven research without much prior knowledge. In this endeavor, one hopes to produce abundant high-quality data to generate testable hypotheses. In most cases, MS will identify the protein no matter what protein is analyzed, this has put great demand on the steps prior to MS analysis. The experimental design that determines what information is to be gained, and the biochemical purification that contains the information to be extracted, will often determine the outcome of the experiment. Whether useful information can be obtained to generate testable hypothesis no longer depends on MS but rather on the experimental design and biochemical purification. As MS is no longer the bottleneck, we first give only a brief account highlighting the most interesting developments in MS that may be useful for the study of protein–protein interaction; next, we discuss some interesting issues that are starting to emerge as we learn more about protein complexes; finally we discuss in some detail one example of using MS to study protein–protein interaction, and propose a new concept that we termed "network analysis proteomics" that aims to identify modular protein interaction networks.

2. MS IDENTIFICATION OF PROTEIN COMPLEXES

The study of protein–protein interaction begins with the identification of interacting proteins. The most convenient way to do this now is perhaps to purify the protein complex and identify the associated proteins with MS. Historically, one-dimensional sodium dodecyl sulfate-polyacrylamide gel electrophoresis (SDS-PAGE) is often the method of choice to separate the purified protein complex. The protein bands are then stained with Coomassie Brilliant Blue or silver; next, specific bands as compared with a negative control are cut out and in-gel digested, and peptides extracted and analyzed with tandem mass spectrometry. In such an approach, in-gel digestion and peptide extraction are the crucial steps, as some peptides are difficult to extract and thus are lost. An emerging trend now is to digest the protein complex in solution to alleviate this difficulty (Washburn et al., 2001; Sanders et al., 2002). Many more peptides are present when the protein complex is digested without separation than those derived from a SDS-PAGE band containing one or two proteins. Because MS can analyze only a limited numbers of peptides on the time scale of a high-performance liquid chromatography (HPLC) elution peak, many peptides cannot be analyzed so that some proteins will be missed. Two-dimensional liquid chromatography is usually required to separate peptides to reduce the complexity, thus maximizing the number of proteins identified. Either approach has its merit. The use of SDS-PAGE allows the visualization of specific bands and rough quantification of the stoichiometry of associated proteins, which are often lost in the in-solution method. More importantly, in-solution digestion requires a highly purified protein complex that does not contain a predominating component, such as antibodies. This requirement is often met by complexes purified using a tandem tag purification (TAP) method (Rigaut et al., 1999). When complexes are purified with antibodies, the large amounts of immunoglobulin G (IgG) that need to be used often present a problem. This, however, can be somewhat alleviated when SDS-PAGE is used to separate IgG from associated proteins. Antibodies can be crosslinked to beads to avoid elution when the associated proteins are eluted. Usually, however, a substantial amount of antibodies leak out to be problematic for the subsequent MS analysis. Perhaps more importantly, in many cases, crosslinking of antibodies to beads may reduce the binding efficiency (presumably by destroying or obscuring the antigen binding site), resulting in reduced amount of associated proteins purified. To purify a sufficient amount of protein that is not abundant in the cell, one is often in a position that he or she cannot afford losing any efficiency in the purification. In such a case, antibody crosslinking may not be an option. Therefore, both in-gel and in-solution digestion should be considered to identify associated proteins.

3. THE USE OF MS IN PROTEIN COMPLEX CHARACTERIZATION

3.1. Specific versus Nonspecific Interaction

After interacting proteins are identified, substantial work is needed to character-ize the interaction to understand the biological significance. The first question to be addressed is the specificity. This is typically determined by an immunoprecipitation (IP) followed by Western blotting, which is still preferred to MS, as Western blotting is still much more sensitive. The use of quantitative MS may also help to establish the specificity. In this scheme, the complex of interest and the control are processed with light and heavy isotope coded affinity tags (ICAT) reagents (Gygi et al., 1999), so that peptides from the complex of interest or the control can be differentiated on the basis of their light and heavy isotope patterns. The nonspecific binding proteins, which are common to the complex of interest and the control, should have a similar intensity of the light and heavy isotopic patterns, whereas the specific proteins will have a higher intensity in the light isotopic patterns. This has been applied to a large RNA polymerase II (Pol II) preinitiation complex (PIC) (Ranish et al., 2003). The bone fide PIC is assembled with recombinant TATA-box binding protein (TBP) using extracts from a yeast strain that is devoid of endogenous TBP on a piece of DNA containing a promoter, and the control PIC is assembled without TBP. The two complexes are then processed with different ICAT reagents and the specific proteins unique to the PIC for-mation can be distinguished from those that bind to DNA nonspecifically. In this way, three proteins are identified as potential new core Pol II components. Such an approach can also be applied to dynamic protein complex, in which association with a particular subunit depends on a signaling event that can be manipulated experimentally. Similar to using ICAT reagents, stable isotope labeling using different isotope (C12 vs. C13) substituted medium to grow cells can also be used to label proteins. This technique has been applied to Crb2 SH2 domain binding proteins in response to epidermal growth factor (EGF) signaling. Two novel proteins were identified whose binding to Crb2 SH2 domain *in vitro*: depends on stimulation by EGF (Blagoev et al., 2003). Although one can argue that more efficient biochemical purification may reduce non-specific protein binding, this approach may still be beneficial for the identification of weakly associated components that may be washed away in more stringent biochemi-cal purifications. The question of how to keep weakly associated proteins intact while reducing the nonspecific binding proteins is a paradox without an obvious solution.

3.2. Direct Interaction and Organization of the Protein Complex

Having established a specific interaction, one often wishes to know whether the interaction is direct. This can be addressed only by *in vitro*: binding using highly purified recombinant proteins. It is also important to know the organization of the protein complex—which molecule is the nearest neighbor of the subunit and how the complex is assembled. In theory, it is possible to crosslink the nearest neighbor using a bidirectional chemical crosslinking reagent and to identify the nearest neighbor using MS; even better, if an affinity moiety (e.g., biotin) is incorporated in the crosslinker, the crosslinked peptides generated by a proteolytic digestion of the crosslinked complex can be purified by streptavidin beads, and then identified with MS. It is, therefore, possible to map the region of contact between two subunits. In practice, many technical issues need to be resolved, the most important of which is how to control the condition in which there is only one crosslinking per complex. Multiple crosslinks turn the complex into an aggregate that is refractory to further MS analysis.

A productive way to obtain information about the organization of a complex is perhaps to figure out the dependence of the complex formation on each individual subunit. This is mostly conveniently done in an organism in which deletion mutants are readily generated, for example, yeast (Shen et al., 2003). If the deletion mutant is viable, one can purify the same protein complex and identify the rest of the associated proteins to obtain the information concerning the dependence of complex association on a particular subunit. Such information will give a rough picture about the organization of the complex. If a subunit serves as an organizer to which most of the other subunits bind, one obtains the most dramatic effect: in such a deletion mutant, the protein complex cannot form. This type of experiment can also be envisaged in human cells, in which small RNA interference is used to knock down the expression of one subunit, followed by examination of the integrity of the complex.

3.3. Stoichiometry

Finally, it may be important to determine the relative stoichiometry of a protein complex. Although this can be most conveniently estimated by a staining method (such as Coomassie Brilliant Blue) when a highly purified protein complex is resolved on a SDS-PAGE, MS may provide a general method. In this method, an isotope-labeled tryptic peptide is synthesized for each individual subunit, purified, and quantified by amino acid analysis. Then these isotope-labeled tryptic peptides are used as internal standards in the analysis of the in-solution tryptic digest of the protein complex. Because the standard is isotope labeled so that it can be differentiated from the endogenous peptides, the relative MS signal of the standard to the endogenous peptide represents the relative peptide abundance. This method provides an accurate determination of the stoichiometry, and does not require a highly purified protein complex as long as the identities of the subunits are known. When the molecular weight of the complex can be measured, or estimated using gel filtration chromatography, the absolute stoichiometry can then be determined.

4. INTERESTING QUESTIONS ABOUT PROTEIN COMPLEXES

So far we have been using the term protein complex loosely, which refers to proteins that interact directly or indirectly. We are quite liberal in using the term protein complex as long as a coimmunoprecipation is demonstrated. It may be beneficial now to define the term protein complex explicitly and to articulate the meanings of various protein complexes under different conditions. This will help us to better understand their biological meanings.

4.1. The Core Complex and the Regulated Complex

The term protein complex is likely derived from early biochemical work in which a set of proteins was found always to copurify during multiple column fractionations. For example, the Mre11 complex, composed of Rad50, Mre11, and NBS1 proteins can bind to the double-stranded break (DSB) end and is important for DSB repair. The Mre11 subunit has the nuclease activity that requires Rad50 for its protein stability, and NBS1 for stimulation of nuclease activity and substrate selectivity (Paull and Gellert, 1999). These three proteins are complexed together most of the time in the cell and are not separable for function. Biochemically, they copurify over multiple columns and are not disrupted by buffers containing high concentration of salt and detergents or SDS (Trujillo et al., 1998). Therefore, they form a stable complex. This does not mean that other proteins do not interact with them, however. It is often quite the opposite. The MDC1 protein, for example, can be coimmunoprecipitated by an antibody against Mre11 (Goldberg et al., 2003). Thus, we can loosely say that MDC1 is in a complex with the Mre11 complex. This MDC1–Mre11 complex is quite different from the stable Mre11 complex, as the MDC1 protein can be dissociated from the Mre11 complex when stringent washing is applied. We can still loosely use the term: the MDC1 complex and the Mre11 complex, but they have different underlying meanings. When MDC1 comes into the Mre11 complex, it is possible that the MDC1 complex assumes a unique function that the Mre11 complex or the MDC1 protein alone cannot provide, in which MDC1 regulates the activity of the Mre11 complex, or the Mre11 complex facilitates the function of MDC1.

To differentiate them, we should define the Mre11 complex as a core complex, and the MDC1–Mre11 complex as a regulated Mre11 complex. There is perhaps only one core Mre11 complex, but there are many regulated Mre11 complexes in which different proteins can join the core Mre11 complex to assume different functions. It is known that the ATM protein interacts with the Mre11 complex to function in DSB signaling, thus forming another regulated Mre11 complex (Uziel et al., 2003); the telomere protein TIF2–RAP1 also interacts with the Mre11 complex to form another regulated Mre11 complex that presumably functions at the telomere (Zhu et al., 2000). Differentiation of the regulated complex versus the core complex should help us conceptualize the function of the unique protein in the regulated complex. We have listed three regulated Mre11 complexes (MDC1, ATM, and TIF2–RAP1) here. Is there any relationship between them? The answer is probably yes. It is possible to imagine that the core Mre11 complex at the telomere, which is recruited by the TRF2–Rap1 complex, may recruit further the ATM kinase and the mediator MDC1 to set up a signaling cascade that allows ATM to phosphorylate a substrate that is brought in by the MDC1 protein to propagate the signal when telomere uncapping is detected. In this situation, the substrate brought in by the MDC1 protein may be a component of the regulated MDC1 complex, whose identity is not yet known.

In reality, most protein complexes are probably regulated complexes that are built around the core complex, which expand greatly the functional repertoire of the core complex. Delineation of the composition of regulated complexes may provide much more insight regarding function than that of the core complex.

4.2. One Protein or a Set of Proteins in Many Complexes?

One protein can sometimes be found in multiple protein complexes that have different functions. Except for the common protein, the rest of the proteins are different. These complexes are different from the core and regulated complexes. The protein discussed earlier, DNMTI-associated protein 1 (DMAP1) first identified as a protein interacting with the cytosine DNA methyltransferase 1 (DNMT1), can be found in two protein complexes. The first one contains at least DMAP1, DNMT1, and proliferation cell nuclear antigen (PCNA) (Rountree et al., 2000), which functions in DNA replication; the second one is the histone acetyltransferase Tip60 complex, which may function in transcription regulation, DNA replication and repair (Cai et al., 2003). These two complexes in cycling cells can be separated biochemically. The relationship between them is not clear, and the factors that determine whether DMAP1 resides in the DNMT1 complex versus in the Tip60 complex are not fully known. As more is learned about protein complexes, we expect to find more cases in which one protein can be found in many different protein complexes.

A set of proteins can also be found in multiple complexes. The smaller subunits of the replication protein C (RFC) complex, RFC2 to RFC5, can be found in the RFC1 complex (Tsurimoto et al., 1989), the checkpoint protein Rad17 complex (Lindsey-Boltz et al., 2001), the CTF18 complex that functions in chromosome cohesin and checkpoint regulation (Mayer et al., 2001; Naiki et al., 2001; Bermudez et al., 2003; Merkle et al., 2003), and another checkpoint protein ELG1 complex (Bellaoui et al., 2003; Ben Aroya et al., 2003; Kanellis et al., 2003). All of these RFC-like complexes are core complexes; they coexist in the cell and presumably have different functions.

One protein or a set of proteins that are found in many complexes presents a conceptual challenge for the understanding of the function of the protein. The cleanest way to assign a function to a protein is by loss of function studies, in which the gene of interest is knocked out and the observed phenotype is attributed to the function of the gene. If a protein can be found in multiple complexes, the protein complex to which the observed phenotype should be assigned when the gene is knocked out cannot be determined. We often find multiple functions for a protein, which may indicate that the protein exists in multiple complexes. This is analogous to the multifunction of a kinase, in which the diverse functions of the kinase are reflected in its multiple substrates.

4.3. Complex–Complex Interactions

The entire cell can be viewed as a factory that contains an elaborate network of interlocking assembly lines in which assemblies of 10 or more protein molecules carry out nearly every major process in a cell. As it carries out its biological functions, each of these protein assemblies interacts with several other large complexes of proteins (Alberts, 1998). The need to study protein complex–protein complex interaction becomes evident when we go a step further to understand protein–protein interaction. A useful analogy is to consider that in chemistry, one finds a hierarchy of elements, compounds, and reactions, and in biology, a hierarchy of proteins, protein complexes, and protein complex–protein complex interactions exists. Just as different chemical reactions define the chemistry, different protein complex–complex interactions determine the biology. The present discussion focuses largely on protein–protein interaction, but we will soon encounter the problem of complex–complex interaction. One example is described below.

Heterochromatin is a higher-order chromatin structure that is important for transcriptional silencing, chromosome segregation, and genome stability. Establishment and maintenance of heterochromatin is regulated not only by genetic elements, but also by epigenetic elements that include histone tail modification (e.g., acetylation and methylation) and DNA methylation. In one study delineating a pathway that maintains heterochromatin structure during cell division, we found that two interacting protein complexes are important in this pathway (Xin et al., 2003). We found that the p33ING1–Sin3–HDAC complex, as well as DNA methyltransferase I (DNMT1) and its interacting protein DMAP1 are required for maintaining heterochromatin structure. p33ING1 and DMAP1 interact physically and colocalize to heterochromatin in late S phase, and are required for the proper localization of heterochromatin protein 1 (HP1) to heterochromatin. The p33ING1–Sin3–HDAC complex and the DMAP1–DNMT1 complex are recruited independently to pericentric heterochromatin regions, but they are both required for deacetylation of histones and methylation of lysine 9 in histone H3. These data support a cooperative model for histone deacetylation, methylation, and DNA methylation in maintaining pericentric heterochromatin structure through cell division. In this case, the p33ING1–Sin3–HDAC1 complex and the DNMT1–DMAP1 complex cooperate to maintain the heterochromatin structure, through complex–complex protein interactions, in which the p33ING1 subunit in the p33ING1 complex and the DMAP1 subunit in the DNMT1 complex connect these two complexes during a specific point in the cell cycle, the late S phase. In the rest of the cell cycle, these two complexes function independently. Important applications of MS may also be found in identifying complex–complex interactions. If two complexes interact during a specific stage of the cell cycle, or in response to a signal, the interaction will be transient. It is necessary to use mild conditions to purify one complex in the hope of copurifying the other complex. Since the copurified complex most likely will be sub-stoichiometrical, it will also be necessary to use a sensitive MS method for the identification.

5. AN EXAMPLE: APPLICATION OF MS IN PROTEIN COMPLEX PURIFICATION/IDENTIFICATION

Although MS has been used widely as a tool in biological research, it is nonetheless an independent discipline. Most of the MS work in biology is done by collaboration or service. A few laboratories including our own are now experimenting on ways to integrate MS in biological research. We are studying signaling networks of tumor suppressor proteins and have been attempting to integrate protein purification, protein identification, and functional study so that all of the work can be performed in the same laboratory.

There are certain advantages of handling biochemical analyses and MS in the same laboratory. MS is now well recognized for its utility to identify proteins. Unfortunately, for the most part current collaboration or service arrangements to utilize this technology are neither flexible nor optimized. MS and biochemical analysis are usually separated. This hinders the utilization of an important property of MS, in that it provides a rapid feedback at every step for protein purification. Because protein purification guided by MS is an iterative process, it requires a major commitment of time and energy for the MS laboratory to repeat interactive steps in the process. The ability to screen associated proteins rapidly during purification provides important clues for the subsequent experimental design to improve the purification. The power of this integrated approach is best illustrated in our purification of the mega-BRCA1 complex.

5.1. Purification and Identification of the Mega-BRCA1 Protein Complexes

Classic biochemical purification of protein complexes usually involves multiple column fractionations. As protein complexes are often subject to high salt conditions to enable elution from the column, they are often disrupted. Therefore, usually only the most stable core complexes that survive the fractionation are purified. Although this method can purify core protein complexes to homogeneity, providing information for the composition of complexes, it misses some components that do not survive the fractionation. These components may be the regulatory subunits of the core complexes or protein complexes that reflect complex–complex interactions.

To gain the maximum amount of information about BRCA1-associated proteins, it is necessary to identify as many interacting proteins as possible, whether their interaction is strong or not. When we attempted to purify BRCA1 from fractionated find nuclear extract (NE), we could find only the BRCA1–BARD1 (BRCA1-associated RING domain protein) heterodimer. We needed to use a gentler biochemical purification scheme. We found that one-step immunoprecipitation from unfractionated nuclear extracts made it possible to minimize disruption of protein complexes and the loss of sub-stoichiometrical components. The alternative approach of engineering stable cell lines that express an epitope-tagged BRCA1, which is elegant and allows double-affinity purification to obtain pure complexes that are suitable for subsequent biochemical assays, was discovered not to be applicable for BRCA1. Generating stable lines is difficult when working with toxic genes. Many tumor suppressors kill the cells when overexpressed even at levels two- to five fold higher than the endogenous proteins. Most retrovirus systems are also less effective packaging genes whose gene products are larger than 150 kDa. Large, toxic proteins are better purified using primary antibodies.

We used antibodies to isolate BRCA1-associated proteins by one-step IP. We found a group of proteins that function in DNA damage repair and checkpoints (ATM, MSH6–MSH2, Rad50, MLH1, BLM and RFC-p140, p40, and p37). Strikingly, all these proteins have roles in recognition of abnormal DNA structures or damaged DNA, suggesting that they may serve as sensors for DNA damage signaling. It is noteworthy that all these BRCA1-associated proteins can form stable complexes by themselves, such as the Mre11–Rad50–NBS1 (M-R-N) complex, MSH6–MSH2, MLH1–PMS2, and the five-subunit RFC complex. When they associate with BRCA1, they most likely form regulated complexes, which may have different functions from the core complexes.

Such abundant information has allowed us to propose a genome surveillance hypothesis in which the DNA repair proteins, transducer kinases (ATM–ATR), BRCA1, and other BRCA1-associated proteins may function to monitor the status of DNA damage during the cell cycle to elicit cell cycle checkpoint activation, DNA repair, or apoptosis (Wang et al., 2000). This wealth of information was crucial for generating the hypothesis, as the model needs to reconcile with many observations. This is quite different from information obtained with other approaches (e.g., yeast two-hybrid systems or stringent biochemical purification) in which one or two partners or the core complexes are found. Limited information makes it impossible to formulate a hypothesis that may provide a global view of the problem.

5.2. The BRCA1-associated Genome Surveillance Complex (BASC) Hypothesis

Our central hypothesis is that repair proteins in the BASC may function as lesion-specific DNA damage sensors and form constitutive signaling modules with the transducer kinases ATM or ATR. In response to DNA damage, they signal to BRCA1 and may be other components of BASC by phosphorylation. These phosphorylated proteins are effectors, which further amplify the signal or elicit cell cycle checkpoint activation, DNA repair, or apoptosis. In particular, we hypothesize that the M–R–N complex will function as a sensor for DSB; the MSH2–MSH6 complex will signal mismatched DNA during DNA replication or damaged base by methylating agents; and the BLM protein plays a role in signaling stalled replication forks in response to replication stress.

5.3. The Mre11–Rad50–NBS1 Complex as a Sensor for DSB

The breakthrough for testing the genome surveillance hypothesis came from our observation that the structural maintenance of chromosomes protein 1 (SMC1) that associates with BRCA1 is phosphorylated in response to ionizing radiation (IR) (Yazdi et al., 2002). SMC1 and SMC3 are evolutionarily conserved chromosomal proteins that are components of the cohesin complex, necessary for sister chromatid cohesion. These proteins may also function in DNA repair. Using MS, we identified S957 and S966 of SMC1 are phosphorylated *in vivo* in response to IR by ATM. Phosphorylation of these sites is required for the activation of the S-phase checkpoint. Consistent with the view that the M–R–N complex functions upstream in response to DSB, we found that the NBS1 protein is also required for SMC1 phosphorylation *in vivo*. Thus, SMC1 is a downstream effector in the ATM–NBS1 dependent S-phase checkpoint pathway in response to IR.

Interestingly, we discovered that phosphorylation of NBS1 on S278 and S343 by ATM is also required for the phosphorylation of SMC1. Because NBS1 phosphorylation does not disrupt the M–R–N complex, and NBS1 phosphorylation is not required for the M–R–N complex to bind DSB, it is predicted that the mutation of Ser to Ala should not inactivate the activity of the M–R–N complex as a DSB sensor. Thus, the phosphorylated NBS1 may have an additional function besides serving as a component of the sensor complex. We proposed a model in which the M–R–N complex initially serves as a sensor, leading to the activation of ATM. ATM, in turn, phosphorylates its substrates, including NBS1. Phosphorylation of NBS1 effectively terminates the function of NBS1 as a component of the sensor complex and converts phosphorylated NBS1 into an adaptor by conformational change. The adaptor NBS1 then positions NBS1-binding proteins for phosphorylation by ATM. Within this model, the role of NBS1 in the context of the M–R–N complex can be more accurately described as an adaptor that brings the substrate, SMC1, to ATM. Our duo-sensor/adaptor model imposes specificity on DNA damage response, meaning that the transducers can convey signals only to downstream effectors that bind to the duo-sensor/adapter proteins. This specificity can explain why specific forms of DNA damage elicit specific responses, although they all may work through the same transducer kinase (ATM or ATR). The specificity is imposed by the sensor–adaptor–effector combination. This theme turns out to be also true for the MSH2–MSH6 heterodimer.

Recently, a mechanism of ATM activation was identified as autophosphorylation of Ser1981 that leads to dissociation of the inactive dimer in response to DSB (Bakkenist and Kastan, 2003). The Ser1981 phosphorylation specific antibody provides an important reagent for the examination of ATM activation. Amazingly, it was later found that the Mre11 complex is required for ATM autophosphorylation, providing a more direct evidence for a role of the Mre11 complex as a DSB sensor (Uziel et al., 2003). Because previous work on the dependence of phosphorylation of ATM substrates on the Mre11 complex may also be alternatively explained by a role of the Mre11 complex as an adaptor, this finding is important in further establishing the role of the Mre11 protein as a DSB sensor.

5.4. The MSH2–ATR Signaling Module Responding to DNA Methylation Damage and Mismatch Incorporation

To test the role of MSH2 in the context of BASC, we carried out analysis of the MSH2-associated proteins by IP and MS. We found that the MSH2 protein physically interacts with ATR to form a signaling module that is required for the phosphorylation of SMC1 and Chk1 in response to DNA damage generated by N-methyl-N'-nitro-N-nitrosoguanidine MNNG. MSH2–MSH6 binds the O^6-methyl-G·C generated by MNNG. This MSH2-dependent response is lesion specific as it responds primarily to MNNG, not IR (Wang and Qin, 2003). The MSH2 proteins have been implicated as upstream elements in response to MNNG- and cisplatin-induced damage. Our findings further strengthen this notion, and establish that ATR is the transducer kinase that participates in the MSH2-dependent DNA damage response pathway. The MSH2–ATR signaling module is analogous to the established signaling module of the DSB repair M–R–N complex with ATM, which responds primarily to DSB (Lim et al., 2000; Wu et al., 2000; Zhao et al., 2000). Such an arrangement in which the repair proteins physically associate with transducer kinases constitutively is intriguing. Since the M–R–N complex binds DSB, and MSH2–MSH6 binds the O^6-methyl-G·C generated by MNNG, the physical association of these repair proteins with transducer kinases put them in close proximity for the possibility of direct damage signaling.

Intriguingly, the duo-sensor/adaptor model may also apply in the case of MSH2–MSH6 heterodimer. MSH2 interacts with ATR directly *in vitro*: and they interact independent of MSH6 *in vivo*; the ATR substrates, SMC1 and BRCA1, both can interact with MSH6 directly *in vitro*. Thus, it can be envisioned that the MSH2–MSH6 binds to O^6-methyl-G·C, leads to ATR activation (MSH2–MSH6 functions as a sensor), and at the same time, MSH6 brings in substrates for the activated ATR to phosphorylate (now MSH6 functions as an adaptor).

5.5. The BLM Protein as an Upstream Element in Response to DNA Replication Inhibition

In two recent studies, evidence was presented that the gene product of the Bloom Syndrome Protein (BLM) functions upstream in response to DNA replication inhibition (Franchitto and Pichierri, 2002; Davalos and Campisi, 2003). In one study, it was shown that BLM and ATR in cancer cells are specifically required to properly relocalize the M–R–N complex at sites of replication arrest. In another study, it was shown that BLM protein in a hTERT immortalized primary human fibroblast derived from a Bloom syndrome patient is required to form BRCA1–NBS1 foci in response to replication inhibition. Most interestingly, helicase-defective BLM is equally capable of BRCA1–NBS1 recruitment, suggesting catalytic and structural roles for BLM. Thus, the structure-specific DNA binding activity of the helicase-deficient BLM is important for signaling, suggesting a more general role of the BLM protein beyond its helicase activity. These findings suggest that BLM is an early responder to damaged replication forks.

6. NETWORK ANALYSIS PROTEOMICS

The preceding example illustrates the power of MS when integrated with biological research. Our approach, as described in this chapter, however, has its limitations, too. The combination of affinity purification followed by SDS-PAGE still does not offer a sufficiently high dynamic range and resolving power for mega-complexes such as the BRCA1 complexes. If we had sufficiently high resolving power and dynamic range, we would have identified every subunit of the regulated BRCA1 complexes. This is apparently not the case. For example, we identified only Rad50 (not Mre11 or NBS1), MLH1 (not PMS2), and only three subunits of the five-subunit RFC complex.

The missing components probably do not stain well by Coomassie Brilliant Blue or they are masked by a large excess of nonspecific proteins of similar molecular weight, which overwhelm the MS analysis so that proteins of low abundance are missed.

To dissect a signaling network in which a disease protein functions, we need to identify proteins that interact with the disease protein directly and indirectly. The collection of these proteins comprises the signaling network. We propose a new concept that we have termed "network analysis proteomics." We begin network analysis proteomics by identifying the disease protein complex (the primary complex, for example, BRCA1). The disease protein is now the focus of attention. By defining the primary complex, we can identify several key components within a signaling pathway. We then begin to purify protein complexes (secondary complexes) using all the components of the previously identified primary complex as baits. In this second step, all components of the primary disease protein complexes become the focus of attention. In the case of BRCA1, which is of low abundance, the secondary complexes are smaller, less complicated, and more abundant. As a result, they should be resolved better on SDS-PAGE and more easily identifiable. The purified secondary complexes are most likely core complexes and more abundant regulated complexes. The identification of these complexes allows us to define the individual pathways that converge to the central disease protein signaling network. This concept has been applied in our analysis of BASC, in which we purified the MSH2 complex after we identified MSH2 as a component of BASC. The MSH2 complex does not contain enough BRCA1, but it contains other components (e.g., ATR) that were not identified in the analysis of BASC. Here, the simpler composition of the MSH2 complex allowed better identification of missing components. The relationship of BASC and the MSH2 complex is obvious now that BASC can be viewed as a regulated MSH2 complex, in which BRCA1 joins the core MSH2 complex, perhaps to give the core MSH2 complex a new function.

If necessary, the same approach can be taken to establish the tertiary complexes derived from proteins that are unique to the secondary complexes. This allows dissecting a network layer by layer. When one can no longer find new components in the tertiary complexes, one has saturated the screen of identifying interacting proteins. Thus, the static signaling network in which the disease protein functions has been established. The ability to dissect the network layer by layer and then to reconstitute the network is the essence of our network analysis proteomics. This approach obviously provides much more information than a random identification of binary protein interactions as everything identified is interconnected within a signaling network; this approach is also more focused and economical for establishing modular signaling networks, as there is no need to identify every complex in the proteome.

REFERENCES

Aebersold, R., and Mann, M. (2003). Mass spectrometry-based proteomics. *Nature* 422:198–207.
Alberts, B. (1998). The cell as a collection of protein machines: preparing the next generation of molecular biologists. *Cell* 92:291–294.

Andersen, J.S., Lyon, C.E., Fox, A.H., Leung, A.K., Lam, Y.W., Steen, H., Mann, M., and Lamond, A.I. (2002). Directed proteomic analysis of the human nucleolus. *Curr. Biol.* 12:1–11.

Bakkenist, C.J., and Kastan, M.B. (2003). DNA damage activates ATM through intermolecular autophosphorylation and dimer dissociation. *Nature* 421:499–506.

Bellaoui, M., Chang, M., Ou, J., Xu, H., Boone, C., and Brown, G.W. (2003). Elg1 forms an alternative RFC complex important for DNA replication and genome integrity. *EMBO J.* 22:4304–4313.

Ben Aroya, S., Koren, A., Liefshitz, B., Steinlauf, R., and Kupiec, M. (2003). ELG1, a yeast gene required for genome stability, forms a complex related to replication factor C. *Proc. Natl. Acad. Sci. USA* 100:9906–9911.

Bermudez, V.P., Maniwa, Y., Tappin, I., Ozato, K., Yokomori, K., and Hurwitz, J. (2003). The alternative Ctf18-Dcc1-Ctf8-replication factor C complex required for sister chromatid cohesion loads proliferating cell nuclear antigen onto DNA. *Proc. Natl. Acad. Sci. USA* 100:10237–10242.

Blagoev, B., Kratchmarova, I., Ong, S.E., Nielsen, M., Foster, L.J., and Mann, M. (2003). A proteomics strategy to elucidate functional protein–protein interactions applied to EGF signaling. *Nat. Biotechnol.* 21:315–318.

Cai, Y., Jin, J., Tomomori-Sato, C., Sato, S., Sorokina, I., Parmely, T.J., Conaway, R.C., and Conaway, J.W. (2003). Identification of new subunits of the multiprotein mammalian TRRAP/TIP60-containing histone acetyltransferase complex. *J. Biol. Chem.* 278:42733–42736.

Corbin, R.W., Paliy, O., Yang, F., Shabanowitz, J., Platt, M., Lyons, C.E., Jr., Root, K., McAuliffe, J., Jordan, M.I., Kustu, S., Soupene, E., and Hunt, D.F. (2003). Toward a protein profile of *Escherichia coli*: comparison to its transcription profile. *Proc. Natl. Acad. Sci. USA* 100:9232–9237.

Davalos, A.R., and Campisi, J. (2003). Bloom syndrome cells undergo p53-dependent apoptosis and delayed assembly of BRCA1 and NBS1 repair complexes at stalled replication forks. *J. Cell Biol.* 162:1197–1209.

Fenyo, D., Qin, J., and Chait, B.T. (1998). Protein identification using mass spectrometric information. *Electrophoresis* 19:998–1005.

Franchitto, A., and Pichierri, P. (2002). Bloom's syndrome protein is required for correct relocalization of Rad50/Mre11/NBS1 complex after replication fork arrest. *J. Cell Biol.* 157:19–30.

Gavin, A.C., Bosche, M., Krause, R., Grandi, P., Marzioch, M., Bauer, A., Schultz, J., Rick, J.M., Michon, A.M., Cruciat, C.M., Remor, M., Hofert, C., Schelder, M., Brajenovic, M., Ruffner, H., Merino, A., Klein, K., Hudak, M., Dickson, D., Rudi, T., Gnau, V., Bauch, A., Bastuck, S., Huhse, B., Leutwein, C., Heurtier, M.A., Copley, R.R., Edelmann, A., Querfurth, E., Rybin, V., Drewes, G., Raida, M., Bouwmeester, T., Bork, P., Seraphin, B., Kuster, B., Neubauer, G., and Superti-Furga, G. (2002). Functional organization of the yeast proteome by systematic analysis of protein complexes. *Nature* 415:141–147.

Goldberg, M., Stucki, M., Falck, J., D'Amours, D., Rahman, D., Pappin, D., Bartek, J., and Jackson, S.P. (2003). MDC1 is required for the intra-S-phase DNA damage checkpoint. *Nature* 421: 952–956.

Gygi, S.P., Rist, B., Gerber, S.A., Turecek, F., Gelb, M.H., and Aebersold, R. (1999). Quantitative analysis of complex protein mixtures using isotope-coded affinity tags. *Nat. Biotechnol.* 17:994–999.

Ho, Y., Gruhler, A., Heilbut, A., Bader, G.D., Moore, L., Adams, S.L., Millar, A., Taylor, P., Bennett, K., Boutilier, K., Yang, L., Wolting, C., Donaldson, I., Schandorff, S., Shewnarane, J., Vo, M., Taggart, J., Goudreault, M., Muskat, B., Alfarano, C., Dewar, D., Lin, Z., Michalickova, K., Willems, A.R., Sassi, H., Nielsen, P.A., Rasmussen, K.J., Andersen, J.R., Johansen, L.E., Hansen, L.H., Jespersen, H., Podtelejnikov, A., Nielsen, E., Crawford, J., Poulsen, V., Sorensen, B.D., Matthiesen, J., Hendrickson, R.C., Gleeson, F., Pawson, T., Moran, M.F., Durocher, D., Mann, M., Hogue, C.W., Figeys, D., and Tyers, M. (2002). Systematic identification of protein complexes in *Saccharomyces cerevisiae* by mass spectrometry. *Nature* 415:180–183.

Kanellis, P., Agyei, R., and Durocher, D. (2003). Elg1 forms an alternative PCNA-interacting RFC complex required to maintain genome stability. *Curr. Biol.* 13:1583–1595.

Krutchinsky, A.N., Kalkum, M., and Chait, B.T. (2001). Automatic identification of proteins with a MALDI-quadrupole ion trap mass spectrometer. *Anal. Chem.* 73:5066–5077.

Lim, D.S., Kim, S.T., Xu, B., Maser, R.S., Lin, J., Petrini, J.H., and Kastan, M.B. (2000). ATM phosphorylates p95/nbs1 in an S-phase checkpoint pathway. *Nature* 404:613–617.

Lindsey-Boltz, L.A., Bermudez, V.P., Hurwitz, J., and Sancar, A. (2001). Purification and characterization of human DNA damage checkpoint Rad complexes. *Proc. Natl. Acad. Sci. USA* 98:11236–11241.

Mann, M., Hendrickson, R.C., and Pandey, A. (2001). Analysis of proteins and proteomes by mass spectrometry. *Annu. Rev. Biochem.* 70:437–473.

Mayer, M.L., Gygi, S.P., Aebersold, R., and Hieter, P. (2001). Identification of RFC(Ctf18p, Ctf8p, Dcc1p): an alternative RFC complex required for sister chromatid cohesion in *S. cerevisiae. Mol. Cell* 7:959–970.

Merkle, C.J., Karnitz, L.M., Henry-Sanchez, J.T., and Chen, J. (2003). Cloning and characterization of hCTF18, hCTF8, and hDCC1. Human homologs of a *Saccharomyces cerevisiae* complex involved in sister chromatid cohesion establishment. *J. Biol. Chem.* 278:30051–30056.

Naiki, T., Kondo, T., Nakada, D., Matsumoto, K., and Sugimoto, K. (2001). Chl12 (Ctf18) forms a novel replication factor C-related complex and functions redundantly with Rad24 in the DNA replication checkpoint pathway. *Mol. Cell Biol.* 21:5838–5845.

Paull, T.T., and Gellert, M. (1999). Nbs1 potentiates ATP-driven DNA unwinding and endonuclease cleavage by the Mre11/Rad50 complex. *Genes Dev.* 13:1276–1288.

Ranish, J.A., Yi, E.C., Leslie, D.M., Purvine, S.O., Goodlett, D.R., Eng, J., and Aebersold, R. (2003). The study of macromolecular complexes by quantitative proteomics. *Nat. Genet.* 33:349–355.

Rigaut, G., Shevchenko, A., Rutz, B., Wilm, M., Mann, M., and Seraphin, B. (1999). A generic protein purification method for protein complex characterization and proteome exploration. *Nat. Biotechnol.* 17:1030–1032.

Rountree, M.R., Bachman, K.E., and Baylin, S.B. (2000). DNMT1 binds HDAC2 and a new co-repressor, DMAP1, to form a complex at replication foci. *Nat. Genet.* 25:269–277.

Rout, M.P., Aitchison, J.D., Suprapto, A., Hjertaas, K., Zhao, Y., and Chait, B.T. (2000). The yeast nuclear pore complex: composition, architecture, and transport mechanism. *J. Cell Biol.* 148:635–651.

Sanders, S.L., Jennings, J., Canutescu, A., Link, A.J., and Weil, P.A. (2002). Proteomics of the eukaryotic transcription machinery: identification of proteins associated with components of yeast TFIID by multidimensional mass spectrometry. *Mol. Cell Biol.* 22:4723–4738.

Shen, X., Ranallo, R., Choi, E., and Wu, C. (2003). Involvement of actin-related proteins in ATP-dependent chromatin remodeling. *Mol. Cell* 12:147–155.

Trujillo, K.M., Yuan, S.S., Lee, E.Y., and Sung, P. (1998). Nuclease activities in a complex of human recombination and DNA repair factors Rad50, Mre11, and p95. *J. Biol. Chem.* 273:21447–21450.

Tsurimoto, T., Fairman, M.P., and Stillman, B. (1989). Simian virus 40 DNA replication *in vitro*: identification of multiple stages of initiation. *Mol. Cell Biol.* 9:3839–3849.

Uziel, T., Lerenthal, Y., Moyal, L., Andegeko, Y., Mittelman, L., and Shiloh, Y. (2003). Requirement of the MRN complex for ATM activation by DNA damage. *EMBO J.* 22:5612–5621.

Wang, Y., and Qin, J. (2003). MSH2 and ATR form a signaling module and regulate two branches of the damage response to DNA methylation. *Proc. Natl. Acad. Sci. USA.* 100:15387–15392.

Wang, Y., Cortez, D., Yazdi, P., Neff, N., Elledge, S.J., and Qin, J. (2000). BASC, a super complex of BRCA1-associated proteins involved in the recognition and repair of aberrant DNA structures. *Genes Dev.* 14:927–939.

Washburn, M.P., Wolters, D., and Yates, J.R., III (2001). Large-scale analysis of the yeast proteome by multidimensional protein identification technology. *Nat. Biotechnol* 19:242–247.

Wolters, D.A., Washburn, M.P., and Yates, J.R., III (2001). An automated multidimensional protein identification technology for shotgun proteomics. *Anal. Chem.* 73:5683–5690.

Wu, X., Ranganathan, V., Weisman, D.S., Heine, W.F., Ciccone, D.N., O'Neill, T.B., Crick, K.E., Pierce, K.A., Lane, W.S., Rathbun, G., Livingston, D.M., and Weaver, D.T. (2000). ATM phosphorylation of Nijmegen breakage syndrome protein is required in a DNA damage response. *Nature* 405:477–482.

Xin, H., Yoon, H., Singh, S.B., Wong, J., and Qin, J. (2004). Components of a pathway maintaining histone modification and HP1 binding at the pericentric heterochromatin in mammalian cells. *J. Biol. Chem.* 279:9539–9546.

Yates, J.R., III (1998). Database searching using mass spectrometry data. *Electrophoresis* 19:893–900.

Yazdi, P.T., Wang, Y., Zhao, S., Patel, N., Lee, E.Y., and Qin, J. (2002). SMC1 is a downstream effector in the ATM/NBS1 branch of the human S-phase checkpoint. *Genes Dev.* 16: 571–582.

Zhao, S., Weng, Y.C., Yuan, S.S., Lin, Y.T., Hsu, H.C., Lin, S.C., Gerbino, E., Song, M.H., Zdzienicka, M.Z., Gatti, R.A., Shay, J.W., Ziv, Y., Shiloh, Y., and Lee, E.Y. (2000). Functional link between ataxia-telangiectasia and Nijmegen breakage syndrome gene products. *Nature* 405:473–477.

Zhu, X.D., Kuster, B., Mann, M., Petrini, J.H., and de Lange, T. (2000) Cell-cycle-regulated association of Rad50/Mre11/NBS1 with TRF2 and human telomeres. *Nat. Genet.* 25:347–352.

4

Molecular Recognition in the Immune System

Eric J. Sundberg and Roy A. Mariuzza

ABSTRACT

Antibody and T-cell receptor (TCR) molecules may be regarded as products of a protein engineering system for the generation of a virtually unlimited repertoire of complementary molecular surfaces. This extreme structural heterogeneity is required for recognition of the infinite array of antigenic determinants presented in nature. Here we broadly discuss the structures of antibodies and TCRs as well as their specific recognition of antigen, the binding energetics of these interactions, the structural basis of the antibody maturation and TCR selection processes, limitations to affinity and specificity for antigens, and the role of conformational flexibility in antigen recognition. A final section highlights research results from the burgeoning field of natural killer cell receptor biology.

1. THE IMMUNOGLOBULIN FOLD: A STRUCTURAL FRAMEWORK FOR MOLECULAR RECOGNITION

1.1. Structural Overview of the Immunoglobulin (Ig) Domain

The basic building blocks of both antibodies and TCRs are small protein domains, each composed of two antiparallel β-sheets and belonging to the "immunoglobulin

ERIC J. SUNDBERG • Center for Advanced Research in Biotechnology, W.M. Keck Laboratory for Structural Biology, University of Maryland Biotechnology Institute, 9600 Gudelsky Drive, Rockville, MD 20850, and Boston Biomedical Research Institute, 64 Grove Street, Watertown, MA 02472, USA. ROY A. MARIUZZA • Center for Advanced Research in Biotechnology, W.M. Keck Laboratory for Structural Biology, University of Maryland Biotechnology Institute, 9600 Gudelsky Drive, Rockville, MD 20850, USA.

Proteomics and Protein–Protein Interactions: Biology, Chemistry, Bioinformatics, and Drug Design, edited by Waksman. Springer, New York, 2005.

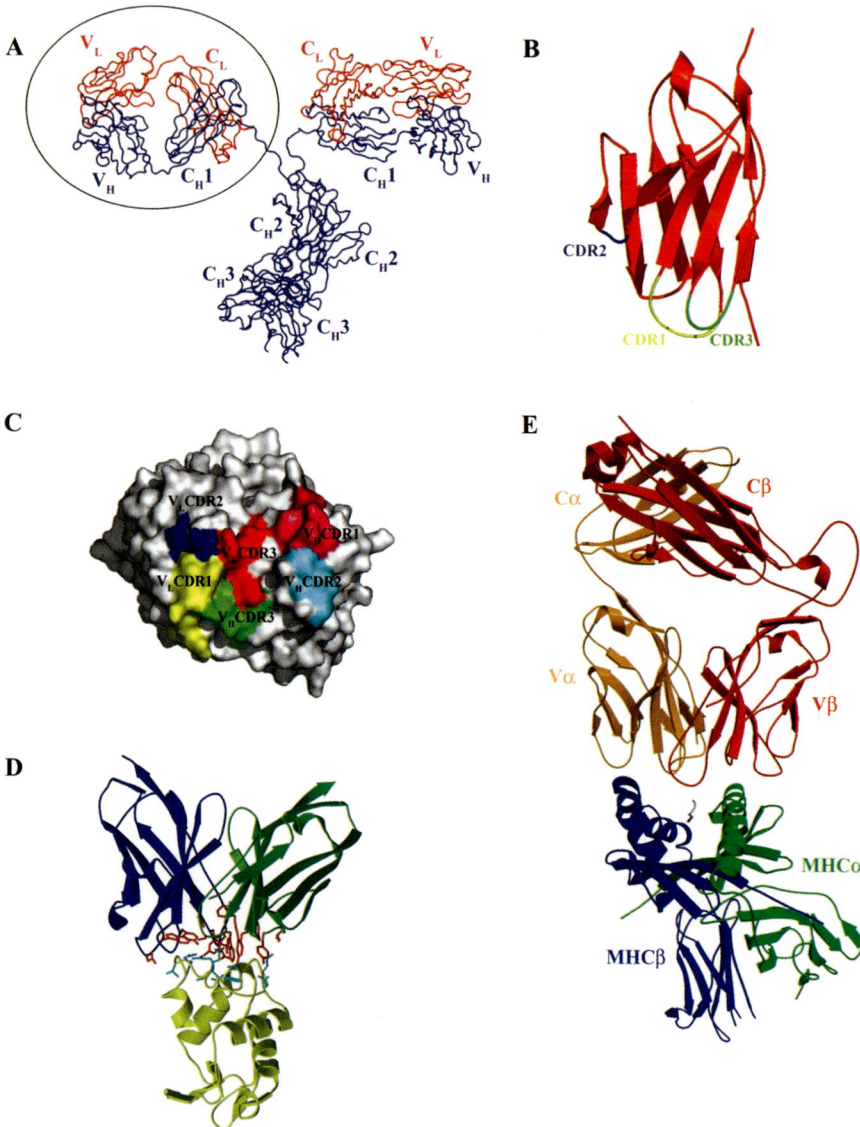

Figure 4.1. Structural overview of antibodies and T cell receptors. (**A**) Structure of the intact murine IgG2a monoclonal antibody, Mab231, including two light chains each composed of a variable (V_L) and a constant (C_L) immunoglobulin (Ig) domain (*red*) and two heavy chains each composed of a variable (V_H) and three constant (C_H1, C_H2, and C_H3) domains (*blue*) (Harris et al., 1998). The two hinge regions are highlighted within the dotted oval, revealing the source of structural asymmetry within the intact antibody. (**B**) Ribbon diagram of a single Ig domain, V_L, of Mab231 highlighting its antiparallel β-sheet secondary structure. The N- and C-termini are marked as well as the complementarity determining region loops, CDR1 (*yellow*), CDR2 (*blue*), and CDR3 (*green*). This structure is representative of all Ig domains, including those of TCRs.

fold" superfamily (Amzel and Poljak, 1979), although the arrangement of these immunoglobulin (Ig) domains differs between these two types of molecules. Figure 4.1 provides an overview of the structural characteristics of Ig domains, how they are assembled to form functional antibodies and TCR and how they generally recognize antigenic molecules. Antibody molecules (Fig. 4.1A–D) are composed of two identical polypeptide chains of approximately 500 amino acids (the heavy or H chains) covalently linked through disulfide bridges to two identical polypeptide chains of roughly 250 residues (the light or L chains). Based on amino acid sequence comparisons, the H and L chains may be divided into N-terminal variable (V) and C-terminal constant (C) portions. Each H chain contains four or five Ig domains (V_H, C_H1, C_H2, $C_H3 \pm C_H4$ depending on the antibody isotype) while each L chain consists of two such domains (V_L, C_L). The V_L and C_L domains are disulfide-linked with the V_H and C_H1 domains, respectively, to form the Fab region of the antibody which is linked through a hinge region to the Fc domain, formed by noncovalent association of the C_H2 to 4 domains from both chains. TCRs are likewise composed of multiple Ig domains and are equivalent to a single Fab domain containing two chains, each comprised of a variable and a constant domain (Fig. 4.1E). Although $\gamma\delta$TCR heterodimers have received varying levels of attention in recent years and one crystal structure of an unliganded $\gamma\delta$TCR is available (Allison et al., 2001), we restrict our discussion to the more well understood $\alpha\beta$TCR heterodimers that recognize antigenic peptides in the context of class I and II major histocompatibility complex (MHC) molecules.

The variable domains of antibodies and TCRs (V_H and V_L, $V\alpha$ and $V\beta$, respectively) each contain three segments, or loops, which connect the β-strands and are highly variable in length and sequence (Wu and Kabat, 1970). These so-called complementarity-determining regions (CDRs) lie in close spatial proximity on the surface of the V domains and determine the conformation of the combining site (Fig. 4.1B–D). In this way, the CDRs confer specific binding activity to apical regions of the Ig domain. The central paradigm of antigen recognition is that the three-dimensional structure formed by the six CDRs recognizes and binds a complementary surface (epitope) on the antigen.

Although CDR loops are hypervariable they adopt a limited number canonical structures in both antibodies (Al-Lazikani et al., 1997) and TCRs (Al-Lazikani et al., 2000). Usage of the six CDR loops that confer antigen binding specificity

---

Figure 4.1. (*Continued*) (**C**) Molecular surface of the antibody combining site of Mab231 formed by the intersection of the apical regions of V_L and V_H. The CDR loops provide a nearly contiguous surface for antigen recognition. Colors are as follows: V_LCDR1 (*yellow*); V_LCDR2 (*blue*); V_LCDR3 (*green*); V_HCDR1 (*magenta*); V_HCDR2 (*cyan*); V_HCDR3 (*red*). This structure is representative of all Ig domains, including those of TCRs. (**D**) Ribbon diagram of the FvD1.3–hen egg lysozyme (HEL) complex. Colors are as follows: HEL (*yellow*), D1.3 V_L domain (*green*), and D1.3 V_H domain (*blue*). Residues of HEL and D1.3 involved in interactions in the antigen–antibody interface are cyan and red, respectively. (**E**) Structure of the extracellular portion of the HA1.7 TCR, including α (*orange*) and β (*red*) chains each composed of a variable ($V\alpha$/$V\beta$) and constant domain ($C\alpha$/$C\beta$), in complex with the MHC molecule HLA-DR1 (α subunit, *green*; β-subunit, *blue*) displaying the hemaglutinnin 306–318 peptide (*gray*) (Hennecke et al., 2000).

varies, especially for antibodies. Antibodies to smaller antigens, such as haptens and peptides, commonly do not utilize all six CDRs (Chitarra et al., 1993; Wilson and Stanfield, 1993), while antiprotein antibodies generally do. Camelid antibodies that have no light chains (Hamers-Casterman et al., 1993) can nonetheless bind protein antigens with nanomolar affinities using as few as two CDR loops (Decanniere et al., 1999). Indeed, some of the contacts to various mammalian antibody CDR loops by protein antigens, while confirmed as structurally belonging to the molecular interface, are energetically meaningless (see below for a discussion of the binding energetics of antigen recognition). In addition, both polyclonal and monoclonal antibodies rasied against small (8- to 15-mer) peptides often bind to both the peptide and to the whole correlate protein, sometimes with higher affinity than antibodies raised directly against the latter (Chersi et al., 2002; Metaxas et al., 2002; Hewer and Meyer, 2003). Framework regions are commonly invoked in antigen recognition to varying degrees, and can comprise up to 15% of the buried surface area of an antibody-antigen complex (Wilson and Stanfield, 1994). The V_HCDRs, and V_HCDR3 in particular, generally make more extensive contacts than V_LCDRs, and the geometrical center of the interface tends to lie near V_HCDR3. A strong correlation exists between residues that do not form contacts with antigen and those residues that are important in defining the canonical backbone structures of the CDR loops (Chothia et al., 1989). These residues tend to pack internally and are therefore less exposed on the antibody combining site surface.

1.2. Structural Characteristics of Antigen-Associated Molecular Interfaces

Antibody–antigen complexes exhibit a high degree of both shape and chemical complementarity at their interacting surfaces (Conte et al., 1999). The application of an algorithm to quantitate shape complementarity in protein–protein interfaces (Lawrence and Colman, 1993) to oligomeric proteins or protease–protease inhibitor complexes gives shape correlation (S_c) values ranging from 0.70 to 0.76 on a scale of 0 (topologically uncorrelated) to 1 (perfect geometrical fit). For antigen–antibody interfaces, S_c values of 0.64 to 0.74 are obtained, indicating poorer average shape correlation—albeit a better topological correlation than for other classes of nonobligatory heterocomplexes (Jones and Thornton, 1996). The combined solvent-accessible surfaces buried in antiprotein antibody–antigen complexes range from approximately 1400 Å^2 to 2300 Å^2, with roughly equal contributions from antigen and antibody, while smaller antigens, such as haptens and peptides, generally bury less overall surface area when bound to antibody. The surface topography of the antigen-contacting surface, as well as other general structural features, of antibodies can vary significantly according to antigen size (MacCallum et al., 1996). While the percentage of the antigen surface buried in the interface with antibody is always high and their surfaces complementary, the antibody contact surface becomes more concave as the antigen becomes smaller. Thus, although the combining sites of antibodies that recognize large protein antigens are generally planar, and are often more planar than a number of other types of protein–protein interfaces (Jones and Thornton,

1996), antibodies that recognize medium-sized antigens, such as peptides, DNA, and carbohydrates, often have a grooved antigen-contacting surface, while even smaller antigens (haptens) are recognized by antibodies with distinct cavities (Webster et al., 1994). A common feature of antipeptide antibody–antigen interactions is a β-turn motif of the peptide buried deeply into the combining site (Stanfield et al., 1990; Garcia et al., 1992; Rini et al., 1992). The amount of surface area on the antibody molecule buried by antigen decreases with antigen size, as less of the antibody surface is utilized to envelop the smaller antigens. Large antigens often contact antibody residues at the edge of the combining site and interact with the more apical portions of the CDR loops, while the interactions of smaller antigens are more restricted to the central portion of the antibody combining site (MacCallum et al., 1996).

Relative to those bound by antibodies, the antigens recognized by TCRs are much more homogeneous, consisting of extended peptide chains of generally a dozen or less amino acid residues within the context of conserved MHC molecules (pMHC). TCRs thus have a dual recognition role, with required specificity for both self and foreign molecular surfaces in the forms of MHC and antigenic peptide molecules, respectively (Fig 4.1E). TCRs recognize pMHC with a relatively conserved orientation (diagonal to the peptide main chain axis $\pm 35°$) with CDR3 loops primarily responsible for molecular readout of the antigenic peptide and the CDR1 and 2 loops interacting with regions of the MHC molecule surrounding the peptide binding groove (Rudolph and Wilson, 2002). Distribution of shape complementarity in the TCR–pMHC interface is broader than in antibody–antigen complexes ranging in S_c values from approximately 0.4 to 0.7. These generally lower values for the former might be expected of macromolecular complexes with substantially lower affinities, although there is no correlation between shape complementarity and affinity within the set of TCR–pMHC complexes for which crystal structures exist (Ding et al., 1998, 1999; Garcia et al., 1998; Degano et al., 2000; Reiser et al., 2000, 2002, 2003; Luz et al., 2002; Buslepp et al., 2003; Stewart-Jones et al., 2003) Buried surface areas in TCR–pMHC complexes, from approximately 1250 Å^2 to 1900 Å^2, are also generally lower than for interfaces formed between antibodies and protein antigens.

1.3. Roles for Interfacial Water Molecules in Antigen Recognition

The structures of both antibody–antigen and TCR–pMHC complexes illustrate the importance of bound water molecules in mediating these interactions. Indeed, with only one exception (Muller et al., 1998), water molecules have been localized in the interfaces of each of the antigen–antibody complexes whose crystal structures have been determined at sufficiently high resolution (<2.5 Å) to allow the identification of ordered waters with a reasonable degree of accuracy (Bhat et al., 1994; Fields et al., 1995; Mylvaganam et al., 1998; Kondo et al., 1999; Li et al., 2000; Faelber et al., 2001). In detailed structural analyses of the FvD1.3–hen egg lysozyme (HEL) complex (Bhat et al., 1994; Braden et al., 1995, 1998) it has been shown that many water molecules from the free antibody and antigen structures are positionally conserved in the complex; that there is a recruitment of water molecules from the bulk solvent to the complex interface as demonstrated by a net gain of water molecules in the complex structures relative to the individual component structures; and that water molecules in the interface, regardless of their positional origin, increase the shape and chemical complementarity of the interacting surfaces.

The paucity of high-resolution TCR–pMHC crystal structures has limited the assessment of the role of water molecules in these complexes. One recent structure of the immunodominant JM22 TCR in complex with a virus matrix protein in the context of the class I MHC HLA-A2 resolved at 1.4 Å resolution, however, has revealed an intricate network of water molecules between the TCR and pMHC molecular surfaces. Other moderate resolution ($\sim$2.5 Å) TCR–pMHC crystal structures (Reiser et al., 2000; Hennecke and Wiley, 2002; Luz et al., 2002) have revealed from 5 to 39 ordered waters positioned within the binding interface with from 6 to 12 of these waters acting as direct intermolecular bridges between the TCR and pMHC components of the complex. These waters appear to be dependent on the particular TCR and pMHC sequences, as none of them are conserved amongst these structures.

2. BINDING ENERGETICS OF ANTIGEN RECOGNITION

2.1. Energetic Characteristics of Antigen-Associated Molecular Interfaces

There exists a functional affinity window for antigen recognition by both antibodies and TCR molecules. Antibodies undergo affinity maturation on encountering their specific antigens (addressed later in this chapter); here the binding properties of fully matured antibodies are discussed. TCRs, conversely, are selected in the thymus prior to exposure to antigen, and thus their affinities for pMHC complexes remain constant. Their high density in the periphery (T cells comprise approximately 10^{11} of the 10^{14} total cells in the human body), cross-reactivity, and clonal expansion of antigen-specific T cells ensure an efficient immune response.

Most mature antibodies have affinities for their specific antigens in the range of 10^7 to 10^8 M^{-1}, although many antibodies that recognize carbohydrates and bacterial polysaccharides fail to reach affinity levels of 10^6 M^{-1}. It has been proposed (Foote and Eisen, 1995) that, because of diffusion rates and the residence time required for antibody internalization controlling on- and off-rates, there exists an affinity ceiling for antibody–antigen interactions of approximately 10^{10} M^{-1}. Antibodies with antigen affinities above this threshold, presumably, would not be further advantaged over their lower affinity counterparts in the antibody selection process in vivo. The existence of this affinity ceiling has been demonstrated for antigen-specific B-cell transfectants, and more importantly, an affinity window for effective B-cell response has been revealed for which a minimum affinity of 10^6 M^{-1} and half-life of 1 s were required for detectable B- cell triggering that reached a plateau for affinities beyond 10^{10} M^{-1} (Batista and Neuberger, 1998). Not surprisingly, when primary response antibodies exhibit affinities for their specific antigens approaching this affinity ceiling, they neither require nor undergo further affinity maturation (Roost et al., 1995). Throughout the effective affinity window, the efficiency of antibody-mediated presentation of antigens to T cells is controlled by the off-rate of the antibody–antigen interaction with slower off-rates correlated to increased signaling (Guermonprez et al., 1998). Beyond this quantitative correlation between affinity and response, there exists qualitative variability in the B- cell response in which some signaling responses are significantly affinity-dependent while others are not (Kouskoff et al., 1998). This effective affinity window, however, appears to shift to a range of lower affinities, with an affinity ceiling of approximately 10^6 M^{-1}, when the antigen is in particulate form, presumably owing to avidity effects. Conversely, the range of the affinity window for extraction of antigen from a noninternalizable surface remains quite broad with an affinity ceiling similar to that of soluble antigens (Batista and Neuberger, 2000). Antigens in these nonsoluble forms are thought to more closely mimic the properties of antigens in vivo.

As the overall affinity of antibody–antigen interactions can vary by several orders of magnitude, so too can the kinetics of these interactions. In a number of kinetic analyses of antiprotein antibodies (England et al., 1997; Xavier et al., 1999; Rajpal and Kirsch, 2000; Gerstner et al., 2002), both association and dissociation rates vary by greater than 2 log-fold. Thermodynamically, the formation of many antibody–antigen complexes reflects an enthalpically driven process with some compensating negative entropy component, alluding to an important role for water. In fact, a strong correlation between decreases in water activity and association constants in the D1.3–HEL complex has been observed by calorimetric binding analyses performed in the presence of cosolutes with polarities lower than that of water (Goldbaum et al., 1996). Although other antibody–protein antigen (Kelley et al., 1992) and antibody–carbohydrate antigen (Sigurskjold et al., 1991) interactions also appear enthalpically driven, this may not be the general rule for antibody–antigen associations owing to the limited number of such systems whose thermodynamics have been rigorously determined. In accordance with the significance of water activity on antigen recognition, antibodies binding to both protein and hapten antigens have exhibited a thermodynamic dependence on the solvent pH and ionic strength (Omelyanenko et al., 1993; Gibas et al., 1997; Xavier et al., 1999; De Genst et al., 2002).

TCR–pMHC interactions generally fall within a lower and broader affinity range, 10^3 to 10^7 M^{-1}, than do antibody–antigen interactions, and exhibit slow association and fast dissociation rates (Davis et al., 1998) that correspond to half-lives of 70 to 0.1 s (Margulies, 1997). There seem to exist roughly as many TCR–pMHC systems that are exceptions to any observed correlation between complex half-lives and levels of T-cell activation (Garcia et al., 1997; Kersh et al., 1998; Boniface et al., 1999; Baker et al., 2000) as there are that adhere to it (Alam et al., 1996, 1999; Lyons et al., 1996; Kersh et al., 1998; Ding et al.,1999). This is likely due, at least in part, to the complexity of the immunological synapse (Bromley et al., 2001), the intercellular environment in which these molecules interact that involves the junction of two cellular membranes and numerous costimulatory molecules. The role of water in TCR–pMHC interactions appears less clear than for antibody–antigen interactions. Binding analysis in the presence of water-disrupting solvents has revealed substantial discrepancies in the importance of water molecules in pMHC engagement by various

TCR molecules (Anikeeva et al., 2003). The release of up to 15 ordered waters bound to the pMHC on TCR engagement has been observed in one system, presumably contributing favorably to the binding energy via an increase in the entropic term (Luz et al., 2002).

Although assessing the binding energetics of numerous associations relevant to antigen recognition can reveal some of the generalities of the binding phenomena for this particular class of interactions, a more complete description of the antigen recognition energetic landscape is revealed through perturbation of these interactions and quantification of the resulting structural, energetic, and functional outcomes. These types of experiments can reveal a more detailed understanding (in the atomic sense) of antigen recognition in particular and the mosaic nature of the binding energetics of macromolecular interactions in general.

2.2. Mutagenesis as a Tool for Better Understanding Binding Energetics

Alanine substitutions into the combining site of the D1.3 antibody have revealed that while residues from each of the six CDRs participate in the molecular interface, only three residues in V_LCDR1 and V_HCDR3 (V_HW92, V_HD100, and V_HY101) make significant energetic contributions to the binding of its cognate antigen, HEL (Dall'Acqua et al., 1996). Conversely, residues located in each of its CDR loops contribute significantly to binding to its anti-idiotypic antibody, E5.2, with the energetically most important residues coming from V_HCDR2 (V_HW52 and V_HD54) and V_HCDR3 (V_HE98, V_HD100, and V_HY101) (Dall'Acqua et al., 1996). Both energetically important and insignificant residues tend to be juxtaposed in the D1.3–HEL and D1.3–E5.2 interfaces (Dall'Acqua et al., 1996, 1998; Goldman et al., 1997), a complementarity of functional epitopes that has been observed in other protein–protein interactions (Clackson and Wells, 1995). The functional surfaces of D1.3 involved in binding HEL and E5.2 can be mapped onto the three-dimensional structures of the complexes (Bhat et al., 1994; Fields et al., 1995; Braden et al., 1996a) (Fig. 4.2A,B). With the exception of V_LW92, which lies at the periphery, the residues of D1.3 most important for binding HEL (V_HY101, V_HD100, V_LY32, and V_HE98) are located in a contiguous patch at the center of the combining site. Residues at the periphery make only a minor contributions to the binding energy. A similar pattern is observed for the D1.3–E5.2 complex, with the most important residues (V_LY32, V_HW52, V_HD54, V_HE98, V_HD100, and V_HY101) forming a central band of key contacts. Only two D1.3 substitutions, V_HD100A and V_HY101A, significantly affect the binding to both HEL and E5.2. Thus, a single set of antibody contact residues on D1.3 can bind two antigens (HEL and E5.2) in energetically distinct ways. Both polar (e.g., D1.3 residues V_HD54, V_HE98 and V_HD100) and nonpolar residues (e.g., D1.3 residues V_LW92 and V_HW52) play a prominent role in stabilization of the D1.3–HEL and D1.3–E5.2 complexes and reveal no clear segregation of polar residues at the periphery of the interface and of nonpolar amino acids at the core.

The interaction between the 2C TCR and the QL9 peptide in complex with the MHC class I molecule H2–L^d has been analyzed by alanine scanning mutagenesis

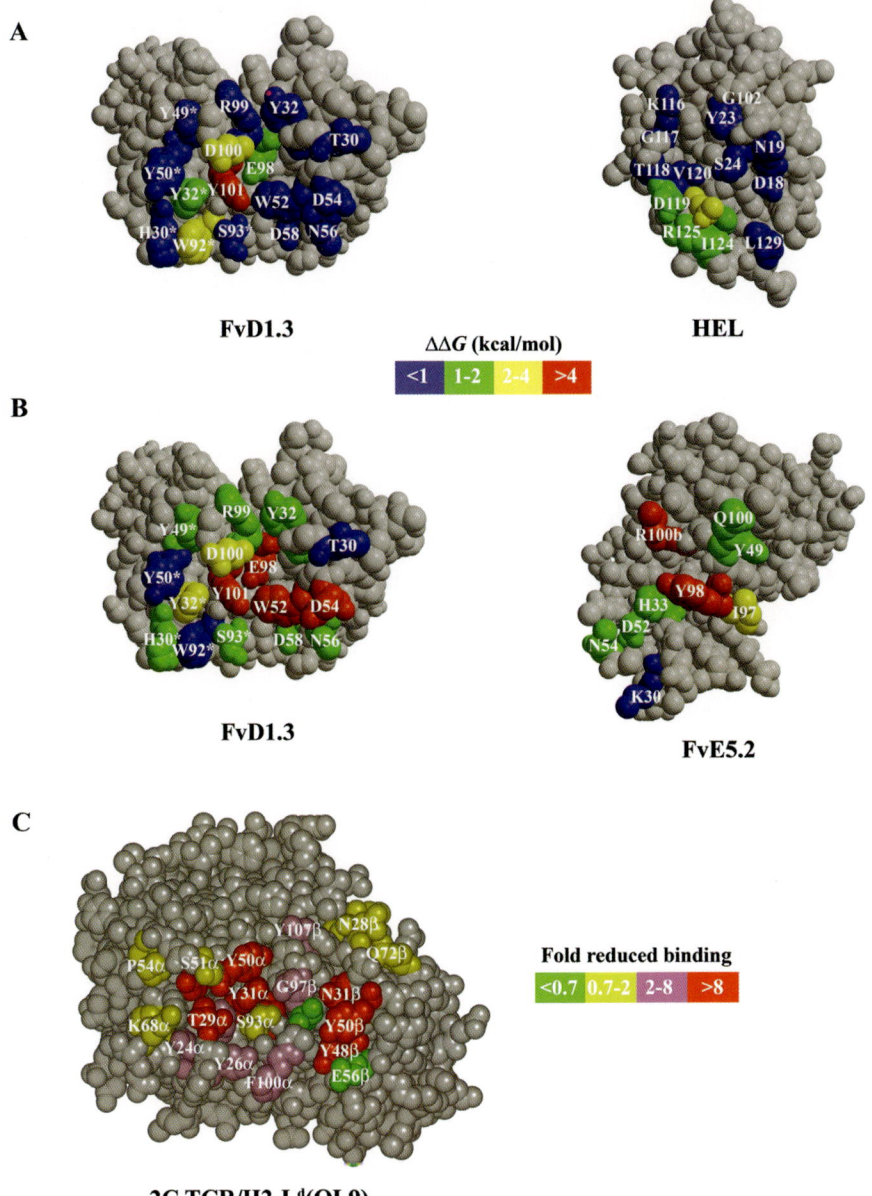

Figure 4.2. Energetic maps of antigen–antibody interfaces. (**A**) Space-filling model of the surface of D1.3 (*left*) in contact with HEL and of the surface of HEL (*right*) in contact with D1.3. The two proteins are oriented such that they may be docked by folding the page along a vertical axis between the components. Residues are color coded according to the loss of binding free energy upon alanine substitution: *red*, >4 kcal/mol; *yellow*, 2–4 kcal/mol; *green*, 1–2 kcal/mol; *blue*, <1 kcal/mol. V_L and V_H residues are labeled in *white* and V_L residues are denoted by an asterisk (*). (**B**) Model of the surface of D1.3 (*left*) in contact with E5.2 and of the surface of E5.2 (*right*) in contact with D1.3. Residues are colored and labeled as in (**A**). (**C**) Model of the surface of the 2C TCR that forms an interface with H2-L^d(QL9) pMHC complex color-coded according to changes in binding energetics derived by alanine scanning mutagenesis (Manning et al., 1998).

(Manning et al., 1998). Nearly all of the residues in each of the CDR1–3 and HV4 loops of the 2C TCR were mutated to alanine and assessed for binding to the QL9– L^d complex by quantitative competition cellular assays and confirmed by surface plasmon resonance analysis. Although some residues from each of the CDR loops were shown to be important for pMHC binding, the most significant binding eneries were attributable to residues residing in the CDR1 and CDR2 loops of both $V\alpha$ and $V\beta$ (Fig. 4.2C). The collective binding energy contributed by the CDR1 and CDR2 loops exceeded that of the CDR3 loops. As the former are generally situated over the MHC surface and the latter over the antigenic peptide (Rudolph and Wilson, 2002), this suggests that positive and negative selection of TCR in the thymus, skewing of TCR toward class I and II MHC, and proper orientation of the TCR on the MHC molecule may all be of paramount importance to the penultimate antigen recognition binding event. Several hypervariable residues that are predicted by homology modeling to not contact pMHC in this complex were nevertheless observed to be energetic hot spots, presumably owing to modulation of flexibility within the CDR loops.

Because of unpredictable disruptions of molecular interactions outside of the interaction of interest, the strength of an interaction between two amino acid residues in a protein or protein–protein complex cannot necessarily be measured by simply mutating one of them (Ackers and Smith, 1985; Fersht, 1988), as in alanine scanning mutagenesis.

A more sophisticated approach to dissecting the energetics of pairwise interactions makes use of double mutant cycles (Ackers and Smith, 1985; Serrano et al., 1990), as have been constructed for amino acid pairs in the D1.3–HEL (Dall'Acqua et al., 1998) and D1.3–E5.2 (Goldman et al., 1997) interfaces in order to measure interaction energies ($\Delta\Delta G_{int}$) for interacting, proximal and distant side chains, as judged from the crystal structures of the complexes (Bhat et al., 1994; Fields et al., 1995; Braden et al., 1996a). In the D1.3–HEL complex, only 3 of the 10 residue pairs in direct contact in the crystal structure exhibited significant coupling energies. Conversely, the broad distribution of energetically important residues in the D1.3–E5.2 interface, as revealed by alaninescanning mutagenesis (Dall'Acqua et al., 1996; Goldman et al., 1997), is mirrored in the results of double mutant cycle analysis of this interface (Goldman et al., 1997). All residue pairs tested exhibited significant coupling energies, including both electrostatic and hydrophobic atomic interactions. Even proximal and distant residue pairs in the D1.3–E5.2 complex exhibited coupling energies significantly greater than experimental error. Small magnitude energetic coupling between amino acid residues separated by large distances has likewise been observed in protein folding (Green and Shortle, 1993; LiCata and Ackers, 1995) and protein–protein interaction (Schreiber and Fersht, 1995) systems. One possible explanation for this phenomenon is that the mutations may introduce solvent rearrangements in the D1.3–E5.2 interface, such as described for complexes between mutants of D1.3 and HEL (Fields et al., 1995; Braden et al., 1996a; Sundberg et al., 2000) and that these localized molecular changes may result in global perturbations in electrostatic fields or vibrational modes within the interface. It is clear from double mutant cycle analysis of the D1.3–HEL and D1.3–E5.2 complexes that neither direct atomic contacts within an antibody–antigen interface necessarily imply energetically productive interactions nor do energetically nonproductive interactions always arise from residues separated by some distance within such an interface.

2.3. Tolerance to Mutations in Antigen-Specific Molecular Interfaces

Mutational analysis of the D1.3–HEL interface has demonstrated that it is remarkably tolerant to mutations that, on the basis of the three-dimensional structure of the wild-type complex, might be expected to have pronounced effects on affinity. For example, truncation of HEL residue Asp18 to alanine should result in the loss of a direct hydrogen bond and seven van der Waals contacts to the side chain of D1.3 V_L Tyr50. Nevertheless, the affinity of HEL D18A for D1.3 ($4.5 \times 10^7 M^{-1}$) is nearly identical to that of the wild type ($8.0 \times 10^7 M^{-1}$), corresponding to a $\Delta\Delta G_{int}$ of only 0.3 kcal/mol. The crystal structure of the FvD1.3–HEL D18A complex at 1.5 Å resolution (Dall'Acqua et al., 1998) reveals that the loss of complementarity in the D1.3–HEL interface resulting from the mutation is compensated by the stable inclusion of additional water molecules and by local rearrangements in solvent structure. Solvent rearrangements, including the incorporation of additional interface waters, have also been observed in structural studies of other site-directed mutants of FvD1.3 in complex with HEL, including V_L Y50S, V_H Y32A, and V_L W92D, but with varying effects on affinity compensation (Ysern et al., 1994; Fields et al., 1996). In other cases, seemingly conservative mutations, such as the substitution of a lysine residue for an arginine in the HyHEL–5/HEL interface concomitant with replacement of the lost guanidinium group with water, have been found to greatly affect antigen–antibody binding (Chacko et al., 1995).

Tolerance to mutation is also apparent in TCR–pMHC interactions. In the A6/HLA–A2/TaxP6A complex resulting from mutation of residue P6 to alanine in the Tax peptide, crystal structures of the wild-type and mutant complexes were nearly indistinguishable (Ding et al., 1998, 1999), even though the former peptide acts as an agonist and the latter as an antagonist. In the mutant complex, T-cell stimulation is lost and the affinity of the mutant TCR–pMHC complex is 100-fold lower than that of the wild-type complex. Altered peptide ligands (APLs), synthetic variants of MHC-presented peptides, were used to repair this packing defect (Baker et al., 2000), and the stepwise filling of the cavity resulted in recovered T-cell stimulatory activity but without correlation to the half-lives of the resulting complexes. In the 2C TCR/H–$2K^b$(dEV8) system, APLs exhibiting relatively small differences in affinity can fulfill a wide range of T-cell activation functions, from antagonism to superagonism. The APL SIYR in the context of H–$2K^b$ acts as a superagonist. When its crystal structure in complex with the 2C TCR is compared to that of the wild-type peptide dEV8, a weak agonist, small conformational changes close to the center of the TCR–pMHC interface were observed (Degano et al., 2000). Likewise, the seemingly inconsequential structural difference of a single methyl group between wild-type and mutant the melanoma peptide antigens TPI_{23-37} displayed by the MHC class II molecule HLA-DR1 (Sundberg et al., 2002b) belie the 5-log difference in stimulation of the same tumor infiltrating T-cell line (Pieper et al., 1999). Wide-ranging functional outcomes resulting from seemingly minor structural differences such as these imply that TCR–pMHC interfaces are highly sensitive to alterations of the pMHC molecular surface engaged by TCR.

3. ANTIGEN CROSS-REACTIVITY AND SPECIFICITY

Although specific recognition of foreign versus self material is tantamount to proper immune function, antibodies and TCRs are frequently involved in spurious interaction events. While antibodies are commonly highly specific for a single antigen, it is not at all uncommon for them to cross-react with many, structurally similar, yet distinct, antigenic molecules. TCR also exhibit broad reactivities. In the most analogous scenario to antibody cross-reactivity, TCR frequently recognize more than one distinct pMHC complex, a phenomenon argued to be critical for antigen recognition (Mason, 1998). TCRs can also be auto-, allo-, or xenoreactive, meaning that they recognize and respond to self peptides or nonself MHC from either the same or different species, respectively.

3.1. Heteroclitic Binding

In some cases, antibodies can bind better to antigens not used in challenging the immune system than to the original immunogen, a phenomenon known as heteroclitic binding. The monoclonal antibody (mAb) D11.15, raised against HEL, interacts with higher affinity with several other avian lysozymes, and the molecular basis for this cross-reactivity has been elucidated (Chitarra et al., 1993). FvD11.15 binding to eight different avian lysozymes was tested, and all of these exhibit high affinity for the antibody, with two, pheasant egg-white lysozyme (PHL) and guinea fowl egg-white lysozyme (GEL), exceeding the affinity of the interaction with HEL. When compared to the crystal structure of the FvD11.15–HEL complex, structures of PHL, GEL, Japanese quail egg-white lysozyme (JEL), and the FvD11.15–PHL complex reveal distinct structural mechanisms for heteroclitic binding. The affinity of JEL for FvD11.15 is slightly lower than that of HEL (1.5×10^9 M^{-1} versus 4.0×10^9 M^{-1}), which is likely derived from two amino acid differences in the loop region from residue 100 to 104 that forms part of the epitope in the FvD11.15–PHL complex. Whereas residue 102 is a Gly in HEL, it is a Val in JEL, while residue 103 is an Asn in HEL and a His in JEL. The result of these changes is a displacement of the 100 to 104 loop region by 7.5 Å into a conformation that would likely clash sterically with the V_HCDR3 loop of FvD11.15 (Fig. 4.3A). Conversely, two amino acid differences between PHL and HEL, at residues 113 (Asn in HEL, Lys in PHL) and 121 (Gln in HEL, Asn in PHL), confer higher affinity to the FvD11.15–PHL complex relative to the complex with HEL by two orders of magnitude. The crystal structure of FvD11.15–PHL reveals that the major structural difference in these two complexes is that Lys113 in PHL makes several nonpolar contacts with V_HTyr57 (Fig. 4.3B).

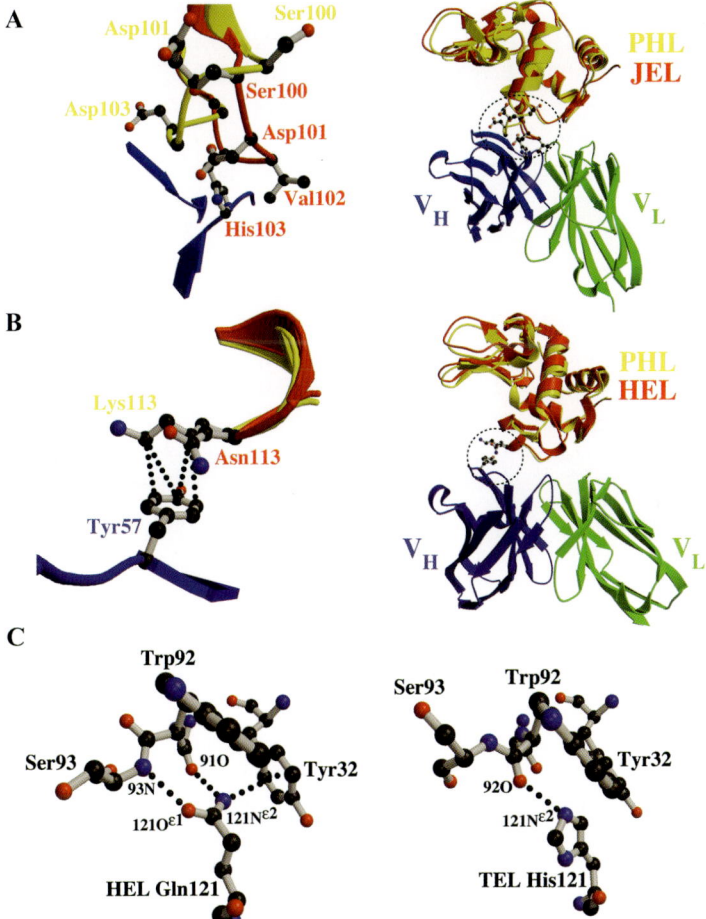

Figure 4.3. Cross-reactivity of antibodies. (**A**) Interaction of FvD11.15 (V$_H$ domain in blue, V$_L$ domain in green) with PHL (*yellow*) and JEL (*red*). The *left* panel is a close-up view of the encircled region in the *right panel*, highlighting the relative displacement of the 100–104 loop region between PHL and JEL resulting in a steric clash between JEL residues Val102 and His103 with the FvD11.15 V$_H$ domain. (**B**) Interaction between FvD11.15 (same color scheme as in [**A**]) with PHL (*yellow*) and HEL (*red*). The *left panel* is a close-up view of the encircled region in the right panel and highlights the productive interactions that are made between FvD11.15 V$_H$Tyr57 and PHL Lys113 (4 hydrogen bonds, indicated by *dotted lines*). Conversely, productive interactions between FvD11.15 V$_H$Tyr57 and HEL Asn113 are largely absent (one hydrogen bond, not shown for clarity) and is likely the reason for the binding affinity discrepancy between the two antigens. (**C**) Hydrogen bonding between FvD1.3 residues V$_L$Tyr32, Phe91, Trp92, and Ser93 with HEL Gln121 (*left panel*) and TEL His121 (*right panel*), the only amino acid difference between these two antigens. HEL Gln121 makes three hydrogen bonds (indicated by *dotted lines*) to the main chain nitrogen atom of Ser93, the main chain oxygen atom of Phe91, and the phenyl ring of Tyr32. All three of these hydrogen bonds are lost in the FvD1.3–TEL complex, however, a peptide flip between FvD1.3 residues Trp92 and Ser93 results in a new hydrogen bond between the TEL His121 side chain and the main chain oxygen atom of Trp92.

Another anti-HEL antibody, D1.3, binds only its immunogen and one other avian lysozyme, bobwhite quail egg-white lysozyme (BEL), with high affinity. Much of the sequence variability between the eight lysozymes tested occurs at HEL residue Gln121. For the highly cross-reactive D11.15, lysozyme residue 121 is located at the periphery of the antigenic epitope. Conversely, for the highly specific D1.3, this residue is located centrally to the binding interface and acts as a hot spot in binding for the D1.3–HEL complex (Dall'Acqua et al., 1996). One of the avian lysozymes that binds poorly to D1.3, turkey egg-white lysozyme (TEL), has been investigated structurally (Braden et al., 1996b). HEL and TEL differ only at residue 121, which is a His in TEL, with a concomitant decrease in affinity by two orders of magnitude, attributable primarily to a reduction in the on-rate of the interaction. Whereas Gln121 of HEL makes two hydrogen bonds to V_L domain main-chain atoms, TEL His121 makes only one hydrogen bond to the antibody light chain and induces a peptide flip between residues V_LW92 and V_LS93, a conformational change that is likely responsible for the slower on-rate of the interaction (Fig. 4.3C).

3.2. Molecular Mimicry in Antibody–Antigen Interactions

Anti-idiotopic antibodies (Poljak, 1994; Pan et al., 1995) recognize an antigenic determinant that is unique to an antibody or group of antibodies, or idiotope. An idiotope is defined functionally by the interaction of an anti-idiotopic antibody (Ab2) with an antibody (Ab1) bearing the idiotope. Conventional Ab2 antibodies recognize idiotopes outside of the antibody combining site paratope, while internal image Ab2 antibodies are able to mimic the molecular surface encountered by Ab1, thereby mimicking stereochemically the antigen specific for Ab1. Numerous efforts have been made to use these molecular mimics as therapeutics, similar to vaccines. As discussed earlier, the D1.3 antibody binds to two structurally distinct ligands—its cognate antigen, HEL, and the anti-idiotypic antibody E5.2—and these interactions exhibit molecular mimicry. The crystal structures of the complexes formed by FvD1.3 with both HEL (Bhat et al., 1994) FvE5.2 (Fields et al., 1995; Braden et al., 1996a) have been determined to high resolution. FvD1.3 contacts HEL and FvE5.2 through essentially the same set of combining site residues and most of the same atoms. Of the 18 FvD1.3 residues that contact FvE5.2 and the 17 that contact HEL, 14 are in contact with both FvE5.2 and HEL. These 14 FvD1.3 residues make up 75% of the total contact area with FvE5.2 and 87% of that with HEL. Furthermore, the positions of the atoms of FvE5.2 that contact FvD1.3 are close to those of HEL that contact FvD1.3, and 6 of the 12 hydrogen bonds in the FvD1.3–FvE5.2 interface are structurally equivalent to hydrogen bonds in the FvD1.3–HEL interface.

3.3. T-Cell Responses to Varied Peptide Antigens

Due to the possible extent of variability of peptide antigens that can be presented MHC molecules, an immune recognition model in which a single TCR is reactive with only one pMHC complex would require a reservoir of T cells that greatly exceeds the actual size of the spleen and the kinetics of a specific immune response would be much too slow (Mason, 1998). Estimations of the number of nonhomologous peptides that can be recognized by a single TCR reach upwards of 10^9 distinct 11-mer peptides (Hemmer et al., 1998). An estimated 1% to 10% of all mature T cells are alloreactive, cross-reacting with nonself MHC molecules (Sherman and Chattopadhyay, 1993). Alloreactivity serves as the primary cause of graft rejection and graft-versus-host disease. While the structural basis of alloresponse induction remains poorly understood, several recent crystal structures have shed light on this type of immune response. A comparison of allogeneic and syngeneic complexes involving the 2C TCR (Luz et al., 2002) reveals that a single MHC mutation imparts alloreactivity (Fig. 4.4A). The mutated residue does not interact directly with TCR but instead modifies the structure of the C-terminal end of the peptide, resulting in a significantly more intimate interaction between the TCR β-chain and pMHC, as evidenced by a relative increase in shape complementarity, buried surface area, and intermolecular contacts. The crystal structure of a xenoreactive complex between a mouse TCR and human pMHC (Buslepp et al., 2003) has revealed that the gross structural requirements for xenoreactivity are similar to that of TCR recognition of self- and allo-pMHC complexes. One caveat regarding xenoreactive complexes is that their functionality is independent of the costimulatory molecules CD4 and CD8, as transspecies MHC class II and I molecules, respectively, do not recognize these coreceptors. In this structure, the orientation of the TCR on top of the pMHC complex is more orthogonal than observed for other TCR/self-pMHC interactions owing to a unique positioning of the V domain on the pMHC surface. Reanalysis of previously determined TCR–pMHC complexes revealed a correlation between $V\alpha$ domain position on the pMHC and a functional dependence on CD8.

At the other end of the specificity spectrum from cross-reactivity lies immunodominance, in which there exists a strong bias in TCR selection such that certain TCRs are used widely in individuals with shared MHC alleles. Recent structures of TCR-pMHC complexes involving imunodominant TCRs have begun to shed some structural insight into this type of skewed immune response. In HLA-A2–positive adults, commonly more than 85% of the circulating T cells that respond to the influenza matrix protein epitope (MP_{58-66}) display $V\beta$ 17 domains with a highly conserved CDR3 loop (Moss et al., 1991; Lehner et al., 1995). In the crystal structure of the representative TCR JM22 in complex with HLA-A2(MP_{58-66}) (Stewart-Jones et al., 2003), the two most highly conserved CDR3 residues are observed inserted into a notch formed between the surfaces of the peptide and MHC moieties (Fig. 4.4B). In another structural analysis immunodominance, tthe LC13 TCR in complex with an antigen from the Epstein–Barr virus latent antigen EBNA 3A displayed by HLA-B8 utilizes a distinct structural mechanism, one of CDR loop conformational rearrangements including the disruption of canonical structures to enhance shape complementarity to the pMHC surface (Kjer-Nielsen et al., 2003).

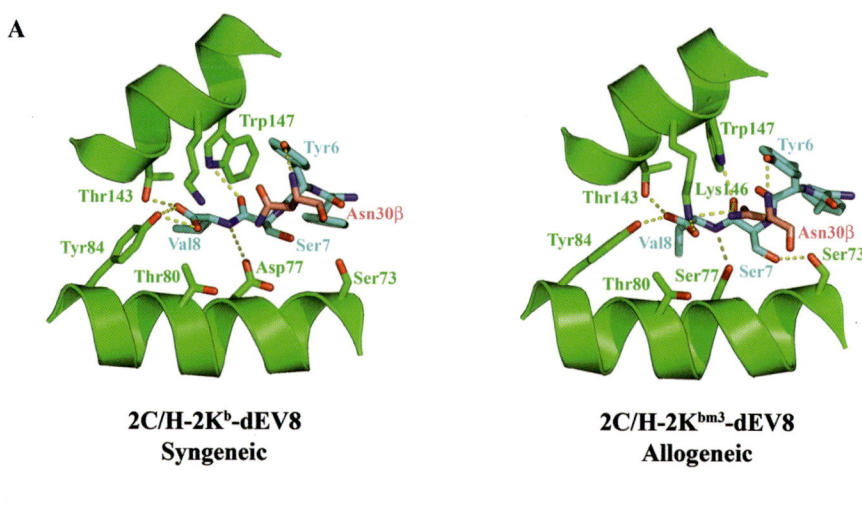

Figure 4.4. Cross-reactivity and immunodominance of TCRs. (**A**) Structural effects of the alloreactive mutation Asp77Ser include rearrangement of the peptide-MHC (*cyan* and *green*, respectively) hydrogen bonding network allowing a more intimate interaction with residue Asn30 of the 2C TCR β-chain (*pink*) between the syngeneic 2C/H-2K^b–dEV8 complex (*left*) and the allogeneic 2C/H-2K^bm3–dEV8 complex (*right*). (**B**) Immunodominance of the JM22 TCR in the recognition of MP(58–66) peptide (*cyan*, side chains removed for clarity) in the context of HLA-A2 (*green*) is due to the filling of a deep pocket by the invariable CDR3β residues of JM22, Arg98, and Ser99 (*pink*).

4. MECHANISTIC ROLES FOR CONFORMATIONAL FLEXIBILITY

4.1. Protein Plasticity in Antibodies

The kinetics of antibody–antigen interactions are commonly temperature dependent. In some cases this may be indicative of the structural plasticity involved in antigen binding. Indeed, the binding kinetics of several anti-HEL antibodies have been shown to conform to a two-state model describing induced fit, with distinct association steps for molecular encounter and docking (Lipschultz et al., 2000; Li et al., 2001b). Although numerous hypotheses concerning the correlation between antibody flexibility and signaling have been proposed over the years, the establishment of molecular flexibility as a component of signaling, beyond the antigen recognition event, remains elusive.

For smaller antigens, notably peptides and DNA, antibody plasticity is generally more pronounced than for protein antigens, although associations with the latter commonly involve a nominal degree of molecular flexibility and cannot necessarily be classified as "lock-and-key" interactions. Two types of backbone movements within the antibody combining site have commonly been observed on antibody–antigen complex formation concerted movements of multiple residue segments of CDR loops and more heterogeneous rearrangements of CDR residues. On binding antigen, heavy chain CDR loops in the antipeptide Fab8F5 undergo essentially rigid-body movements in which the unliganded loop conformations are conserved, while changes in the main-chain conformation of the light chain are not significant (Tormo et al., 1994) (Fig. 4.5D). The largest backbone displacement, greater than 7 Å, occurs for the V_HCDR3 residue Tyr102. The culmination of concerted heavy-chain CDR movements toward the light chain reduces the volume of the antigen binding site by some 3% relative to the unbound Fab8F5. Other examples of segments of CDR loops moving en masse toward antigen have been observed (Stanfield et al., 1990). In Fab17/9, a significant rearrangement of the V_HCDR3 loop is induced by binding of its peptide antigen (Fig. 4.5E), for which the largest backbone changes are 5 Å (Rini et al., 1992). Restructuring of CDR loop regions from both the heavy and light chains of the anti-DNA antibody FabBV04-01 have also been observed (Herron et al., 1991). Induced CDR loop movements on antigen binding seem to be less extreme for antiprotein antibodies. Generally, these are small, concerted displacements of less than 3 Å (Prasad et al., 1993; Bhat et al., 1994; Braden et al., 1994; Mylvaganam et al., 1998; Li et al., 2000; Faelber et al., 2001); (Fig. 5.5F).

4.2. Antigen Flexibility on Recogntion by Antibodies

Molecular flexibility is not limited to a single side of the interface, as a number of structural studies have shown varying degrees of protein plasticity for antigens on

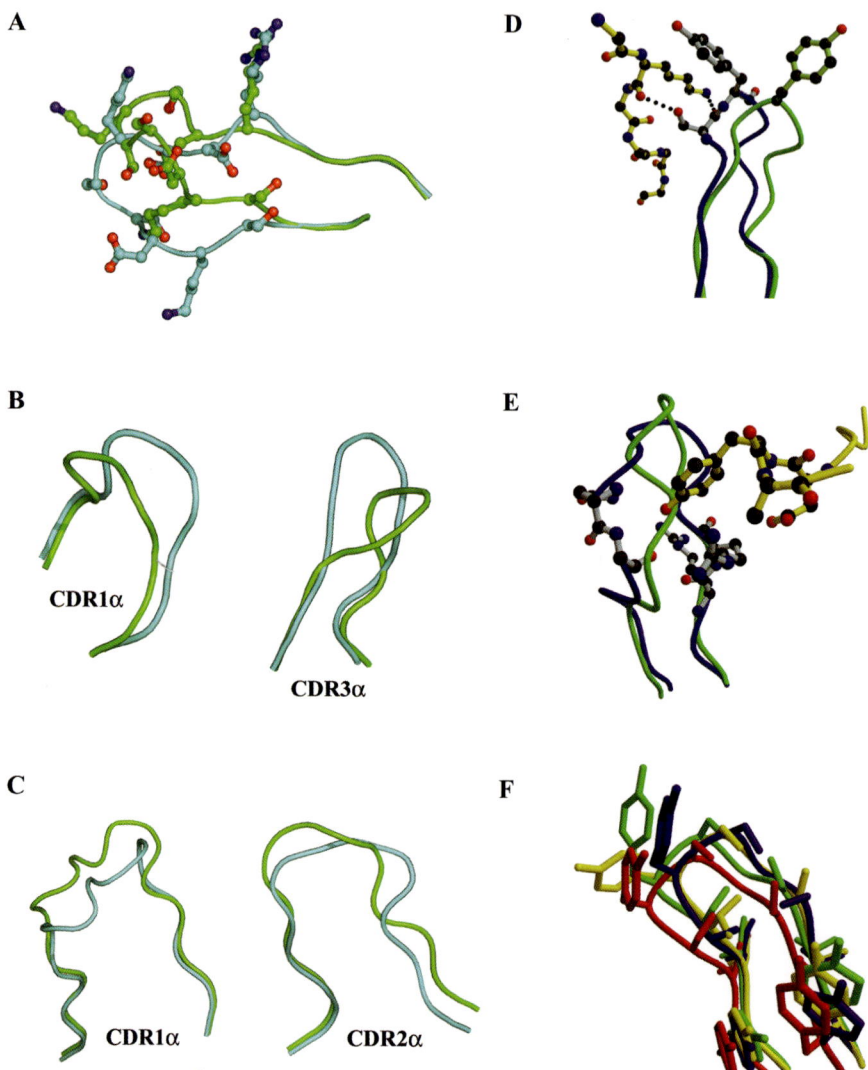

Figure 4.5. Antibody and T-cell receptor conformational changes induced by antigen binding. (**A**) Restructuring of the AB loop of the Cα domain of the LC13 TCR in the putative interface formed with CD3ε, a critical T-cell signaling molecule. The unliganded structure is in *cyan*; the conformation when bound to HLA-B8(EBNA 3A) is in *green*. (**B**) Rigid-body movements in the CDR1α and CDR3α loops in the 2C TCR from its unliganded structure (*cyan*) to its structure when bound to H-2K^b(dEV8) (*green*). (**C**) Disruption of canonical structures of the CDR1α and CDR2α loops of the LC13 TCR between its free (*cyan*) and bound (*green*) states. (**D**) Concerted movement of the Fab8F5 CDR H3 loop induced on binding its peptide antigen. The unbound Fab structure is in green, the bound Fab structure in blue and the peptide antigen in *yellow*. V$_H$Ser101 in the bound form makes two hydrogen bonds to Lys157 of the peptide and V$_H$Tyr102 is displaced by more than 7 Å between the unbound and bound Fab8F5 molecules. (**E**) Atomic rearrangement of the CDR H3 loop of the anti-peptide Fab17/9. The color scheme is the same as in (**D**). Side chains of residues in contact between the antibody and antigen in the bound complex are shown.

recognition by both antibodies and TCR. HEL can be crystallized in several space groups (Ramanadham et al., 1990; Harata, 1994; Kurinov and Harrison, 1995). Comparison of the structures reveals significant flexibility of several loops at the molecular surface, including a number of $C\alpha$ atom displacements greater than 3 Å between HEL molecules from different space groups. Between crystal structures of HEL bound to different antibodies, some main chain movements become more pronounced. Relative to the D1.3–HEL complex (Bhat et al., 1994), Gly102 and Asn103 of HEL are displaced some 8 Å in complexes with the anti-HEL antibodies HyHEL-10 (Padlan et al., 1989) and D11.15 (Chitarra et al., 1993). In the HyHEL–63/HEL complex (Li et al., 2000), residues 99 to 104 of this same loop region of HEL have a root mean square deviation of 6.8 Å relative to their positions in HEL complexed with D1.3 (Bhat et al., 1994). Molecular movement in this HyHEL–63 complex is highlighted by a peptide flip at residue Asp101 that allows the formation of five hydrogen bonds between this residue and the antibody. Smaller conformational changes are seen in the HIV-1 capsid protein p24 on binding Fab13B5 (Berthet-Colominas et al., 1999; Monaco-Malbet et al., 2000). Localized to the turn portion of the helix-turn-helix motif, the flexibility of the antigen is highlighted by a 4 Å displacement of the carbonyl oxygen of Pro207 that points in a direction opposite its unbound form to adapt to the molecular environment of the antibody. Owing to the high degree of flexibility of uncomplexed peptides (Dyson and Wright, 1995), the Fab13B5–p24 complex, with its continuous peptide-like epitope, may present the best current measure for the role of peptide antigen plasticity in antibody recognition. Increased antigen flexibility, however, is not always beneficial to epitope recognition by antibodies. To produce mimics of the N-terminal sequence of transforming growth factor alpha epitope recognized by the monoclonal antibody tAb2, peptides required cyclization to constrain their conformations to ones that are suitable for binding (Hahn et al., 2001).

4.3. Protein Plasticity of TCR and Its Potential Signaling Role

For TCR–pMHC complexes, on the contrary, evidence is mounting that flexibility in the binding interface directly contributes to T-cell activation. Binding studies utilizing surface plasmon resonance and isothermal titration calorimetric analyses have established that, at least in some cases, conformational changes are required for complex formation (Boniface et al., 1999; Willcox et al., 1999). A two-step mechanism has been observed for complex formation between the 2B4 TCR and MCC–IE^k

---

Figure 4.5. (*Continued*) (**F**) Contrary to many antipeptide antibodies, antiprotein antibodies generally exhibit relatively small conformational changes upon binding antigen as shown for the anti-HEL antibody FabHyHEL63. Superposition of the CDR H2 loops of FabHyHEL63 bound to HEL (*blue*) and three different unbound forms: solved in the C2 spacegroup (*green*); one molecule from the asymmetric unit of the free antibody solved in the P1 spacegroup (*red*); and the second molecule of the asymmetric unit of the free antibody solved in the P1 spacegroup (*yellow*).

pMHC complex in which MHC contacts dominate initial association while peptide contacts control stabilization (Wu et al., 2002). When changes in heat capacities, an indicator of conformational change induced on binding, is coupled with the half-life of TCR–pMHC complex stabilization, the degree of T-cell activation could be accurately predicted for a single TCR binding to a series of pMHC ligands (Krogsgaard et al., 2003). An intriguing structural link between TCR–pMHC complex formation and intracellular signaling has been suggested to explain conformational changes observed in the region of the $C\alpha$ domain of the LC13 TCR at the putative interface with the signaling molecule $CD3\epsilon$ on binding to HLA–B8(EBNA 3A) (Kjer-Nielsen et al., 2003) (Fig. 4.5A). It remains to be seen whether these flexibility-signaling correlations will hold as the analysis of distinct TCR–pMHC complexes expands rapidly in the coming years.

TCR recognition of pMHC can also be described by an induced fit mechanism. While thermodynamic analysis has shown that these interactions commonly exhibit slow association rates with large entropic barriers (Davis et al., 1998; Boniface et al., 1999; Willcox et al., 1999), indicative of the ordering of the molecular interface during the binding event, recent sructural comparisons of TCR in its unliganded state and in complex pMHC have cemented conformational rearrngements as a hallmark of TCR–pMHC recognition. The 2C TCR exhibits distinct conformations of the $V\alpha$CDR1 and $V\alpha$CDR3 loops between its free (Garcia et al., 1996) and liganded (Garcia et al., 1998) crystal structures. These structural changes are largely en masse hinge motions of the CDR loops in the absence of rearrangements of the canonical loop structures (Fig. 4.5B). Conversely, the free (Kjer-Nielsen et al., 2002b) and bound (Kjer-Nielsen et al., 2003) crystal structures of the immunodominant LC13 TCR has revealed the disruption of canonical $V\alpha$CDR1 and $V\alpha$CDR2 loop structures as a mechanism for producing an enhanced fit to the HLA–B8(EBNA 3A) complex (Fig. 5.5C). TCR recognition of APLs exhibits varying degrees of conformational rearrangements in the $V\beta$CDR3 loops depending on the TCR–APL/MHC system investigated (Ding et al., 1999; Degano et al., 2000).

4.4. pMHC Flexibility Induced by TCR Binding

Conformational changes in the pMHC moiety have been shown to play a role in formation of functional TCR–pMHC complexes. The HLA–B8 residue Gln155 forms a hydrogen bond with the P7 Tyr residue of the EBNA 3A peptide in the unbound structure (Kjer-Nielsen et al., 2002a), while in the complex between the LC13 TCR and HLA–B8(EBNA 3A), Gln155 adopts a different rotamer conformation in order to form a hydrogen bond with Thr30 of the $V\alpha$CDR1 loop (Kjer-Nielsen et al., 2003). A small rigid-body shift of the bound peptide was also observed in these structures. Peptide mobility may prove to be a common feature in antigen recognition by TCR, especially for relatively long peptides that must form a bulge above the center of the peptide-binding groove in MHC class I molecules on account of their ends being fixed in conserved pockets of the MHC (Speir et al., 2001).

5. FUNCTIONAL MATURATION OF ANTIBODIES AND T CELL RECEPTORS

5.1. Different Mechanisms for Producing Functional Antibodies and TCR

The function of the immune system is dependent on the recognition of essentially any antigenic material, yet the structural diversity of antigens greatly outweighs the genetic diversity encoded by immune system genes. Thus, molecular recognition of diverse antigens is accomplished by producing antibodies and TCRs with specificity for almost any antigen via recombination and imprecise joining of antibody or TCR gene segments. This focuses molecular diversity at the contiguous molecular surface formed by the CDR loops, the combining site for antigen recognition. This results in germline antibodies and TCRs of relatively low affinity and specificity (Tonegawa, 1983). This junctional diversity in the primary repertoire can produce CDR loops of different lengths and varying structures (Chothia et al., 1992; Tomlinson et al., 1995).

It has been estimated that these somatic recombination and joining events can produce up to 10^{18} different TCRs (Janeway et al., 1999), which would exceed the total number cells in the human body. T cells undergo thymic selection to pare down to roughly 10^{11} clonal T cells. During development, T cells that react with self proteins are negatively selected, while those that can react with self MHC molecules are positively selected, resulting in the deletion of 95% to 98% of all T cells (Palmer, 2003). These T cells bear TCR with the structural properties required for efficient recognition of nearly any foreign, antigenic peptide presented by self MHC molecules, while largely avoiding recognition of self peptides and thus autoimmunity.

The differential affinity requirements for antibodies and TCR (approximately nanomolar and micromolar, respectively) necessitates, in the case of antibodies, a secondary process for improving affinity and specificity once diversity has been established. The somatic hypermutation of antibody V regions spreads structural diversity generated by gene segment recombination to regions at the periphery of the binding site (Tomlinson et al., 1996). Selective expansion of antibody clones on the basis of antigen affinity produce mature antibodies that are high in both affinity and specificity (Rajewsky, 1996). Somatic hypermutation is primarily a point mutation process in gene regions that are highly conserved in the primary repertoire that can result, at times, in codon insertions or deletions (Tomlinson et al., 1996). It has been shown that the presence or absence of certain V_HCDR3 junctional amino acids can determine the affinity maturation pathway of an antibody by biasing subsequent amino acid replacements by somatic hypermutation (Furukawa et al., 1999) and that these affects are correlated to the structure and flexibility of the V_HCDR3 loop in the germline antibodies (Furukawa et al., 2001).

5.2. Maturation of Antibodies via Somatic Hypermutation

Structural and energetic studies comparing germline and mature antibodies bound to the same antigen have advanced our understanding of the effects of somatic hypermutation on antibody affinity maturation. The mature Fab48G7 and its germline counterpart, Fab48G7g, both bind a nitrophenyl phosphonate transition-state analog, but with a 30,000-fold difference in affinity, primarily due to a decrease in the dissociation rate (Wedemayer et al., 1997a). The sequence differences between the Fabs are limited to nine somatic hypermutations, six in V_H and three in V_L, located up to 15 Å from the bound hapten. Crystal structures of the unliganded germline Fab48G7g and its complex with hapten (Wedemayer et al., 1997a) reveal large conformational changes induced upon antigen binding, while crystal structures of the mature Fab48G7 (Patten et al., 1996; Wedemayer et al., 1997b) in its free and hapten-bound forms exhibit very few conformational changes on complex formation. The conformational changes induced upon binding antigen by Fab48G7g are later observed in the mature Fab structure even in the absence of antigen (Fig. 4.6), and thus it appears, at least in the case of the Fab48G7 system, that the affinity maturation process is driven in large part by a mechanism of preorganizing the antibody combining site into a conformation that is favorable for binding its hapten antigen.

Through the introduction of forward and back site-directed mutations in the germline and mature Fabs and measurements of binding affinities, the effects of the nine somatic hypermutations on the affinity maturation pathway of Fab48G7 have been dissected (Yang and Schultz, 1999). In this system, the effect on binding of the individual mutations was either positive or neutral, yet their additive changes in affinity were not equal to the overall change in affinity between the germline and mature Fabs. Double mutations revealed a high degree of cooperativity between mutations, not only between individually neutral mutations but also between even the two most positive individual mutations. Cooperativity between somatic hypermutations, however, does not appear to be a required mechanism for affinity maturation. For Fab39-A11, which catalyzes a Diels–Alder reaction, only two somatic mutations exist between the germline and mature counterparts, of which only one contributes the majority of binding affinity to mature Fab (Romesberg et al., 1998). Another catalytic antibody, AZ-28, which catalyzes an oxy-Cope rearrangement, has six somatic mutations, five of which contribute to differences in affinity between germline and mature antibodies in a strictly additive way (Ulrich et al., 1997). In the affinity maturation of an antiprotein antibody, FvD1.3, the five somatic hypermutations have also been shown to be energetically additive (England et al., 1999). In this system, changes in antigen affinity are dominated by the only mutated amino acid that is in direct contact with the antigen, HEL.

The quantity and cooperativity of somatic hypermutations may be dependent on the affinity differences between the germline and mature antibodies. While the affinity discrepancy between Fab48G7 and Fab48G7g is 30,000-fold (Wedemayer et al., 1997a), FabAZ-28, with only five significant somatic mutations, has an antigen affinity only 40-fold greater than its germline counterpart (Ulrich et al., 1997). Furthermore, Fab39-A11 and Fab39-A11g, with only one significant amino acid difference, both bind nine haptens, for most of which the difference in affinity is within an order of magnitude (Romesberg et al., 1998). Germline and mature FvD1.3 also differ by only five amino acids and by 60-fold in affinity (England et al., 1999). If one considers

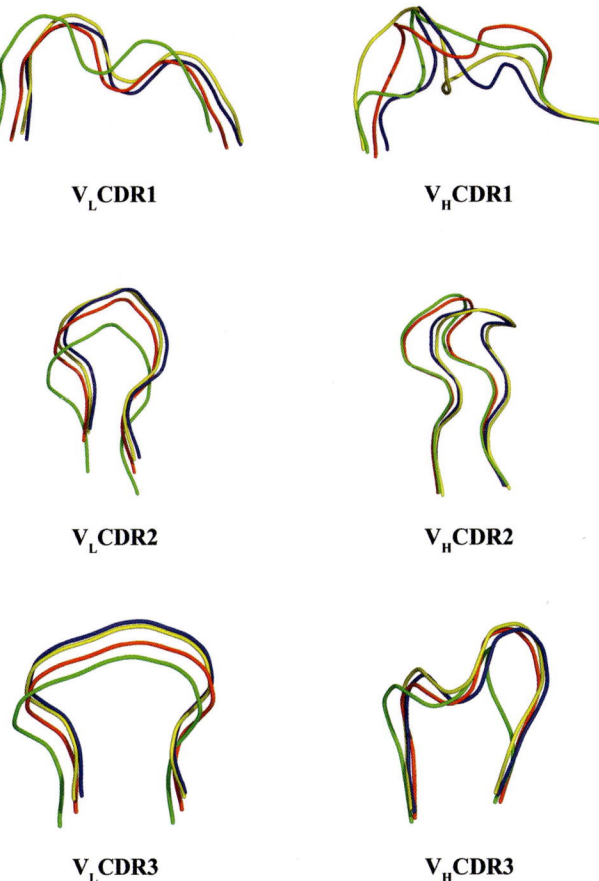

V_LCDR1 V_HCDR1

V_LCDR2 V_HCDR2

V_LCDR3 V_HCDR3

Figure 4.6. Affinity maturation via preorganization of the antibody combining site in Fab48G7. Superposition of the CDR loops of the free germline Fab48G7 (*green*), the antigen bound germline Fab48G7 (*red*), the unliganded mature Fab48G7 (*blue*), and the liganded mature Fab48G7 (*yellow*). The conformational changes invoked on antigen binding by the germline antibody are commonly replicated in the mature antibody, especially for V_LCDR1, V_LCDR2, V_LCDR3, and V_HCDR3.

that mature antibodies must break a minimum affinity threshold for antigen binding through a limited number of somatic mutations to be functional in vivo, then it follows that the number of somatic mutations will increase as the difference in affinities between germline and mature antibodies gets larger and cooperativity between the somatic mutations will be utilized in cases where the affinity maturation process must overcome extreme germline-mature affinity discrepancies. Precise affinity ranges for the lack or presence of cooperativity associated with somatic hypermutation may or may not actually exist. One complicating factor is the observation of negative cooperativity in the affinity maturation pathway of a protein–protein interaction with an affinity difference of 1500-fold (Yang et al., 2003).

Recently, the crystal structures of four closely related anti-HEL antibodies (HyHEL8, HyHEL10, HyHEL26, and HyHEL63), representing different stages of affinity maturation, were determined bound to the same site on HEL (Li et al., 2003), revealing that enhanced binding is achieved by the burial of increasing amounts of apolar surface, at the expense of polar surface, accompanied by improved shape complementarity (Fig. 4.7A,B). The increase in hydrophobic interactions, which can fully account for the 30-fold affinity improvement in these anti-HEL antibodies according to an experimental estimate of the hydrophobic effect in protein–protein interactions (Sundberg et al., 2000), is the consequence of subtle, yet highly correlated, structural rearrangements in antibody residues at the periphery of the interface with antigen, adjacent to the central energetic hot spot, whose structure remains unaltered (Fig. 4.7C,D). While increasing hydrophobic interactions and improving the fit at peripheral sites that have not been optimized for binding, and whose plasticity and ability to accommodate mutations render them permissive to such optimization, constitute effective strategies for maturing antiprotein antibodies, other, as yet unobserved mechanisms may be utilized by various antibodies for affinity maturation.

Some of the energetic factors involved in the preorganization of mature antibodies through somatic hypermutation of germline antibodies have been elucidated recently using surface plasmon resonance techniques in which different binding characteristics at various temperatures of the same complex provide information relative to the enthalpic and entropic contributions to the interaction. The affinities of panels of early primary and secondary response monoclonal antibodies (mAbs) for a model synthetic 40-mer peptide were determined at two temperatures (Manivel et al., 2000). While the effects of temperature on the dissociation step of the interaction was similar for mAbs in both panels, opposite temperature effects on association were observed for each panel of mAbs. For primary mAbs, complex association was enthalpically highly favorable but entropically unfavorable, while dissociation was enthalpically unfavorable and entropically favorable. The equilibrium binding for primary mAbs was enthalpically driven with a large entropic cost of complex formation, resulting in relatively low affinity. Conversely, in secondary mAbs, association was enthalpically unfavorable but the entropic costs had been reduced dramatically. Because the dissociation step of the reaction was similar to that for primary mAbs, equilibrium binding in the seceondary mAbs was essentially independent of enthalpy effects, and instead, was driven by entropic changes. Thus, the relative high affinity of the secondary mAbs is derived exclusively from the nearly complete abolishment of any entropic costs of complex association in comparison to the primary mAbs. While these experiments seem to confirm the idea of antibody affinity maturation through paratope preorganization, at least for an antipeptide antibody, it is intriguing to note that the increased affinities in the antihapten Fab48G7 and the antiprotein FvD1.3 systems derive nearly entirely from decreases in the dissociation phases of the reactions (Wedemayer et al., 1997a; England et al., 1999). Although similar experiments examining enthalpy and entropy effects on antigen binding to germline and mature Fab48G7 and FvD1.3 have not been performed, it is likely that these types of experiments would reveal that these complexes are stabilized due to large entropic barriers to dissociation in the mature versus germline antibodies.

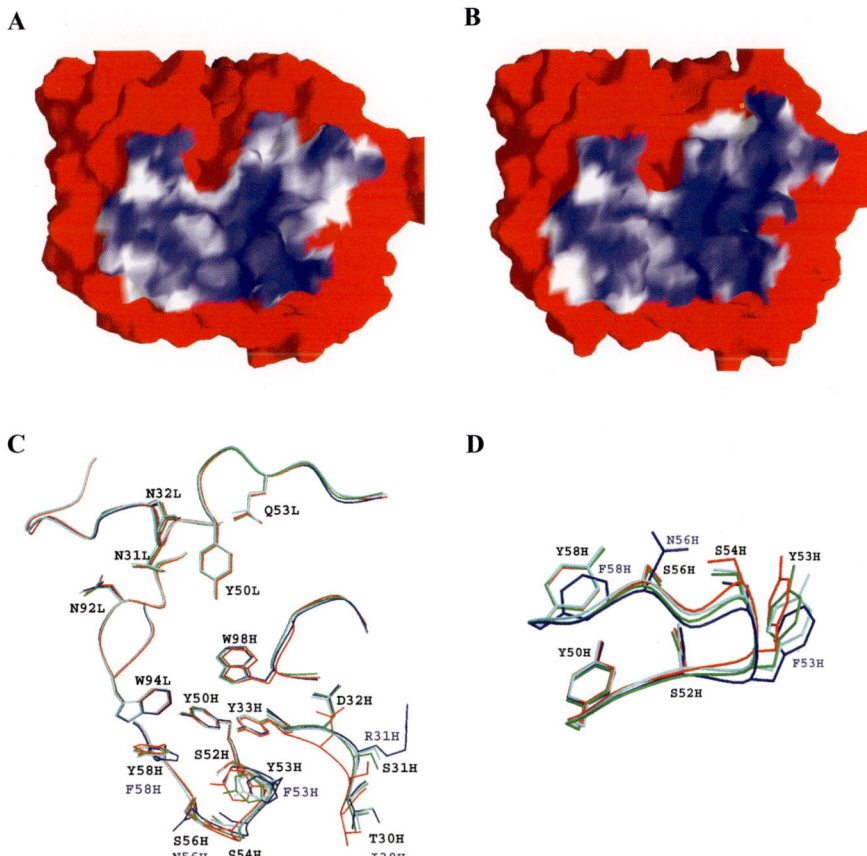

Figure 4.7. Shape complementarity and conformational differences in antibody/HEL interfaces. (A) Molecular surface of germline antibody HyHEL26 viewed at the site that interacts with HEL in the HyHEL26–HEL complex. Regions with higher S_c values, indicating closer topological match with HEL, are more blue; regions with topologically uncorrelated surfaces ($S_c = 0$) are white. (B) Molecular surface of affinity-matured antibody HyHEL8 viewed at the binding site for HEL in the HyHEL8–HEL complex. HyHEL8 binds HEL with 35-fold higher affinity than HyHEL26. As in (A), regions with better geometric fits to the antigen are more blue. The higher-affinity HyHEL8/HEL interface is also the more complementary (133). (C) Comparisons of the combining sites of antibodies HyHEL26 (*red*), HyHEL63 (*green*), HyHEL10 (light blue) and HyHEL8 (dark blue) in their complexes with HEL. These antibodies recognize HEL with relative affinities HyHEL26 < HyHEL63 < HyHEL10 < HyHEL8. (D) Close-up view of the V_HCDR2 loops (residues V_H50–58) in (C), showing the progressive shift of the Tyr–Phe53 side chain from its position in the the lowest-affinity (*red*) to the highest-affinity (*dark blue*) complex. These concerted movements in V_HCDR2 improve shape complementarity at the V_H–HEL interface. The shifts also increase the amount of apolar surface buried in the interfaces, concomitant with tighter binding and a reduction in polar buried surface (Li et al., 2003).

6. NATURAL KILLER CELL RECEPTORS: FURTHER INSIGHTS INTO IMMUNE RECOGNITION

6.1. Two Structurally Distinct Families of Natural Killer Cell Receptors

Natural killer (NK) cells are an essential component of innate immunity toward tumors and virally infected cells (McQueen and Parham, 2002; Yokoyama and Plougastel, 2003). NK cell function is regulated by a dynamic balance between positive signaling receptors (resulting in target cell lysis) and negative signaling receptors (preventing lysis) that ultimately determines the outcome of NK cell–target cell encounters. The downregulation of surface MHC class I molecules, which is often a hallmark of the morbid cell, triggers NK cell-mediated cytolysis by failing to engage MHC class I specific inhibitory receptors on the NK cell. Several receptor families have been identified on primate and rodent NK cells that monitor MHC class I expression on surrounding cells (McQueen and Parham, 2002; Natarajan et al., 2002; Yokoyama and Plougastel, 2003). These include members of the Ly49 family (Ly49A through W), KIRs, LIRs, and CD94/NKG2 receptors. In addition, the activating receptor NKG2D recognizes distant homologs of MHC class I molecules, such as MIC-A and RAE-I, that are selectively upregulated in stressed tissues (Vivier et al., 2002).

NK receptors belong to two structurally distinct groups, the Ig superfamily (KIRs, LIRs) and the C-type lectin superfamily (Ly49s, NKG2D, CD94–NKG2) (McQueen and Parham, 2002; Natarajan et al., 2002; Yokoyama and Plougastel, 2003). NK receptors of the Ig superfamily are type I transmembrane glycoproteins containing one or more Ig-like extracellular domains, whereas the C-type lectin receptors are homo- (Ly49s, NKG2D) or heterodimeric (CD94–NKG2) type II transmembrane glycoproteins, with each chain containing a single, extracellular C-type lectin domain. A remarkable property of certain Ly49s (e.g., Ly49C), and of NKG2D, is their ability to recognize multiple MHC class I, or MHC class I-like, ligands. Thus, the broadly MHC-reactive Ly49C receptor, which binds H-2K^b, H-2K^d, H-2D^b, H-2D^d, and H-2D^k, is considerably more promiscuous than Ly49A, whose specificity is largely confined to H-2D^d and H-2D^k (Anderson et al., 2001). Similarly, NKG2D recognizes an array of distinct MHC class I-like molecules, including MIC-A, MIC-B, ULBP, MULT1, RAE-1, and H60 (Vivier et al., 2002). X-ray crystallographic studies of Ly49s (Tormo et al., 1999; Dam et al., 2003) and NKG2D (Li et al., 2001a, 2002; Radaev et al., 2001) have revealed that these NK receptors employ entirely different strategies to achieve multispecific immune recognition.

6.2. Basis for MHC Specificity of Ly49 Receptors

In the crystal structure of mouse Ly49C bound to H-2K^b (Dam et al., 2003), the Ly49C homodimer engages H-2K^b bivalently, such that each C-type lectin domain makes identical interactions with MHC class I at a cavity beneath the peptide-binding platform formed by the heavy chain α_1/α_2 and α_3 domains and β_2-microglobulin (β_{2m}) (Fig. 4.8A). Ly49A binds to a very similar site on MHC class I in its complex with

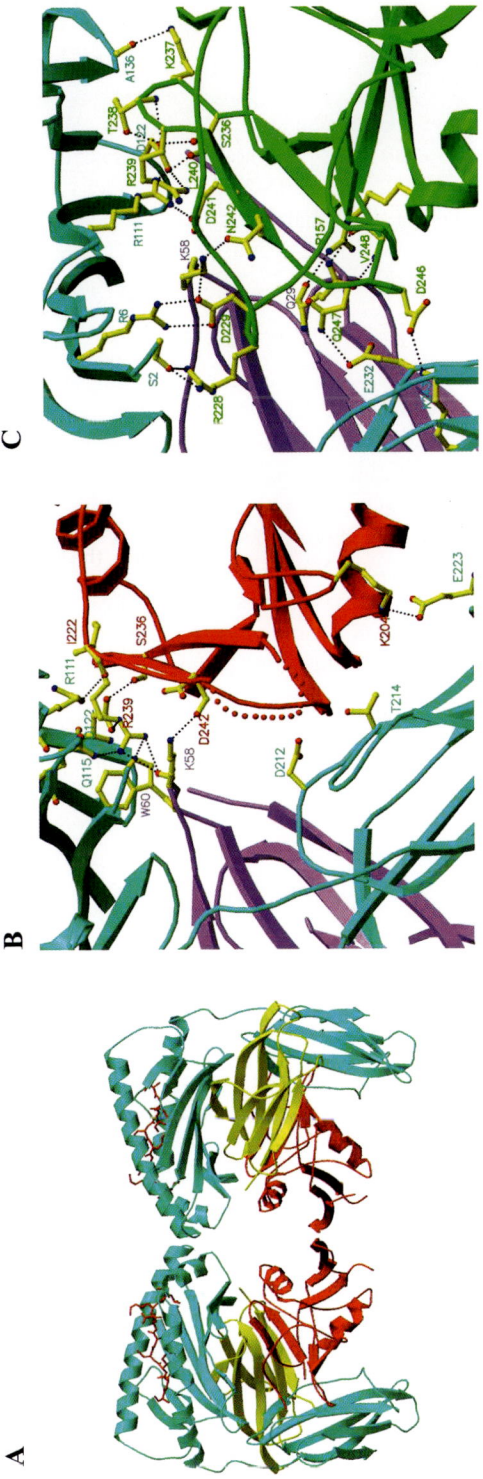

Figure 4.8. Structure of Ly49–MHC class I complexes. (**A**) Ribbon diagram of the Ly49C–H-2K[b] complex (Dam et al., 2003), in which the Ly49C dimer (*red*) crosslinks two MHC class I molecules. The H-2K[b] heavy chain is *blue*, the MHC-bound peptide is *red*, and b2m is *yellow*. (**B**) Detailed view of the Ly49C–H-2K[b] interface showing hydrogen bonds (*dotted black lines*) between Ly49C (*red*) and the H-2K[b] heavy chain (*blue*) or b2m (*violet*); the side chains of interacting residues are *yellow*. The portion of the Ly49C L3 loop, encompassing residues 226–231, which is disordered in the crystal structure, is drawn arbitrarily (*red balls*). With the exception of H-2K[b] residues Asp212 and Thr214, whose side chains could potentially interact with the disordered L3 loop of Ly49C, only residues forming hydrogen bonds are shown. (**C**) Stereo diagram showing hydrogen bonds between Ly49A (*green*) and the H-2D[d] heavy chain (*blue*) or b2m (*violet*) in the Ly49A/H-2D[d] complex (Tormo et al., 1999).

H-2D^d (Tormo et al., 1999), where it contacts β_{2m} and mostly nonpolymorphic residues of the heavy chain. The Ly49C–H-2K^b and Ly49A–H-2D^d interfaces are both characterized by poor shape complementarity, based on S_c values of 0.58 and 0.63, respectively, similar to those for TCR-peptide–MHC complexes (Rudolph and Wilson, 2002), but less than the S_c of the NKG2D–MIC-A interface (Li et al., 2001a). Moreover, the Ly49C–H-2K^b interface, like that of Ly49A–H-2D^d, is very hydrophilic and dominated by polar interactions. Hot spot residues of Ly49A, whose mutation to alanine abolished binding to H-2D^d (Wang et al., 2002), are identical, or conservatively substituted, in Ly49C.

Although the general features of the Ly49C–H-2K^b and Ly49A–H-2D^d interfaces are very similar, the network of intermolecular electrostatic interactions is far more extensive in the Ly49A–H-2D^d than the Ly49C–H-2K^b complex (Fig. 4.8B,C) (Tormo et al., 1999; Dam et al., 2003). Whereas Ly49A forms 20 hydrogen bonds with H-2D^d, the Ly49C–H-2K^b interface comprises only 10 such bonds, along with correspondingly fewer van der Waals contacts and salt bridges. In addition, the Ly49A–H-2D^d complex buries a total solvent-accessible surface of 2820 Å^2, compared to only 2160 Å^2 for the Ly49C–H-2K^b complex. These differences are largely attributable to a particular loop (L3) of Ly49C, comprising residues 226 to 231, that is disordered in the crystal structure of the Ly49C–H-2K^b complex (Fig. 4.8B). Site-directed mutations in this apparently flexible loop did not reduce binding to H-2K^b, indicating that L3, which shows high sequence variability among Ly49s, contributes little to complex stabilization (Dam et al., 2003). By contrast, L3 is well defined in the Ly49A–H-2D^d structure (Tormo et al., 1999), where it forms numerous specific interactions with ligand, including several hydrogen bonds through hot spot residue 229 (Fig. 4.8C).

These factors indicate that Ly49 receptors have evolved a two-tiered strategy for recognizing MHC class I, which explains the substantially broader MHC specificity of Ly49C compared to Ly49A. Primary recognition is mediated by a small number of conserved hot spot residues in structurally conserved regions of Ly49s that together contribute most of the binding free energy. Overlayed onto these primary interactions are secondary ones involving more variable portions of the receptors that confer varying levels of MHC specificity. Thus, the promiscuous Ly49C receptor binds H-2K^b almost exclusively through structurally conserved regions, with little participation by variable elements that could restrict this receptor's broad MHC cross-reactivity. By contrast, the more selective Ly49A receptor, in addition to contacting H-2D^d through the same conserved regions as Ly49C, utilizes the variable L3 loop to impose a narrower binding specificity. An analogous mechanism has been evoked to explain differences in the TCR Vβ-binding specificities of bacterial superantigens (Sundberg et al., 2002a).

6.3. Basis for Multispecific Ligand Recognition by NKG2D

The homodimeric C-type lectin-like receptor NKG2D binds multiple protein ligands that are distant structural homologs of MHC class I, including MIC-A, ULBP3, and RAE-1β (Natarajan et al., 2002; Vivier et al., 2002). However, unlike true MHC class I molecules, NKG2D ligands bind neither antigenic peptides (or any other small molecule) nor β_{2m}, and ULBP3 and RAE-1β even lack the heavy chain α_3 domain, existing on the cell surface as isolated, GPI-linked α_1/α_2 platform domains. Crystal structures have been reported for mouse and human NKG2D in free form (Wolan et al., 2001; McFarland et al., 2003), for mouse NKG2D in complex with RAE-1β (Radaev et al., 2001), and for human NKG2D bound to MIC-A and ULBP3 (Li et al., 2001a, 2002). In the complex structures (Fig. 4.9A–C), the NKG2D homodimer binds orthogonally to the axes of the helices of the α_1/α_2 platform domain of the ligands in a manner resembling the docking mode of TCR onto MHC class I, but distinct from that of Ly49C, each of whose subunits comprises an independent binding site for MHC (Fig. 4.8A). This recognition of asymmetric ligands by a symmetric receptor is mediated by very similar surfaces on the NKG2D monomers that specifically interact with two distinct surfaces on MIC-A, ULBP3, or RAE-1β. For example, in the NKG2D–MIC-A complex (Li et al., 2001a), 7 of the 11 contact residues contributed by each NKG2D monomer are common to both MIC-A binding surfaces, although 6 are engaged in different interactions at the two interfaces. The single NKG2D binding site has therefore evolved to recognize six different surfaces on the α_1/α_2 domains of MIC-A, ULBP3, and RAE-1β, which are distantly related in terms of both sequence (~25% identity) and detailed structure.

Two mechanisms have been proposed to explain multispecific ligand recognition by NKG2D (Li et al., 2001a, 2002; Radaev et al., 2001). NKG2D may possess a degree of conformational flexibility, or plasticity, that allows a single receptor to re-configure its binding site to accommodate diverse ligands (induced fit). Alternatively, an essentially rigid binding site on NKG2D may make different interactions with structurally distinct ligand surfaces, without significant conformational changes in the receptor (rigid adaptation). To distinguish between these possibilities, Strong and colleagues (McFarland et al., 2003) performed in silico alanine-scanning mutagenesis of the complexes formed by NKG2D with MIC-A, ULBP3, and RAE-1β to quanti-tate the relative contributions different residues make to the binding energetics. The computational method employed (Chevalier et al., 2002; Kortemme and Baker, 2002) considers shape complementarity, polar interactions involving hydrogen bonds and ion pairs, and protein–solvent interactions, including a penalty for unsolvated buried polar groups. This analysis suggested that the free energy of binding is unevenly dis-tributed across the NKG2D–ligand interfaces, resulting in obvious hot spots and the energetic dominance of one of the NKG2D subunits. Importantly, the predicted hot spots were associated with structurally conserved receptor elements that interact with relatively conserved residues of the ligands, which argues in favor of a rigid adaptation recognition mechanism. Arguing against induced fit is the absence of large confor-mational changes in NKG2D on ligand binding, although some structural differences were noted between free and bound forms of the receptor (McFarland et al., 2003).

To unambiguously resolve whether induced fit or rigid adaptation better describes multispecific recognition by NKG2D, detailed thermodynamic and kinetic analyses were carried out on four different NKG2D–ligand pairs (McFarland and Strong, 2003). Induced-fit interactions, as exemplified by TCR binding peptide–MHC, typically

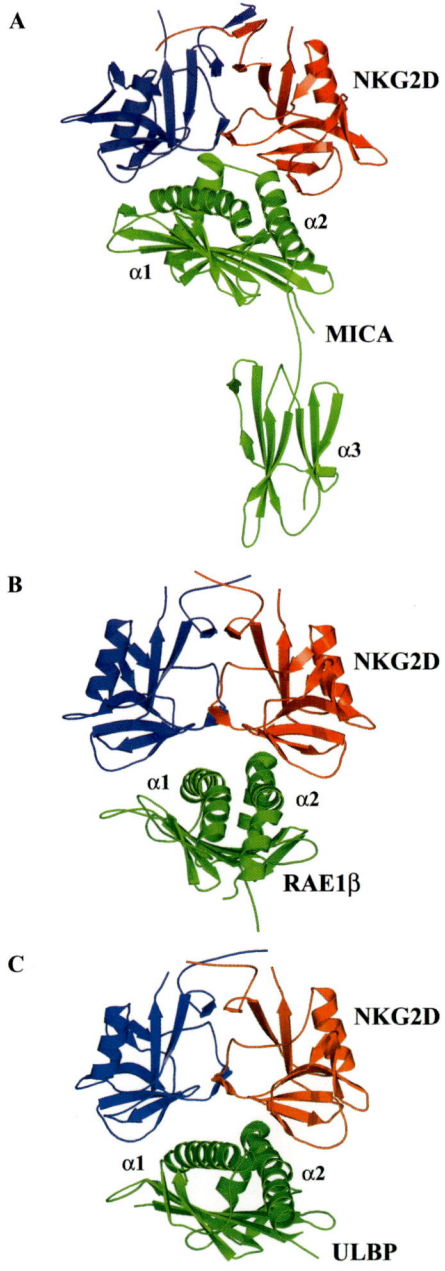

Figure 4.9. Interaction of NKG2D with the MHC class I homologs MIC-A, RAE-1β, and ULBP3. (**A**) Ribbon representation of the human NKG2D/MIC-A complex (Li et al., 2001a). (**B**) The mouse NKG2D–RAE-1β complex (Li et al., 2002). (**C**) The mouse NKG2D–ULBP3 complex (Radaev et al., 2001). MIC-A, RAE-1b, and ULBP3 are *green*; the NKG2D monomers are *blue* and *red*.

exhibit slow on-rates resulting from high entropic barriers to forming an ordered transition state (Davis et al., 1998); the resulting unfavorable entropy changes on binding are compensated by favorable enthalpic terms. In addition, TCR–peptide/MHC interactions are characterized by large heat capacity changes (ΔC_p) (Willcox et al., 1999), presumably as a result of the ordering of flexible binding sites, concomitant with the burial of protein surface and release of ordered water molecules. However, the binding of NKG2D to MIC-A, MIC-B, ULBP1, and RAE-1β displayed fast on-rates and activation energies for association close to the diffusion limit (McFarland and Strong, 2003). Moreover, entropy drove the reactions more than enthalpy, while the experimentally measured heat capacity changes closely matched values calculated using empirical relationships between ΔC_p and buried surface area that assume no significant conformational changes in the interacting species on complex formation. Thus, the binding energetics of NKG2D–ligand interactions are inconsistent with induced-fit recognition, but are fully compatible with rigid protein–protein association.

REFERENCES

Ackers, G.K., and Smith, F.R. (1985). Effects of site-specific amino acid modification on protein interactions and biological function. *Annu. Rev. Biochem.* 54: 597–629.

Al-Lazikani, B., Lesk, A.M., and Chothia, C. (1997). Standard conformations for the canonical structures of immunoglobulins. *J. Mol. Biol.* 273:927–948.

Al-Lazikani, B., Lesk, A.M., and Chothia, C. (2000). Canonical structures for the hypervariable regions of T cell alphabeta receptors. *J. Mol. Biol.* 295:979–995.

Alam, S.M., Travers, P.J., Wung, J.L., Nasholds, W., Redpath, S., Jameson, S.C., and Gascoigne, N.R. (1996). T-cell-receptor affinity and thymocyte positive selection. *Nature* 381:616–620.

Alam, S.M., Davies, G.M., Lin, C.M., Zal, T., Nasholds, W., Jameson, S.C., Hogquist, K.A., Gascoigne, N.R., and Travers, P.J. (1999). Qualitative and quantitative differences in T cell receptor binding of agonist and antagonist ligands. *Immunity* 10:227–237.

Allison, T.J., Winter, C.C., Fournie, J.J., Bonneville, M., and Garboczi, D.N. (2001). Structure of a human gammadelta T-cell antigen receptor. *Nature* 411:820–824.

Amzel, L.M., and Poljak, R.J. (1979). Three-dimensional structure of immunoglobulins. *Annu. Rev. Biochem.* 48:961–997.

Anderson, S.K., Ortaldo, J.R., and McVicar, D.W. (2001). The ever-expanding Ly49 gene family: repertoire and signaling. *Immunol. Rev.* 181:79–89.

Anikeeva, N., Lebedeva, T., Krogsgaard, M., Tetin, S.Y., Martinez-Hackert, E., Kalams, S.A., Davis, M.M., and Sykulev, Y. (2003). Distinct molecular mechanisms account for the specificity of two different T-cell receptors. *Biochemistry* 42:4709–4716.

Baker, B.M., Gagnon, S.J., Biddison, W.E., and Wiley, D.C. (2000). Conversion of a T cell antagonist into an agonist by repairing a defect in the TCR/peptide/MHC interface: implications for TCR signaling. *Immunity* 13:475–484.

Batista, F.D., and Neuberger, M.S. (1998). Affinity dependence of the B cell response to antigen: a threshold, a ceiling, and the importance of off-rate. *Immunity* 8:751–759.

Batista, F.D., and Neuberger, M.S. (2000). B cells extract and present immobilized antigen: implications for affinity discrimination. *Embo J.* 19:513–520.

Berthet-Colominas, C., Monaco, S., Novelli, A., Sibai, G., Mallet, F., and Cusack, S. (1999). Head-to-tail dimers and interdomain flexibility revealed by the crystal structure of HIV-1 capsid protein (p24) complexed with a monoclonal antibody Fab. *Embo J.* 18:1124–1136.

Bhat, T.N., Bentley, G.A., Boulot, G., Greene, M.I., Tello, D., Dall'Acqua, W., Souchon, H., Schwarz, F.P., Mariuzza, R.A., and Poljak, R.J. (1994). Bound water molecules and conformational stabilization help mediate an antigen-antibody association. *Proc. Natl. Acad. Sci. USA* 91:1089–1093.

Boniface, J.J., Reich, Z., Lyons, D.S., and Davis, M.M. (1999). Thermodynamics of T cell receptor binding to peptide-MHC: evidence for a general mechanism of molecular scanning. *Proc. Natl. Acad. Sci. USA* 96:11446–11451.

Braden, B.C., Souchon, H., Eisele, J.L., Bentley, G.A., Bhat, T.N., Navaza, J., and Poljak, R.J. (1994). Three-dimensional structures of the free and the antigencomplexed Fab from monoclonal anti-lysozyme antibody D44.1. *J. Mol. Biol.* 243, 767–781.

Braden, B.C., Fields, B.A., and Poljak, R.J. (1995). Conservation of water molecules in an antibody-antigen interaction. *J. Mol. Recognit.* 8:317–325.

Braden, B.C., Fields, B.A., Ysern, X., Dall'Acqua, W., Goldbaum, F.A., Poljak, R.J., and Mariuzza, R.A. (1996a). Crystal structure of an Fv-Fv idiotope-anti-idiotope complex at 1.9 A resolution. *J. Mol. Biol.* 264:137–151.

Braden, B.C., Fields, B.A., Ysern, X., Goldbaum, F.A., Dall'Acqua, W., Schwarz, F.P., Poljak, R.J., and Mariuzza, R.A. (1996b). Crystal structure of the complex of the variable domain of antibody D1.3 and turkey egg white lysozyme: a novel conformational change in antibody CDR-L3 selects for antigen. *J. Mol. Biol.* 257:889–894.

Braden, B.C., Goldman, E.R., Mariuzza, R.A., and Poljak, R.J. (1998). Anatomy of an antibody molecule: structure, kinetics, thermodynamics and mutational studies of the antilysozyme antibody D1.3. *Immunol. Rev.* 163:45–57.

Bromley, S.K., Burack, W.R., Johnson, K.G., Somersalo, K., Sims, T.N., Sumen, C., Davis, M.M., Shaw, A.S., Allen, P.M., and Dustin, M.L. (2001). The immunological synapse. *Annu. Rev. Immunol.* 19:375–396.

Buslepp, J., Wang, H., Biddison, W.E., Appella, E., and Collins, E.J. (2003). A correlation between TCR Valpha docking on MHC and CD8 dependence: implications for T cell selection. *Immunity* 19:595–606.

Chacko, S., Silverton, E., Kam-Morgan, L., Smith-Gill, S., Cohen, G., and Davies, D. (1995). Structure of an antibody-lysozyme complex unexpected effect of conservative mutation. *J. Mol. Biol.* 245:261–274.

Chersi, A., Galati, R., Ogino, T., Butler, R.H., and Tanigaki, N. (2002). Anti-peptide antibodies that recognize conformational differences of HLA class I intracytoplasmic domains. *Hum. Immunol.* 63:731–741.

Chevalier, B.S., Kortemme, T., Chadsey, M.S., Baker, D., Monnat, R.J., and Stoddard, B.L. (2002). Design, activity, and structure of a highly specific artificial endonuclease. *Mol. Cell.* 10:895–905.

Chitarra, V., Alzari, P.M., Bentley, G.A., Bhat, T.N., Eisele, J.L., Houdusse, A., Lescar, J., Souchon, H., and Poljak, R. J. (1993). Three-dimensional structure of a heteroclitic antigen-antibody cross-reaction complex. *Proc. Natl. Acad. Sci. USA* 90:7711–7715.

Chothia, C., Lesk, A.M., Tramontano, A., Levitt, M., Smith-Gill, S.J., Air, G., Sheriff, S., Padlan, E.A., Davies, D., Tulip, W.R., Colman, P.R., Spinelli, S., Alzari, P.M. and Poljak, R.J. (1989). Conformations of immunoglobulin hypervariable regions. *Nature* 342:877–883.

Chothia, C., Lesk, A.M., Gherardi, E., Tomlinson, I.M., Walter, G., Marks, J.D., Llewelyn, M.B., and Winter, G. (1992). Structural repertoire of the human VH segments. *J. Mol. Biol.* 227:799–817.

Clackson, T. and Wells, J.A. (1995). A hot spot of binding energy in a hormonereceptor interface. *Science* 267:383–386.

Conte, L.L., Chothia, C., and Janin, J. (1999). The atomic structure of protein–protein recognition sites. *J. Mol. Biol.* 285:2177–2198.

Dall'Acqua, W., Goldman, E.R., Eisenstein, E., and Mariuzza, R.A. (1996). A mutational analysis of the binding of two different proteins to the same antibody. *Biochemistry* 35:9667–9676.

Dall'Acqua, W., Goldman, E.R., Lin, W., Teng, C., Tsuchiya, D., Li, H., Ysern, X., Braden, B.C., Li, Y., Smith-Gill, S.J., and Mariuzza, R.A. (1998). A mutational analysis of binding interactions in an antigen-antibody protein-protein complex. *Biochemistry* 37:7981–7991.

Dam, J., Guan, R., Natarajan, K., Dimasi, N., Chlewicki, L.K., Kranz, D.M., Schuck, P., Margulies, D.H., and Mariuzza, R.A. (2003). Variable MHC class I engagement by Ly49 natural killer cell receptors demonstrated by the crystal structure of Ly49C bound to H-2K(b). *Nat. Immunol.* 4:1213–1222.

Davis, M.M., Boniface, J.J., Reich, Z., Lyons, D., Hampl, J., Arden, B., and Chien, Y. (1998). Ligand recognition by alpha beta T cell receptors. *Annu. Rev. Immunol.* 16:523–544.

Decanniere, K., Desmyter, A., Lauwereys, M., Ghahroudi, M.A., Muyldermans, S., and Wyns, L. (1999). A single-domain antibody fragment in complex with RNase A: noncanonical loop structures and nanomolar affinity using two CDR loops. *Struct. Fold. Des.* 7:361–370.

Degano, M., Garcia, K.C., Apostolopoulos, V., Rudolph, M.G., Teyton, L., and Wilson, I.A. (2000). A functional hot spot for antigen recognition in a superagonist TCR/MHC complex. *Immunity* 12:251–261.

De Genst, E., Areskoug, D., Decanniere, K., Muyldermans, S., and Andersson, K. (2002). Kinetic and affinity predictions of a protein-protein interaction using multivariate experimental design. *J. Biol. Chem.* 277:29897–29907.

Ding, Y.H., Smith, K.J., Garboczi, D.N., Utz, U., Biddison, W.E., and Wiley, D.C. (1998). Two human T cell receptors bind in a similar diagonal mode to the HLAA2/Tax peptide complex using different TCR amino acids. *Immunity* 8:403–411.

Ding, Y.H., Baker, B.M., Garboczi, D.N., Biddison, W.E., and Wiley, D.C. (1999). Four A6-TCR/peptide/HLA-A2 structures that generate very different T cell signals are nearly identical. *Immunity* 11:45–56.

Dyson, H.J., and Wright, P.E. (1995). Antigenic peptides. *Faseb J.* 9:37–42.

England, P., Bregegere, F., and Bedouelle, H. (1997). Energetic and kinetic contributions of contact residues of antibody D1.3 in the interaction with lysozyme. *Biochemistry* 36:164–172.

England, P., Nageotte, R., Renard, M., Page, A.L., and Bedouelle, H. (1999). Functional characterization of the somatic hypermutation process leading to antibody D1.3, a high affinity antibody directed against lysozyme. *J. Immunol.* 162:2129–2136.

Faelber, K., Kirchhofer, D., Presta, L., Kelley, R.F., and Muller, Y.A. (2001). The 1.85 A resolution crystal structures of tissue factor in complex with humanized Fab D3h44 and of free humanized Fab D3h44: revisiting the solvation of antigen combining sites. *J. Mol. Biol.* 313:83–97.

Fersht, A.R. (1988). Relationships between apparent binding energies measured in sitedirected mutagenesis experiments and energetics of binding and catalysis. *Biochemistry* 27:1577–1580.

Fields, B.A., Goldbaum, F.A., Ysern, X., Poljak, R.J., and Mariuzza, R.A. (1995). Molecular basis of antigen mimicry by an anti-idiotope. *Nature* 374:739–742.

Fields, B.A., Goldbaum, F.A., Dall'Acqua, W., Malchiodi, E.L., Cauerhff, A., Schwarz, F.P., Ysern, X., Poljak, R.J., and Mariuzza, R.A. (1996). Hydrogen bonding and solvent structure in an antigen-antibody interface. Crystal structures and thermodynamic characterization of three Fv mutants complexed with lysozyme. *Biochemistry* 35:15494–15503.

Foote, J., and Eisen, H.N. (1995). Kinetic and affinity limits on antibodies produced during immune responses. *Proc. Natl. Acad. Sci. USA* 92:1254–1256.

Furukawa, K., Akasako-Furukawa, A., Shirai, H., Nakamura, H., and Azuma, T. (1999). Junctional amino acids determine the maturation pathway of an antibody. *Immunity* 11:329–338.

Furukawa, K., Shirai, H., Azuma, T., and Nakamura, H. (2001). A role of the third complementarity-determining region in the affinity maturation of an antibody. *J. Biol. Chem.* 276:27622–27628.

Garcia, K.C., Ronco, P.M., Verroust, P.J., Brunger, A.T., and Amzel, L.M. (1992). Three-dimensional structure of an angiotensin II-Fab complex at 3 A: hormone recognition by an anti-idiotypic antibody. *Science* 257:502–507.

Garcia, K.C., Degano, M., Stanfield, R.L., Brunmark, A., Jackson, M.R., Peterson, P.A., Teyton, L., and Wilson, I.A. (1996). An alphabeta T cell receptor structure at 2.5A and its orientation in the TCR-MHC complex. *Science* 274:209–219.

Garcia, K.C., Tallquist, M.D., Pease, L.R., Brunmark, A., Scott, C.A., Degano, M., Stura, E.A., Peterson, P.A., Wilson, I.A., and Teyton, L. (1997). Alphabeta T cell receptor interactions with syngeneic and allogeneic ligands: affinity measurements and crystallization. *Proc. Natl. Acad. Sci. USA* 94:13838–13843.

Garcia, K.C., Degano, M., Pease, L.R., Huang, M., Peterson, P.A., Teyton, L., and Wilson, I.A. (1998). Structural basis of plasticity in T cell receptor recognition of a self peptide-MHC antigen. *Science* 279:1166–1172.

Gerstner, R.B., Carter, P., and Lowman, H. B. (2002). Sequence plasticity in the antigenbinding site of a therapeutic anti-HER2 antibody. *J. Mol. Biol.* 321:851–862.

Gibas, C.J., Subramaniam, S., McCammon, J.A., Braden, B.C., and Poljak, R.J. (1997). pH dependence of antibody/lysozyme complexation. *Biochemistry* 36:15599–15614.

Goldbaum, F.A., Schwarz, F.P., Eisenstein, E., Cauerhff, A., Mariuzza, R.A., and Poljak, R.J. (1996). The effect of water activity on the association constant and the enthalpy of reaction between lysozyme and the specific antibodies D1.3 and D44.1. *J. Mol. Recognit.* 9:6–12.

Goldman, E.R., Dall'Acqua, W., Braden, B.C., and Mariuzza, R.A. (1997). Analysis of binding interactions in an idiotope-antiidiotope protein–protein complex by double mutant cycles. *Biochemistry* 36:49–56.

Green, S.M., and Shortle, D. (1993). Patterns of nonadditivity between pairs of stability mutations in staphylococcal nuclease. *Biochemistry* 32:10131–10139.

Guermonprez, P., England, P., Bedouelle, H., and Leclerc, C. (1998). The rate of dissociation between antibody and antigen determines the efficiency of antibodymediated antigen presentation to T cells. *J. Immunol.* 161:4542–4548.

Hahn, M., Winkler, D., Welfle, K., Misselwitz, R., Welfle, H., Wessner, H., Zahn, G., Scholz, C., Seifert, M., Harkins, R., et al. (2001). Cross-reactive binding of cyclic peptides to an anti-TGFalpha antibody Fab fragment: an X-ray structural and thermodynamic analysis. *J. Mol. Biol.* 314:293–309.

Hamers-Casterman, C., Atarhouch, T., Muyldermans, S., Robinson, G., Hamers, C., Songa, E.B., Bendahman, N., and Hamers, R. (1993). Naturally occurring antibodies devoid of light chains. *Nature* 363:446–448.

Harata, K. (1994). X-ray structure of a monoclinic form of hen egg-white lysozyme crystallized at 313 K. Comparison of two independent molecules. *Acta Crystallogr.* D50:250–257.

Harris, L.J., Skaletsky, E., and McPherson, A. (1998). Crystallographic structure of an intact IgG1 monoclonal antibody. *J. Mol. Biol.* 275:861–872.

Hemmer, B., Vergelli, M., Pinilla, C., Houghten, R., and Martin, R. (1998). Probing degeneracy in T-cell recognition using peptide combinatorial libraries. *Immunol. Today* 19:163–168.

Hennecke, J., and Wiley, D.C. (2002). Structure of a complex of the human alpha/beta T cell receptor (TCR) HA1.7, influenza hemagglutinin peptide, and major histocompatibility complex class II molecule, HLA-DR4 (DRA*0101 and DRB1*0401): insight into TCR cross-restriction and alloreactivity. *J. Exp. Med.* 195:571–581.

Hennecke, J., Carfi, A., and Wiley, D.C. (2000). Structure of a covalently stabilized complex of a human alphabeta T-cell receptor, influenza HA peptide and MHC class II molecule, HLA-DR1. *Embo J.* 19:5611–5624.

Herron, J.N., He, X.M., Ballard, D.W., Blier, P.R., Pace, P.E., Bothwell, A.L., Voss, E.W., Jr., and Edmundson, A.B. (1991). An autoantibody to single-stranded DNA: comparison of the three-dimensional structures of the unliganded Fab and a deoxynucleotide-Fab complex. *Proteins* 11:159–175.

Hewer, R., and Meyer, D. (2003). Peptide immunogens based on the envelope region of HIV-1 are recognized by HIV/AIDS patient polyclonal antibodies and induce strong humoral immune responses in mice and rabbits. *Mol. Immunol.* 40:327–335.

Janeway, C., Travers, P., Mark, W., and Capra, J. (1999). *Immunobiology: The Immune System in Health and Disease*, 4th edit Current Biology Publications, New York.

Jones, S., and Thornton, J. M. (1996). Principles of protein-protein interactions. *Proc. Natl. Acad. Sci. USA* 93:13–20.

Kelley, R. F., O'Connell, M.P., Carter, P., Presta, L., Eigenbrot, C., Covarrubias, M., Snedecor, B., Bourell, J.H., and Vetterlein, D. (1992). Antigen binding thermodynamics and antiproliferative effects of chimeric and humanized antip185HER2 antibody Fab fragments. *Biochemistry* 31:5434–5441.

Kersh, E.N., Shaw, A.S., and Allen, P.M. (1998). Fidelity of T cell activation through multistep T cell receptor zeta phosphorylation. *Science* 281:572–575.

Kjer-Nielsen, L., Clements, C.S., Brooks, A.G., Purcell, A.W., Fontes, M.R., McCluskey, J., and Rossjohn, J. (2002a). The structure of HLA-B8 complexed to an immunodominant viral determinant: peptide-induced conformational changes and a mode of MHC class I dimerization. *J. Immunol.* 169:5153–5160.

Kjer-Nielsen, L., Clements, C.S., Brooks, A.G., Purcell, A.W., McCluskey, J., and Rossjohn, J. (2002b). The 1.5 A crystal structure of a highly selected antiviral T cell receptor provides evidence for a structural basis of immunodominance. *Structure (Camb)* 10:1521–1532.

Kjer-Nielsen, L., Clements, C.S., Purcell, A.W., Brooks, A.G., Whisstock, J.C., Burrows, S.R., McCluskey, J., and Rossjohn, J. (2003). A structural basis for the selection of dominant alphabeta T cell receptors in antiviral immunity. *Immunity* 18:53–64.

Kondo, H., Shiroishi, M., Matsushima, M., Tsumoto, K., and Kumagai, I. (1999). Crystal structure of anti-Hen egg white lysozyme antibody (HyHEL-10) Fv-antigen complex. Local structural changes in the protein antigen and water-mediated interactions of Fvantigen and light chain-heavy chain interfaces. *J. Biol. Chem.* 274:27623–27631.

Kortemme, T., and Baker, D. (2002). A simple physical model for binding energy hot spots in protein-protein complexes. *Proc. Natl. Acad. Sci. USA* 99:14116–14121.

Kouskoff, V., Famiglietti, S., Lacaud, G., Lang, P., Rider, J. E., Kay, B. K., Cambier, J. C., and Nemazee, D. (1998). Antigens varying in affinity for the B cell receptor induce differential B lymphocyte responses. *J. Exp. Med.* 188:1453–1464.

Krogsgaard, M., Prado, N., Adams, E.J., He, X.L., Chow, D.C., Wilson, D. B., Garcia, K.C., and Davis, M.M. (2003). Evidence that structural rearrangements and/or flexibility during TCR binding can contribute to T cell activation. *Mol. Cell.* 12:1367–1378.

Kurinov, I.V., and Harrison, R.W. (1995). The influence of temperature on lysozyme crystals. Structure and dynamics of protein and water. *Acta Crystallogr.* D51:98–109.

Lawrence, M.C., and Colman, P.M. (1993). Shape complementarity at protein/protein interfaces. *J. Mol. Biol.* 234:946–950.

Lehner, P.J., Wang, E.C., Moss, P.A., Williams, S., Platt, K., Friedman, S.M., Bell, J.I., and Borysiewicz, L.K. (1995). Human HLA-A0201-restricted cytotoxic T lymphocyte recognition of influenza A is dominated by T cells bearing the V beta 17 gene segment. *J. Exp. Med.* 181:79–91.

Li, Y., Li, H., Smith-Gill, S.J., and Mariuzza, R.A. (2000). Three-dimensional structures of the free and antigen-bound Fab from monoclonal antilysozyme antibody HyHEL-63(,). *Biochemistry* 39:6296–6309.

Li, P., Morris, D.L., Willcox, B.E., Steinle, A., Spies, T., and Strong, R.K. (2001a). Complex structure of the activating immunoreceptor NKG2D and its MHC class I-like ligand MICA. *Nat. Immunol.* 2:443–451.

Li, Y., Lipschultz, C.A., Mohan, S., and Smith-Gill, S.J. (2001b). Mutations of an epitope hot-spot residue alter rate limiting steps of antigen-antibody protein-protein associations. *Biochemistry* 40:2011–2022.

Li, P., McDermott, G., and Strong, R.K. (2002). Crystal structures of RAE-1beta and its complex with the activating immunoreceptor NKG2D. *Immunity* 16.77–86.

Li, Y., Li, H., Yang, F., Smith-Gill, S.J., and Mariuzza, R.A. (2003). X-ray snapshots of affinity maturation in an antibody–antigen protein–protein interface. *Nat. Struct. Biol.* 10:482–488.

LiCata, V.J., and Ackers, G.K. (1995). Long-range, small magnitude nonadditivity of mutational effects in proteins. *Biochemistry* 34:3133–3139.

Lipschultz, C.A., Li, Y., and Smith-Gill, S. (2000). Experimental design for analysis of complex kinetics using surface plasmon resonance. *Methods* 20:310–318.

Luz, J.G., Huang, M., Garcia, K.C., Rudolph, M.G., Apostolopoulos, V., Teyton, L., and Wilson, I.A. (2002). Structural comparison of allogeneic and syngeneic T cell receptor-peptide-major histocompatibility complex complexes: a buried alloreactive mutation subtly alters peptide presentation substantially increasing V(beta) Interactions. *J. Exp. Med.* 195:1175–1186.

Lyons, D.S., Lieberman, S.A., Hampl, J., Boniface, J.J., Chien, Y., Berg, L.J., and Davis, M.M. (1996). A TCR binds to antagonist ligands with lower affinities and faster dissociation rates than to agonists. *Immunity* 5:53–61.

MacCallum, R.M., Martin, A.C., and Thornton, J.M. (1996). Antibody-antigen interactions: contact analysis and binding site topography. *J. Mol. Biol.* 262:732–745.

Manivel, V., Sahoo, N.C., Salunke, D.M., and Rao, K.V. (2000). Maturation of an antibody response is governed by modulations in flexibility of the antigen-combining site. *Immunity* 13:611–620.

Manning, T.C., Schlueter, C.J., Brodnicki, T.C., Parke, E.A., Speir, J.A., Garcia, K.C., Teyton, L., Wilson, I.A., and Kranz, D.M. (1998). Alanine scanning mutagenesis of an alphabeta T cell receptor: mapping the energy of antigen recognition. *Immunity* 8:413–425.

Margulies, D.H. (1997). Interactions of TCRs with MHC-peptide complexes: a quantitative basis for mechanistic models. *Curr. Opin. Immunol.* 9:390–395.

Mason, D. (1998). A very high level of crossreactivity is an essential feature of the T-cell receptor. *Immunol. Today* 19:395–404.

McFarland, B.J., and Strong, R.K. (2003). Thermodynamic analysis of degenerate recognition by the NKG2D immunoreceptor: not induced fit but rigid adaptation. *Immunity* 19:803–812.

McFarland, B.J., Kortemme, T., Yu, S.F., Baker, D., and Strong, R.K. (2003). Symmetry recognizing asymmetry: analysis of the interactions between the C-type lectin-like immunoreceptor NKG2D and MHC class I-like ligands. *Structure (Camb)* 11:411–422.

McQueen, K.L., and Parham, P. (2002). Variable receptors controlling activation and inhibition of NK cells. *Curr. Opin. Immunol.* 14:615–621.

Metaxas, A., Tzartos, S., and Liakopoulou-Kyriakide, M. (2002). The production of antihexapeptide antibodies which recognize the S7, L6 and L13 ribosomal proteins of Escherichia coli. *J. Pept. Sci.* 8:118–124.

Monaco-Malbet, S., Berthet-Colominas, C., Novelli, A., Battai, N., Piga, N., Cheynet, V., Mallet, F., and Cusack, S. (2000). Mutual conformational adaptations in antigen and antibody upon complex formation between an Fab and HIV-1 capsid protein p24. *Struct. Fold. Des.* 8:1069–1077.

Moss, P.A., Moots, R.J., Rosenberg, W.M., Rowland-Jones, S.J., Bodmer, H.C., McMichael, A.J., and Bell, J.I. (1991). Extensive conservation of alpha and beta chains of the human T-cell antigen receptor recognizing HLA-A2 and influenza A matrix peptide. *Proc. Natl. Acad. Sci. USA* 88:8987–8990.

Muller, Y.A., Chen, Y., Christinger, H.W., Li, B., Cunningham, B. C., Lowman, H. B., and de Vos, A. M. (1998). VEGF and the Fab fragment of a humanized neutralizing antibody: crystal structure of the complex at 2.4 A resolution and mutational analysis of the interface. *Structure* 6:1153–1167.

Mylvaganam, S.E., Paterson, Y., and Getzoff, E.D. (1998). Structural basis for the binding of an anti-cytochrome c antibody to its antigen: crystal structures of FabE8-cytochrome c complex to 1.8 A resolution and FabE8 to 2.26 A resolution. *J. Mol. Biol.* 281:301–322.

Natarajan, K., Dimasi, N., Wang, J., Mariuzza, R.A., and Margulies, D.H. (2002). Structure and function of natural killer cell receptors: multiple molecular solutions to self, nonself discrimination. *Annu. Rev. Immunol.* 20:853–885.

Omelyanenko, V.G., Jiskoot, W., and Herron, J. N. (1993). Role of electrostatic interactions in the binding of fluorescein by anti-fluorescein antibody 4-4-20. *Biochemistry* 32:10423–10429.

Padlan, E.A., Silverton, E.W., Sheriff, S., Cohen, G.H., Smith-Gill, S.J., and Davies, D.R. (1989). Structure of an antibody-antigen complex: crystal structure of the HyHEL-10 Fab-lysozyme complex. *Proc. Natl. Acad. Sci. USA* 86:5938–5942.

Palmer, E. (2003). Negative selection—clearing out the bad apples from the T-cell repertoire. *Nat. Rev. Immunol.* 3:383–391.

Pan, Y., Yuhasz, S.C., and Amzel, L.M. (1995). Anti-idiotypic antibodies: biological function and structural studies. *FASEB J.* 9:43–49.

Patten, P.A., Gray, N.S., Yang, P.L., Marks, C.B., Wedemayer, G.J., Boniface, J.J., Stevens, R.C., and Schultz, P.G. (1996). The immunological evolution of catalysis. *Science* 271:1086–1091.

Pieper, R., Christian, R.E., Gonzales, M.I., Nishimura, M.I., Gupta, G., Settlage, R.E., Shabanowitz, J., Rosenberg, S.A., Hunt, D.F., and Topalian, S.L. (1999). Biochemical identification of a mutated human melanoma antigen recognized by CD4(+) T cells. *J. Exp. Med.* 189:757–766.

Poljak, R.J. (1994). An idiotope–anti-idiotope complex and the structural basis of molecular mimicking. *Proc. Natl. Acad. Sci. USA* 91:1599–1600.

Prasad, G.S., Earhart, C.A., Murray, D.L., Novick, R.P., Schlievert, P.M., and Ohlendorf, D.H. (1993). Structure of toxic shock syndrome toxin 1. *Biochemistry* 32:13761–13766.

Radaev, S., Rostro, B., Brooks, A.G., Colonna, M., and Sun, P.D. (2001). Conformational plasticity revealed by the cocrystal structure of NKG2D and its class I MHC-like ligand ULBP3. *Immunity* 15:1039–1049.

Rajewsky, K. (1996). Clonal selection and learning in the antibody system. *Nature* 381:751–758.

Rajpal, A., and Kirsch, J.F. (2000). Role of the minor energetic determinants of chicken egg white lysozyme (HEWL) to the stability of the HEWL.antibody scFv-10 complex. *Proteins* 40:49–57.

Ramanadham, M., Sieker, L.C., and Jensen, L.H. (1990). Refinement of triclinic lysozyme: II. The method of stereochemically restrained least squares. *Acta Crystallogr B* 46 (Pt 1):63–69.

Reiser, J.B., Darnault, C., Guimezanes, A., Gregoire, C., Mosser, T., Schmitt-Verhulst, A.M., Fontecilla-Camps, J.C., Malissen, B., Housset, D., and Mazza, G. (2000). Crystal structure of a T cell receptor bound to an allogeneic MHC molecule. *Nat. Immunol.* 1:291–297.

Reiser, J.B., Gregoire, C., Darnault, C., Mosser, T., Guimezanes, A., Schmitt-Verhulst, A.M., Fontecilla-Camps, J.C., Mazza, G., Malissen, B., and Housset, D. (2002). A T cell receptor CDR3beta loop undergoes conformational changes of unprecedented magnitude upon binding to a peptide/MHC class I complex. *Immunity* 16:345–354.

Reiser, J.B., Darnault, C., Gregoire, C., Mosser, T., Mazza, G., Kearney, A., van der Merwe, P.A., Fontecilla-Camps, J.C., Housset, D., and Malissen, B. (2003). CDR3 loop flexibility contributes to the degeneracy of TCR recognition. *Nat. Immunol.* 4:241–247.

Rini, J.M., Schulze-Gahmen, U., and Wilson, I.A. (1992). Structural evidence for induced fit as a mechanism for antibody-antigen recognition. *Science* 255:959–965.

Romesberg, F.E., Spiller, B., Schultz, P.G., and Stevens, R.C. (1998). Immunological origins of binding and catalysis in a Diels-Alderase antibody. *Science* 279:1929–1933.

Roost, H.P., Bachmann, M.F., Haag, A., Kalinke, U., Pliska, V., Hengartner, H., and Zinkernagel, R.M. (1995). Early high-affinity neutralizing anti-viral IgG responses without further overall improvements of affinity. *Proc. Natl. Acad. Sci. USA* 92:1257–1261.

Rudolph, M.G., and Wilson, I.A. (2002). The specificity of TCR/pMHC interaction. *Curr. Opin. Immunol.* 14:52–65.

Schreiber, G., and Fersht, A.R. (1995). Energetics of protein–protein interactions: analysis of the barnase-barstar interface by single mutations and double mutant cycles. *J. Mol. Biol.* 248:478–486.

Serrano, L., Horovitz, A., Avron, B., Bycroft, M., and Fersht, A.R. (1990). Estimating the contribution of engineered surface electrostatic interactions to protein stability by using double-mutant cycles. *Biochemistry* 29:9343–9352.

Sherman, L.A., and Chattopadhyay, S. (1993). The molecular basis of allorecognition. *Annu. Rev. Immunol.* 11:385–402.

Sigurskjold, B.W., Altman, E., and Bundle, D.R. (1991). Sensitive titration microcalorimetric study of the binding of Salmonella O-antigenic oligosaccharides by a monoclonal antibody. *Eur. J. Biochem.* 197:239–246.

Speir, J.A., Stevens, J., Joly, E., Butcher, G.W., and Wilson, I.A. (2001). Two different, highly exposed, bulged structures for an unusually long peptide bound to rat MHC class I RT1-Aa. *Immunity* 14:81–92.

Stanfield, R.L., Fieser, T.M., Lerner, R.A., and Wilson, I.A. (1990). Crystal structures of an antibody to a peptide and its complex with peptide antigen at 2.8 A. *Science* 248:712–719.

Stewart-Jones, G.B., McMichael, A.J., Bell, J.I., Stuart, D.I., and Jones, E.Y. (2003). A structural basis for immunodominant human T cell receptor recognition. *Nat. Immunol.* 4:657–663.

Sundberg, E.J., Urrutia, M., Braden, B.C., Isern, J., Tsuchiya, D., Fields, B.A., Malchiodi, E.L., Tormo, J., Schwarz, F.P., and Mariuzza, R.A. (2000). Estimation of the hydrophobic effect in an antigen–antibody protein-protein interface. *Biochemistry* 39:15375–15387.

Sundberg, E.J., Li, Y., and Mariuzza, R.A. (2002a). So many ways of getting in the way: diversity in the molecular architecture of superantigen-dependent T-cell signaling complexes. *Curr. Opin. Immunol.* 14:36–44.

Sundberg, E.J., Sawicki, M.W., Southwood, S., Andersen, P.S., Sette, A., and Mariuzza, R.A. (2002b). Minor structural changes in a mutated human melanoma antigen correspond to dramatically enhanced stimulation of a CD4+ tumor-infiltrating lymphocyte line. *J. Mol. Biol.* 319:449–461.

Tomlinson, I.M., Cox, J.P., Gherardi, E., Lesk, A.M., and Chothia, C. (1995). The structural repertoire of the human V kappa domain. *EMBO J.* 14:4628–4638.

Tomlinson, I.M., Walter, G., Jones, P.T., Dear, P.H., Sonnhammer, E.L., and Winter, G. (1996). The imprint of somatic hypermutation on the repertoire of human germline V genes. *J. Mol. Biol.* 256:813–817.

Tonegawa, S. (1983). Somatic generation of antibody diversity. *Nature* 302:575–581.

Tormo, J., Blaas, D., Parry, N.R., Rowlands, D., Stuart, D., and Fita, I. (1994). Crystal structure of a human rhinovirus neutralizing antibody complexed with a peptide derived from viral capsid protein VP2. *EMBO J.* 13:2247–2256.

Tormo, J., Natarajan, K., Margulies, D.H., and Mariuzza, R.A. (1999). Crystal structure of a lectin-like natural killer cell receptor bound to its MHC class I ligand. *Nature* 402:623–631.

Ulrich, H.D., Mundorff, E., Santarsiero, B.D., Driggers, E.M., Stevens, R.C., and Schultz, P.G. (1997). The interplay between binding energy and catalysis in the evolution of a catalytic antibody. *Nature* 389:271–275.

Vivier, E., Tomasello, E., and Paul, P. (2002). Lymphocyte activation via NKG2D: towards a new paradigm in immune recognition? *Curr. Opin. Immunol.* 14:306–311.

Wang, J., Whitman, M.C., Natarajan, K., Tormo, J., Mariuzza, R.A., and Margulies, D. H. (2002). Binding of the natural killer cell inhibitory receptor Ly49A to its major histocompatibility complex class I ligand. Crucial contacts include both H-2Dd AND beta 2-microglobulin. *J. Biol. Chem.* 277:1433–1442.

Webster, D.M., Henry, A.H., and Rees, A.R. (1994). Antibody–antigen interactions. *Curr. Opin. Struct. Biol.* 4:123–129.

Wedemayer, G.J., Patten, P.A., Wang, L.H., Schultz, P.G., and Stevens, R.C. (1997a). Structural insights into the evolution of an antibody combining site. *Science* 276:1665–1669.

Wedemayer, G.J., Wang, L.H., Patten, P.A., Schultz, P.G., and Stevens, R.C. (1997b). Crystal structures of the free and liganded form of an esterolytic catalytic antibody. *J. Mol. Biol.* 268:390–400.

Willcox, B.E., Gao, G.F., Wyer, J.R., Ladbury, J.E., Bell, J.I., Jakobsen, B.K., and van der Merwe, P. A. (1999). TCR binding to peptide-MHC stabilizes a flexible recognition interface. *Immunity* 10:357–365.

Wilson, I.A., and Stanfield, R.L. (1993). Antibody-antigen interactions. *Curr. Opin. Struct. Biol.* 3:113–118.

Wilson, I.A., and Stanfield, R.L. (1994). Antibody-antigen interactions: new structures and new conformational changes. *Curr. Opin. Struct. Biol.* 4:857–867.

Wolan, D.W., Teyton, L., Rudolph, M.G., Villmow, B., Bauer, S., Busch, D.H., and Wilson, I.A. (2001). Crystal structure of the murine NK cell-activating receptor NKG2D at 1.95 A. *Nat. Immunol.* 2:248–254.

Wu, T.T., and Kabat, E.A. (1970). An analysis of the sequences of the variable regions of Bence Jones proteins and myeloma light chains and their implications for antibody complementarity. *J. Exp. Med.* 132:211–250.

Wu, L.C., Tuot, D.S., Lyons, D.S., Garcia, K.C., and Davis, M.M. (2002). Two-step binding mechanism for T-cell receptor recognition of peptide MHC. *Nature* 418:552–556.

Xavier, K.A., McDonald, S.M., McCammon, J.A., and Willson, R.C. (1999). Association and dissociation kinetics of bobwhite quail lysozyme with monoclonal antibody HyHEL-5. *Protein Eng.* 12:79–83.

Yang, J., Swaminathan, C.P., Huang, Y., Guan, R., Cho, S., Kieke, M.C., Kranz, D.M., Mariuzza, R.A., and Sundberg, E.J. (2003). Dissecting cooperative and additive binding energetics in the affinity maturation pathway of a protein-protein interface. *J. Biol. Chem.* 278:50412–50421.

Yang, P.L., and Schultz, P.G. (1999). Mutational analysis of the affinity maturation of antibody 48G7. *J. Mol. Biol.* 294:1191–1201.

Yokoyama, W.M., and Plougastel, B.F. (2003). Immune functions encoded by the natural killer gene complex. *Nat. Rev. Immunol.* 3:304–316.

Ysern, X., Fields, B.A., Bhat, T.N., Goldbaum, F.A., Dall'Acqua, W., Schwarz, F.P., Poljak, R.J., and Mariuzza, R.A. (1994). Solvent rearrangement in an antigen–antibody interface introduced by site-directed mutagenesis of the antibody combining site. *J. Mol. Biol.* 238:496–500.

5

Computational Methods for Predicting Protein–Protein Interactions

A. Walker-Taylor and D.T. Jones

ABSTRACT

Protein–protein interactions perform an integral role in a diverse range of cellular and extracellular processes. These interactions provide a means for cells to communicate both internally and externally; they facilitate the anabolic and catabolic reactions of metabolism; they are important for transcriptional and translational control; and they are also important in maintaining cell structure. These examples are by no means exhaustive, however. Understanding these protein–protein interactions is a very important area of research in light of their global implications. Moreover, the knowledge of how to predict these interactions would be extremely useful and could be exploited to block these interactions as a therapeutic intervention, for example, and also to implicate novel functional interactions. However, prediction of these interactions is by no means trivial, as many layers of complexity must be considered, but without any definitive rules that can be applied. This chapter highlights some of these complexities, describes available databases and repositories of protein–protein interactions, and discusses some of the main methods that are currently used to predict these interactions.

ALICE WALKER-TAYLOR • Computer Science Department, University College London, Gower Street, London, WC1E 6BT, UK.
DAVID T. JONES • Computer Science Department, University College London, Gower Street, London, WC1E 6BT, UK.

Proteomics and Protein–Protein Interactions: Biology, Chemistry, Bioinformatics, and Drug Design, edited by Waksman. Springer, New York, 2005.

1. BACKGROUND

The first two complete drafts of the human genome sequence were published in 2001 (Venter et al., 2001). Although minor differences were noted between the two drafts, the overall conclusions concerning gene numbers, repeated sequences, and chromosomal organization were remarkably similar. Furthermore, both groups identified 30,000 to 35,000 genes, far fewer than the 100,000 expected from an earlier (admittedly "back of the envelope") calculation (Hood and Galas, 2003). One fitting hypothesis for this is that molecular complexity arises as a result of the multitude of interactions between these proteins, and not simply because of the sheer number of proteins involved. It is these interactions that are described in this chapter.

Many other genomes have now been fully sequenced as a result of various sequencing projects, and the impact that all these projects has had on bioinformatics research is enormous. The sequencing of these genomes is necessary but requires further interpretation to provide a more complete understanding of these living systems. The useful pieces of information of these genome sequences are the genes embedded within the noise from the intervening ("junk") DNA. As these genes have the potential to be transcribed and translated, it follows that there is a vast amount of information on protein sequences that can be potentially extracted.

Now is undoubtedly the time to be making the most of the available information and preparing for the future influx of these gene and protein sequences. In isolation of other transcription and translation information, gene sequences are also of limited value, but they provide the necessary starting point that paves the way for the next steps of a research pathway. Thus, the next logical step is to determine the biochemical functions of these proteins and their interactions (proteomics), two factors that are often tightly linked.

1.1. The Importance and Variety of Protein–Protein Interactions

Most proteins within cells do not exist in isolation, but interact with other proteins either directly or indirectly, via mechanisms that ultimately involve some form of binding. A protein's cellular function is its role within a living cell (in vivo). This relies entirely on the repertoire of the other cellular components with which the protein must interact. Most proteins in a cell function cooperatively in highly ordered networks, in order for the cellular metabolism to run smoothly. They may interact with a cell's nucleic acids, in which case the function may be to regulate gene transcription or control DNA replication. Another example is the binding to proteins of small nonprotein ligands, as seen in the case of enzyme binding to prevent catalytic activity. An enzyme's catalytic function may also be otherwise modified by binding of a highly reactive metal ion to the active site, such as for electron transfer reactions.

Finally, proteins may interact with other proteins. These interactions may provide a variety of functions such as catalytic, structural, localization, cleavage, transferral, or inhibitory functions. Protein localization occurs to position a protein in a specific cellular location. This allows the cellular contents to remain in their highly ordered compartments, and prevents any mixing that may produce potentially undesirable reactions. Cleavage may convert a proenzyme into its catalytic form, and ligand binding may put the partner protein in a desirable orientation for the cleavage site to be exposed. Transferring a particular chemical group from one protein to another may alter its function, or enable it to further propagate a cellular pathway. As with small protein inhibitors, these large proteins may also bind in order to provide an inhibitory function (blocking undesirable reactions).

The presence of the highly ordered pathways and networks in cells illustrates the importance of protein–protein interactions. The positive and negative feedback mechanisms involved in all these interaction networks enables cells to regulate their activities with immense accuracy. The rate of protein synthesis, translocation, and degradation define the cellular concentration of a particular protein. It may be a critical concentration of this protein that is needed as the trigger for a particular interaction.

Alongside the repertoire of essential "housekeeping" proteins, specialized cells possess cell-specific proteins that allow the cells to carry out their unique function. The ability to predict protein–protein interactions allows potential targeted regulation of the protein in these specialized cells. This in turn, gives potential to regulate the cell's activity within the organism. In addition, prediction of these interactions may provide clues as to which gene to regulate at the DNA level.

The ability to predict protein–protein interactions and develop certain rules or hypotheses concerning such interactions has obvious implications in elucidating biological pathways. There is also the obvious implication of this work for the design of therapeutic drugs, which could be targeted against specific proteins if their interaction "mechanism" was solved.

1.2. Protein–Protein Interaction Prediction Using Known Structures

Some prominent methods of protein interaction prediction rely on having experimentally determined protein structures and complexes to work with, and these can be obtained by way of their crystallization. A protein's structure can then be solved from the synthesized crystals using X-ray diffraction or neutron diffraction analysis. On the other hand, nuclear magnetic resonance (NMR) is a technique that is generally used for proteins in solution, usually used for proteins that cannot be crystallized. The use of this NMR technique for solving many protein structures is restricted, however, as it is limited to small structures or complexes (of about 30 KDa). To generate the output, NMR spectra are generated by placing a sample in a magnetic field and applying radiofrequency pulses.

Solved protein structures, as well as theoretical models, are deposited in a databank of protein structures called the Protein Data Bank (PDB) (Bernstein et al., 1977) and this databank is freely accessible on the Web (http://www.rsbc.org). It is a redundant databank, largely populated by uncomplexed protein crystal structures. As of January 13, 2004, it contained 23,914 deposited structures (Fig. 5.1) but only a very small percentage of these represent protein–protein complexes. The reason for this is that crystallization is a very elusive and time-consuming procedure, involving a great deal of experimental trial and error to find the optimal crystallization conditions. The challenge of crystallization is multiplied when looking to crystallize a

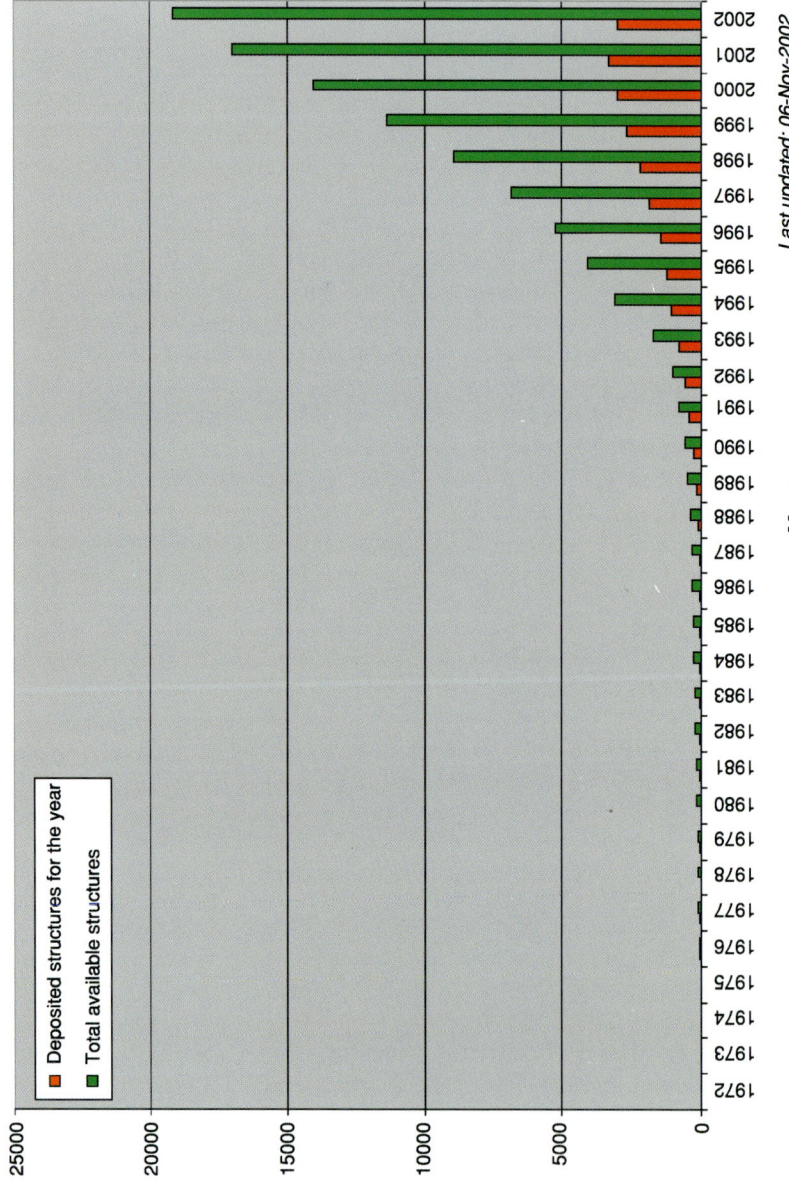

Figure 5.1. Data on the growth in the number of structures in the PDB.

protein complex rather than just an uncomplexed structure. Indeed, it may be of value to find a method that may be able to somehow standardize these conditions, or find some kind of correlation between the conditions required for crystallization and the nature of the protein structure.

The crystallization procedure generates many individual structures bound in a lattice formation of crystal contacts. Thus, it is important to be able to distinguish these crystal contacts from biologically relevant contacts. The ability to recognize the difference between these two kinds of interface by means of conservation has been studied by Elcock and McCammon (2001) and Valdar and Thornton (2001a,b). The comparison of these interfaces and ability to distinguish them by size was investigated by Janin and Rodier (1995).

The protein–protein structures present in the PDB for which the monomeric structures have been solved individually are of particular value for the development of protein–protein interaction prediction techniques. In these cases the final target structure is available, but the native unbound proteins are also available for work in trying to predict this structure. These unbound structures may show subtle conformational differences to the proteins present in their complexed form. The ultimate goal is to find interactions of these unbound monomeric proteins. Docking is perhaps the primary method used to predict interactions from these protein structures. Many docking procedures have been created and are being continually refined to enable accurate predictions (Halperin et al., 2002). These methods involve protein structures that have been solved to a high resolution, in order to enable interaction prediction to use atomic detail.

1.3. Prediction in the Absence of Protein Structures

Having addressed the idea of using solved protein structures in the prediction of protein–protein interactions, it should be mentioned that numerous methods have also been designed to predict interactions without these structural data. Such methods are sequence based, and include the use of interaction motifs and use of evolutionary information that is discussed in more detail later in this chapter.

1.4. Databases of Protein–Protein Interactions

As previously mentioned, the major source of data for protein–protein interactions remains the Protein Data Bank (PDB) (Bernstein et al., 1977), which provides crystallographic data on protein–protein complexes. At the end of 2002 there were just over 450 structures of complexes in the PDB. If, however, we remove theoretical models and close sequence homologs, these 450 structures reduce to approximately 130, which is a very small dataset on which to base any bioinformatics approaches to identifying protein–protein interactions. A further problem with PDB is that in many cases the data bank entries do not contain the biologically relevant multimer. For this reason, the Protein Quarternary Structure (PQS) database was set up at the European Bioinformatics Institute (EBI) (http://www.pqs.ebi.ac.uk). The files of the

PQS database are created from a program that generates the probable whole structure representing the natural biological state of the complex. The program that generates the PQS files uses the crystal symmetry operations recorded in the PDB file to build up the complete unit cell from the given asymmetric unit. The interfaces generated are then assessed as to whether they are likely to be crystal contacts or true biological interfaces. This is achieved by using a set of empirical tests, including evaluating the degree of sequence conservation of the interfacial residues (Ponstingl et al., 2003). Thus, these inconsistencies with the PDB files are all taken into account when the files are parsed to find and analyze those residues found in contact.

Although atomic resolution structural data remains the gold standard for predicting and modeling protein–protein interactions, the recent development of other experimental techniques for experimentally determining interacting pairs of proteins has resulted in the development of a number of other protein–protein interaction databases.

A number of databases of protein–protein interactions have been developed. Probably the most widely known of these are Biomolecular Interaction Network Database (BIND; Bader et al., 2003), Database of Interacting Proteins (DIP; Xenarios et al., 2000) and a Molecular Interactions Database (MINT; Zanzoni et al., 2002). However, a number of more specific databases, such as Curagen's PathCalling yeast interaction database (Uetz et al., 2000), are available maintained.

Although the various interaction databases provide valuable resources, the lack of a standard data model, lack of quality control, and redundancy reduces their overall overall value. In answer to these problems, the IntAct project (Hermjakob et al., 2004) was started, and aims to define a standard for the representation and annotation of protein–protein interaction data, and to provide a public repository of experimental data on protein–protein interactions (http://www.ebi.ac.uk/intact/index.html).

If we look at the DIP database as a typical protein–protein interaction database, we see that more than 30 different experimental techniques are in fact represented in DIP (see Fig. 5.2). In April 2000 the most frequently used methods included affinity column (96 entries), in vitro binding (100 entries), immunoblotting (109 entries), coimmunoprecipitation (248 entries), and immunoprecipitation (574 entries). However, by far the most popular technique in DIP comes from one large study based on the yeast two-hybrid test (1903 entries). In some cases, the interactions documented are determined with many different experiments. As such, it is possible to evaluate the confidence of an interaction being a true positive by the number and the reliability of the particular experiments performed. These confidence values are given as an association constant for the interacting proteins, but in February 2002 this totaled only 32 complexes.

2. METHOD FOR PREDICTING PROTEIN–PROTEIN INTERACTIONS

2.1. Global Interface Properties

A study by Jones and Thornton (1996) characterized interfaces in terms of residue propensities, conservation, interaction propensities, protrusion, and planarity, to name

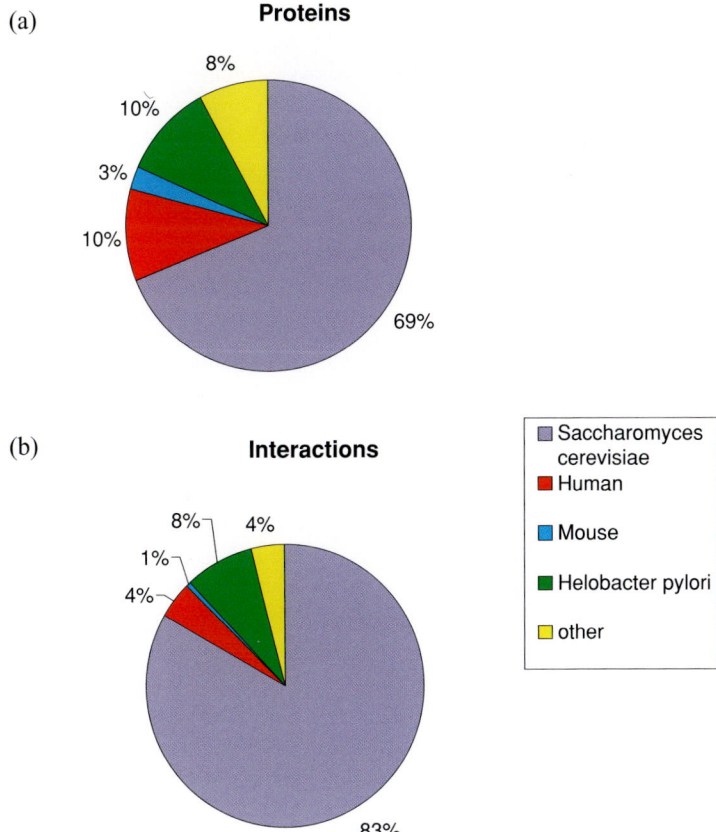

Figure 5.2. DIP organism protein (**a**) and interaction (**b**) distribution (February 2002 and April 2000). Percentages are of total proteins and interactions present.

a few, and found them distinguishable from the rest of the protein surface. It is possible, however, that in some cases most of the protein surface contributes to a protein–protein interaction with one or multiple interaction partners (binding with distinct interfaces on the surface of the same protein). Thus, the problem with defining the rest of the protein surface is that this surface may form part of other interfaces, which would invalidate a proper statistical comparison.

Establishing differences between the interface and the rest of the protein surface is also useful in order to distinguish real biological interactions from crystal contacts (artificial interfaces). The crystallization of a protein complex or monomer involves the association of individual biologically active units into a lattice formation of these units, and the interfaces that contribute to the lattice structure are the crystal contacts. Often in the PDB files the atomic coordinates of the smallest asymmetric part of the crystal are documented, but this does not necessarily represent the full biologically active structure. However, more of the crystal structure can easily be generated by symmetry building operations in the files. Factors such as the size of the interface and the level of residue conservation have been used as a way to distinguish between these two different types of interface (Valdar and Thornton, 2001a) when the biologically active state is in question. However, use has been made of these crystal contacts to derive protein–protein interaction potentials—in order to rank docked solutions (Robert and Janin, 1998)—relaxing the criteria that defines a protein–protein interface. There seems to be little justification for the use of these contacts in terms of biological reasoning that they should reflect functional interfaces however.

2.2. Classes of Interface

From an evolutionary perspective, protein–protein interfaces have evolved over time to optimize the interface to suit their individual biological functions. This function may have required the evolution of specific binding strength. Thus, this example shows that an attempt to characterize or predict the interactions observed between proteins should ideally be carried out in light of their biological function.

It is also important to distinguish between the different types of complexes, in terms of their type of physical interaction, when analyzing the intermolecular interfaces. Nonobligatory complexes can be placed in a separate class to obligatory complexes as, for molecules to exist as independent entities, additional constraints are imposed on those structures (Lo Conte et al., 1999). Thus, complexes can be classified as obligatory or nonobligatory, but also as being either transient or permanent. All obligatory complexes must be permanent by definition, whereas nonobligatory complexes can be either transient or permanent. The component proteins of an obligatory complex are unable to exist in isolation, whereas the component proteins of nonobligatory complexes are able to exist as independent entities prior to their association. The obligatory proteins form the subunits of oligomeric proteins, such as hemoglobin that consists of two identical α-subunits and two identical β-subunits and forms an $\alpha_2\beta_2$ tetramer.

The difference between obligatory and nonobligatory complexes is qualitative and, as such, obvious and definable differences can be seen between the two, such as increased hydrophobicity (or decreased polarity) at the interface of obligatory complexes (Jones et al., 2000). Indeed, obligatory interfaces were found by Jones et al. (2000) to be similar to protein domain interfaces. On the other hand, the transience of a nonobligatory complex is quantitative, and there is a sliding scale (dependent on the value of the dissociation constant), which represents the degree of transience. There is no energy of dissociation cutoff that can define this value. The transience of a complex may also be tightly linked to the environmental conditions surrounding the interaction (such as the temperature, salt concentration, etc). For example, the pH of a solution may affect the number of charged groups at the interface and may greatly affect the dynamics of the interaction. Thus, the surrounding environment of interacting proteins is also an important factor to consider. Most serine protease inhibitor complexes are fairly distinct, however, and can be taken as examples of nonobligatory permanent complexes. The component proteins of the complex are able to exist in isolation, but on association they are unable to dissociate (or do so only under very high temperatures, which may in fact cause the protein to first denature). However, there are many other examples that show varying degrees of transience. Thus, it makes sense to separate nonobligate from obligate complexes to investigate protein–protein interactions, but it is usually difficult and less useful to separate transient complexes from permanent complexes. Nevertheless, it is of interest to discover which are the important factors determining dissociation constants, and how these factors are reflected by the nature of the protein interactions themselves.

Different functional classes of protein–protein interaction may have very different modes of binding that are specifically adapted to their specialized functions. For example, a fundamental difference exists between serine protease inhibitors and antibody–antigen complexes (Jackson, 1999). This study showed that protease inhibitors use "main-chain–main-chain" interactions, whereas antigen–antibody complexes use "side-chain–side-chain" interaction. The difference can be explained by different biological roles, and at the chemical level the differences may partly explain the large difference in binding affinity of the two classes of complex. Serine proteases must bind tightly to their target proteases. This may be best achieved using constrained "main-chain–main-chain" conformation, and means that the inhibitor will be highly committed to the enzyme. For the antibody–antigen complex to be a general utility binding molecule, however, the antibody main-chain conformation cannot be committed to any particular main-chain motif (as it must bind all protein motifs).

Another example of different modes of interaction relating to different function is given by Wang (2002), who compares recognition of cell surface receptors by other physiological receptors, as compared to recognition by viral recognition proteins. It was found that recognition between two cell surface molecules has several features distinct from other homo- and heterocomplexes. He demonstrates that cell recognition between receptors is a multivalent, reversible, and avidity-driven process, and that the charge complementarity, rather than shape complementarity, plays an important role in this weak but specific binding. Evidence suggests that viruses can take advantage of this at the structural level by binding much more strongly to these physiological receptors. This is further evidence of the wide variety of interactions that reflect different functional roles.

2.3. Pattern of the Interface

There is generally considered to be a relationship between size of the interface and size of the proteins involved in the interaction, such that the size of the protein and the size of its interface show a positive correlation (Janin and Chothia, 1990). Although in most cases the interface forms a single patch, there are also cases in which the interaction site is dispersed over the protein surface in multiple patches for a single interaction. This is especially thought to be the case for large interfaces (Chakraborty and Janin, 2002).

Distinction has also been made between interfaces in terms of interface size, based on the nature of the residues present. Glaser et al. (2001) found that hydrophobic residues were more abundant in large interfaces and polar residues more abundant in small interfaces.

2.4. Hydrophobicity

The average hydrophobicity of the protein surface is low, while that of the core is high. One avenue explored is whether the hydrophobic effect, established as a main force that guides protein folding, is also the main driving force in protein–protein association (Tsai et al., 1997a; Tsai and Nussinov, 1997). Analysis indicates that although the hydrophobic effect plays a dominant role in protein binding, it is not as strong as observed in interior of protein monomers—especially for nonobligate interfaces.

Larsen et al. (1998) conducted a visual survey of 136 homodimers. This study showed that the pattern of hydrophobicity over the surface is quite variable and also that most homodimers were stabilized by a number of hydrophobic patches and bridging waters. However, the study did not find a correlation between the presence/absence of a hydrophobic core (which was present in one third of the homodimers) and specific function.

One early study suggested that hydrophobicity is the major factor stabilizing protein–protein association, while complementarity plays a selective role in determining which proteins associate (Chothia and Janin, 1975). However, averaging features over a diverse set of protein–protein interactions blurs the information on how the problem of maintaining structural integrity is solved in individual interfaces.

Xu et al. (1997a,b) found that both hydrophobic and electrostatic/hydrogen bonding interactions are important for stability. There tends to be a high proportion of hydrophobic residues in the protein interior, however, and a high proportion of charged and polar residues buried in the interface. This suggests that hydrogen bonds and ion pairs contribute more to the stability of protein binding than to protein folding. It has been found, however, that for obligatory interfaces there are more hydrophobic and less polar/charged residues at the interface than on the rest of the protein surface (Jones et al., 2000). Nonobligatory interfaces appear on average to have a similar polar to nonpolar distribution to the rest of the protein surface. They have been described by Janin (1999) as having either a hydrophobic core or several small hydrophobic patches distributed throughout the interface.

From this observation, a model was developed describing the nature of interface as being either "wet" or "dry" (Janin, 1999). It has also been documented that a single hydrophobic core yields the same stabilization as a collection of small hydrophobic patches of similar total surface area (Tsai et al., 1997).

Thus, the nonobligatory interface appears to constitute a compromise between the stabilization contributed by the hydrophobic effect, on the one hand, and avoiding large patches on the surface that would be too hydrophobic on the other. Such patches would be unfavorable to the monomers in solution, which is an issue not addressed in the case of obligatory complexes, as by definition they cannot exist alone.

Complex formation therefore usually results in the burial of a number of charged and/or polar residues. This difference between protein folding and protein–protein interaction can be explained in terms of energetics. Because the forces that drive protein–protein association need not be as large as those that drive protein folding, the relative contribution of hydrophobic residues can be reduced.

2.5. Electrostatics

McCoy et al. (1997) performed a quantative study of the charge complementarity (CC), and the electrostatic complementarity (EC) at protein–protein interfaces. They found that all interfaces had significant EC but insignificantly small CC. The results demonstrate the importance of long-range effects of charges. It was shown that the EC value was not related to the number of salt brigdes, and it was shown that other interactions contributed considerably to the EC value.

Lee and Tidor (2001) extensively studied the barnase–barstar complex and found that the complex is electrostatically optimized and does in fact show charge complementarity of the interface. The proteins of this complex have many polar and charged groups at their interface so this is a particularly relevant system for this type of study.

In a review by Sheinerman et al. (2000) it was demonstrated that interfacial charged/polar residues may enhance complex stability, but that the total effects of electrostatics is net destabilising. A later study by Sheinerman and Honig (2002) used continuum electrostatics to investigate the contribution of electrostatics to the binding free energy. In two cases it was found to oppose binding; in one case the net effect was close to zero, and in the last case it provided a significant driving force favoring binding. The extent to which the desolvation of buried charges is compensated for by the formation of hydrogen bonds and ion pairs is an important factor for these differences. The differences also suggest distinct interfaces can be designed to exploit electrostatic and hydrophobic forces in very different ways.

2.6. Interface Bonds

The type of interactions across protein–protein interfaces is largely determined by the nature of the amino acids present. The nature of these interactions also largely contributes to the strength of the protein–protein association. Xu et al. (1997b) showed that the geometry of hydrogen bonds across the interface is generally less optimal and of wider distribution than that observed within chains. Whereas in folding practically all degrees of freedom are available to the chain to attain its optimal configuration, this is not the case for rigid binding. Here the protein molecules are already folded, with only 6 degrees of translational and rotational freedom available to the chains to achieve their most favorable configuration. These constraints cause many polar residues buried in the interface to form weak hydrogen bonds with amino acid residues of a protein, rather than strongly hydrogen bonding to the solvent. Since interfacial hydrogen bonds are weaker than intrachain ones, to compete with binding water, more water molecules are involved in bridging hydrogen bond networks across the protein interface than in the interior (i.e., interfacial water permits better hydrogen-bond geometries to be accepted). The differences between interfacial hydrogen bonding patterns and the intrachain ones highlights that complexes formed by rigid binding may be far removed from the global minimum conformations. From this it could also be speculated that rigid body binding complexes should have a larger number of water molecules bridging the interface.

Averages of interface bond composition have been investigated, although they have limited meaning given the large variation among individual interfaces. One study (Lo Conte et al., 1999) showed the average interface size to be $1600 \pm 400 Å^2$ and that it contributed about 10% of a typical monomer surface. The average interface was also found to possess 10 intermolecular hydrogen bonds, but this result showed a high standard deviation. Every third hydrogen bond was found to involve at least one charged residue, and 13% were formed between the uncharged groups of two charged residues.

According to McDonald and Thornton (1994), it is rarely the case that the number of charges at the interface cancels out. They find that more often than not charge is shared across many polar, uncharged residues on the partner interface. However, it has been proposed that it is these charges and the net dipole of the protein that actually steer the proteins in the correct orientation as they approach prior to an interaction. This has been extensively investigated for the barnase–barstar complex (Camacho et al., 1999).

The importance of the CH–O main chain hydrogen bond at protein–protein interfaces has recently been analyzed (Jiang and Lai, 2002). These are weaker than conventional hydrogen bonds, but their number cannot be neglected as they are recognized to play an important role in the stabilization and function of interactions. The energy contribution of this type of interaction has not yet been fully explored, however. This study showed that the average energy contribution of a conventional hydrogen bond to the interface is 30%, that of a CH–O bond is 17%, and that of hydrophobic interactions is 50%. The CH–O bond contribution was shown to reach as high as 40% to 50% in some cases, however.

2.7. Interface Water

Another study also showed the importance of interfacial water molecules—calculated to contribute to about 25% of the total calculated binding strength. Some mutation studies result in released crystallographic water molecules near the mutation site. These studies lend support to the notion that water molecules bound to a crystallographic complex contribute significantly toward stabilization, and are major contributors to the energetics of protein–protein complexes (Covell and Wallqvist, 1997). This is thought to be especially true of protease inhibitor complexes (Huang et al., 1995). The importance of water at the interface can also be demonstrated by the finding that the same number of hydrogen bonds form by bridging water molecules as by direct hydrogen bonds across the interface.

Vaughan et al. (1999) investigated the structural response to mutation at the protein–protein interface. They crystallized three mutants of the barnase–barstar complex, where interactions across the interface were deleted by simultaneous mutation of both residues involved in the interaction. In all double mutants they found that water molecules fill the created pockets and cavities. These water molecules mimic the deleted side chains by occupying positions close to the noncarbon atoms of truncated side chains and remaking many hydrogen bonds made by the truncated side chains in the wild-type. This suggests that water molecules may account for the plasticity of binding. However, they found that the exact response to mutation is context dependent and that the same mutant can vary depending on the environment within the crystal.

In another study, Fischer and Verma (1999) found that the binding of buried structural water increased the flexibility of binding. This was indicated by the increase in vibrational entropy on binding an initial water molecule to a fully dehydrated bovine trypsin inhibitor. This phenomenon of increased flexibility has also been demonstrated for other larger ligands. A problem with this study is the lack of numerous examples showing cases in which binding of water causes this increase in this proposed binding flexibility.

Larsen (1998) found that the inclusion of a large water-repellent core is an exception rather than the rule. In most interfaces it was found that the water is scattered throughout the interface area. Hubbard and Argos (1994) performed a survey of cavities formed within the interfaces and found that water-sized cavities cover about 10% of a typical interface (more than half of which correlated with crystallographically observed waters).

2.8. Shape

Shape is a simple but nonetheless powerful tool that may be used to predict protein–protein interactions, and is extensively used in docking studies (Norel et al., 1995). Issues raised concern the most effective way one is able to measure the shape complementarity between interacting proteins, and numerous methods exist to measure this parameter (e.g., Connolly, 1983; Norel et al., 1999).

At a global level, interfaces have been shown to be rather flat, with shape complementarity at the more local level of protrusions and cavities formed by residue knobs and holes (Wodak and Janin, 2002). This is in contradiction to traditional teachings that illustrate a convex surface fitting into a concave surface, analogous to a key in a lock. However, in some cases there is a concave surface that fits a convex surface, such as a serine protease inhibitor protruding into the concave surface of the protease enzyme—but this should be seen as the exception rather than the rule.

An early docking study by Norel et al. (1994) used a knob and hole representation of the entire molecular surface with no additional information regarding the binding sites. This surface description was done by using pairs of critical points along with surface normals. They were able to successfully dock 15 of 16 complexes, indicating the importance and distinguishing powers of surface complementarity.

A fast three-dimensional superposition procedure, using the interfaces of known protein structures, has been used to search for geometrically similar surface areas relating to the protein interface (Preißner et al., 1999). Large numbers of structurally similar interfaces were found on the surfaces of unrelated proteins. Interestingly, even patches from different types of secondary structure were found resembling each other. Thus, this indicates convergent evolution of these interfaces, and that there is a limited number of interface geometries used for protein–protein interactions.

2.9. Amino Acid Preferences

It is obviously important to consider the amino acids involved in the interaction— to investigate their type, the number involved in forming bonds, and their order of occurrence in the interface (or footprint). However, as well as the amino acids actually involved in the interaction, other residues lying outside the interface site may play an important in maintaining the three-dimensional structure of the interface. It may be that the surrounding environment has an important role in maintaining a specific binding site shape and determining the interaction.

Thus, another area of research explores whether there are amino acid preferences for the interaction site. Results indicate that some residues are substantially more frequent at interaction sites as compared to their frequency in the protein generally (Viller and Kauver, 1994; Glaser et al., 2001). It may be extrapolated that the resemblance in residue utilization at the binding sites in unrelated proteins leads to an implicit similarity in the characterization of the binding site. A limit to the number of binding motifs results in limited specificity and this could account for cross-reactivity that is sometimes seen.

2.10. Interaction Hot Spots

Interaction hot spots are residues of the interface that contribute to a high proportion of the interface binding energy, because the free energy of binding is not evenly distributed across the interface. A database of alanine mutants (for which change in free energy of binding was measured) was scanned to show the presence of these hot spots of protein–protein interactions (Bogan and Thorn, 1998). These hot spots of binding energy represented only a small subset of residues in the dimer interface. This study also showed little correlation between surface accessibility and the contribution of a residue to the binding energy. Kortemme and Baker (2002) have also recently presented a physical model for binding energy hot spots in protein–protein complexes.

A study involving 11 clustered interface families and three-dimensional superpositioning analysis was used to identify the occurrence of matched residues within a family, by finding conserved residues in spatially similar environments (Hu et al., 2000). This study found these matched residues to be energetic hot spots that were enriched in tyrosine, tryptophan, and arginine, surrounded by energetically less important residues—that most likely serve in the occlusion of bulk solvent from the interface. Occlusion of this bulk solvent is thought to be crucial for binding energetics. The enrichment of specific polar residues in the largely hydrophobic binding site has been extended by other studies to cover other polar residues.

It was found that the polar residues were generally conserved at the interface (Hu et al., 2000), indicating functional importance. It is unclear, however, whether the buried polar interactions are energetically net stabilizing, or whether they merely facilitate specificity (Hendsch and Tidor, 1994). Also, some residues that do not form contacts across the interface contribute significantly to the free energy of binding when analyzed by alanine scanning mutagenesis (Delano, 2002), perhaps by destabilizing the unbound protein. It has also been shown that catalytic and other functionally important residues in proteins can often be mutated to yield more stable proteins (Elcock, 2001).

Because of these hot spots of interaction, intermolecular binding has been shown to be relatively robust to mutational studies (Clackson and Wells, 1995). Mutations of these hot spot residues were shown to have a greater effect than others, and some of these residues were also shown to be absolutely essential for preserving binding specificity. These functionally more important residues form the binding epitope, and contribute, as discussed, to a large fraction of the binding energy. Another study investigating the structural response to mutation at the interface showed that the interface structure is also relatively robust to mutation (Vaughan et al., 1999). This again may highlight the importance of the interaction hot spots described.

Kortemme and Baker (2002) developed a simple physical model to account for the whole range of experimentally measured free energy changes brought about by alanine mutation at protein–protein interfaces. The model is able to predict the results of alanine scanning experiments for 19 protein–protein interfaces with an average unsigned error of 1.06 kcal/mol. It included terms of shape complementarity, polar interactions, and a penalty for desolvation of polar groups in the interface.

2.11. Interaction Energetics

There must be a fine energetic balance between attraction and repulsion that causes two proteins to be temporarily bound before instability causes their dissociation. There may be fundamental differences in the binding sites that can be used to predict the energetics of these complexes. The timing of this dissociation may also be critical to allow the impeccably accurate interaction networks that are observed in living organisms to function. In many cases there is a trigger, such as binding of a metal ion, that disrupts the stability of a pair of bound proteins and causes their dissociation. Discovering how this trigger alters the binding site may provide some insight into the factors contributing to the transience of a complex. Factors such as conformational change (which disrupts the energetic stability) could be envisaged to trigger such dissociation.

The thermodynamic stability is given by the value of the dissociation constant (K_d) or the Gibbs free energy of dissociation. This is a balance of several large terms favoring or opposing complex formation. The major terms opposing protein association are those losses it produces in the translational, rotational, and internal degrees of freedom. Major terms favoring complex formation are the hydrophobic energy (gained from the hydrophobic surface burial), electrostatic energy, and hydrogen bond energy. Although energy terms roughly balance in protein–protein interactions, there is not known to be any simple direct correlation of energies of association with the general structural characteristics of interfaces. This is because of the geometry and environment of the protein that modulate the real values for particular interfaces.

Janin (1995, 1997) examined the kinetic model for protein association for rigid body approximation. He proposed that association starts with random collisions at a specific rate by translational diffusion. This creates an encounter pair, which can then evolve into a stable complex if the molecules are correctly oriented and positioned. Here the surfaces become dehydrated and the internal degrees of freedom relax to optimize short-range interactions. Long-range interactions affect both the random collision rate (affected by net molecular charge) and the rate of complex evolution to the stable state. In addition, electric dipoles are proposed to contribute to the steering effect that preorients the molecules before they collide and affects the probability of evolution to form a stable complex. Janin explains that rigid-body approximation is essential to the analysis of kinetics here, and that this approximation makes recognition simple compared with folding. Without it, protein–protein recognition becomes as complex as protein folding.

However, conformational change on complex formation is an important energy term that must often be considered. Wright and Dyson (1999) investigated intrinsically unstructured proteins and reassessed the structure–function relationship. The found that many proteins are unstructured, forming unfolded or non–globular structures in the cell. They argue that the intrinsic lack of structure can confer a functional advantage on a protein in its ability to bind different targets. They also argue that it allows precise control over the thermodynamics of the binding process, and provides a simple mechanism for inducibility by interaction with components of the cellular machinery. Numerous examples of domains that are unstructured in solution but that become structured on binding to the target have been noted in the areas of cell cycle control and transcriptional regulation.

Noskov and Lim (2001) developed a method to compute the absolute binding free energies of experimental structures using a free energy decomposition scheme. They assumed additivity of three physical processes: desolvation of X-ray structures, isomerization of the X-ray structure to a local energy minimum, and noncovalent complex formation. They found that the binding free energies were in agreement with experimental data. However, errors were incurred when considering proteins that underwent conformational changes on complex formation.

Ma et al. (2002) also developed an empirical model, using three variables to describe free energy of protein associations. These were the number of hydrophobic pairs, the side-chain accessible number and the buried apolar solvent accessible areas. It was found that the side-chain accessible number characterized the loss of side chain conformational entropy of protein interactions. They claim that the method could be used in a rescoring cycle to find the true binding mode of protein–protein docking simulations, as it is quick and simple. However the method does not take into account atomic detail, and is limited to rigid body binding examples so is relatively crude.

Xu et al. (1997a) found that the number of hydrophilic bridges across the interface shows a strong positive correlation to binding free energy, and found that salt bridges across the interface can significantly stabilize complexes in some cases.

There are differences in the contributions of hydrophilic bridges between folding and binding attributable to the different environments to which the hydrophilic groups are exposed before and after bridge formation. On binding these groups are buried in an environment whose residual composition can be much more hydrophilic than one after folding.

2.12. Conservation

Another issue is to find the extent and distribution of conservation of interacting residues in relation to the rest of the protein sequence. It is interesting to examine the pattern of conservation and analyze the distribution of the conserved residues found to be present at the interface, and to find whether this can be related to the protein function in some way. For example, the conserved residues may be localized to a central patch or dispersed over the interface. The nature of the conserved residues could also provide insight into the protein's interaction mechanism. There are several important factors to consider when quantifying the conservation of residues of a protein sequence, and this is a topic that has been extensively studied by Valdar (2002), but there is no single best definition of how to measure conservation.

Conserved residues are likely to fulfil some role such as structural stability, catalysis, or recognition, and may serve as fingerprints to characterize an interface family. The various studies showing conservation of interface residues (Hu et al., 2000; Teichmann, 2002; Noreen and Thornton, 2003) suggest that during evolution, the homodimers, the enzyme inhibitors, and the heterocomplexes have evolved to optimize their interface interactions. In contrast, antibody–antigen complexes tend to be selected principally according to their binding affinity, without being subject to evolutionary optimisation (Decanniere et al., 2001). It is difficult to differentiate between residue conservation conferring binding specificity and conservation owing to the role of residues in energy hot spots. Conserved interface residues have been largely located around the center of the interfaces, protected from the bulk solvent. Other studies have pointed to different conservation of specific residue types, such as an analysis of homodimer interfaces (Valdar, 2002) that showed glycine to be the most invariant residue of these interfaces.

As described previously, conservation has also been used to distinguish biologically relevant interfaces from crystal contacts (Valdar and Thornton, 2001a) on the basis that biologically relevant interfaces will be more conserved than the interfaces of crystal contacts.

2.13. Interaction Motifs

The examination of the possible recurrence of a specific binding motif is another method used to study protein–protein interfaces. Such an interaction motif is a conserved sequence or structure thought to have evolved for specific types of interaction. This motif may become degenerate if the proteins evolve away from one another to perform different functions. On the other hand, a motif may be found in unrelated proteins as a consequence of convergent evolution. Examples of such specialized motifs are the tetratricopeptide motif, PDZ domain network motif, and a class of zinc finger motif.

The tetratricopeptide (TPR) motif is a protein–protein interaction module and is found, in multiple copies, in a number of functionally different proteins that facilitate specific interactions with a partner protein (Blatch and Lassle, 1999). The TPR motif may represent an ancient protein–protein interaction module that has been recruited by different proteins and adapted to specific functions.

The PDZ domain motif is an example of an 80- to 90-amino-acid repeat motif present in 50 unrelated proteins, and mediating a diverse set of interactions (Fanning and Anderson, 1996). The existence of this particular example of motif raises the possibility that competition between different interactions may occur during formation of macromolecular complexes mediated by proteins containing these domains. It was found that single amino acid substitutions alter the specificity and affinity of PDZ domains for their ligands (Gee et al., 2000).

Zinc fingers are extremely common protein domains associated with DNA binding but only recently has a structurally distinct subclass of genuine zinc finger domains been implicated for involvement in protein–protein interactions (Matthews et al., 2001). Little is known about these domains, which have been identified on the basis of sequence homology, rather than their ability to bind zinc. These findings have implications for the prediction of protein function from protein sequences.

Sprinzak and Margalit (2001) used motif combinations (sequence signatures) to find interacting proteins. A statistical analysis performed on all possible combinations of sequence-signature pairs identified those pairs that are overrepresented in the database of yeast interacting proteins. The study demonstrated how the use of the correlated sequence-signatures as identifiers of interacting proteins can reduce significantly the search space, and enable directed experimental interaction screens.

Tsai et al. (1997b) draws attention to different structural motifs at the interface between two-state model complexes (where the chains fold cooperatively), and three-state models (representing the binding of already folded monomers). The origin of this difference can be understood in terms of the different nature of folding and binding involved. This is one good reason to separate oligomers from transient protein interactions when characterizing a protein–protein interface.

Tsai and Nussinov (1997) also show the two state complex formation is the outcome of the hydrophobic effect, and analogous to the formation of a compact hydrophobic nuclei in protein folding. This study shows that hydrophobicity is a critical distinguishing feature between two-state and three-state complexes.

2.14. Evolution

Evolutionary studies use sequence-based methods for the study of protein–protein interactions, and several phylogenetic adaptations using this sequence information have been implemented.

During evolution, functionally linked proteins tend to be preserved or eliminated in a new species. This property of correlated evolution is used in one study (Pellegrini et al., 1999) to characterize each protein by its phylogenetic profile encoding the presence or absence of a protein in every known genome. It is shown that proteins having matching or similar profiles tend to be functionally linked. This method of phylogenetic profiling allows prediction of the function of uncharacterized proteins.

Coevolution of proteins with their interaction partners (Pazos and Valencia, 2001; Cohen and Goh 2002) has been another area of investigation, based on the theory that interacting proteins coevolve to preserve their function. Many proteins have evolved as part of molecular complexes and the specificity of their interaction is essential to function. The network of necessary inter-residue contacts must consequently constrain the protein sequence to some extent. In other words, the sequence of an interacting protein must reflect the consequence of this process of adaptation. It is reasonable to assume that sequence changes accumulated during evolution in one protein must be compensated for by changes in its interacting partner protein. By building phylogenetic trees from the multiple sequence alignments of proteins, a correlation coefficient between two proteins can be calculated that quantifies the coevolution of a protein ligand with its receptor. One application of this particular approach is when applied to orphan ligands and receptors in search for orphan binding partners. Another method, using correlated mutations, leads to the possibility of developing a method for predicting contacting pairs of residues from sequence alone (Pazos et al., 1997).

The evolutionary trace method attempts to quantify the contribution of individual binding residues to the overall free energy of binding (Lichtarge et al., 1996). Since active sites are under evolutionary pressure to maintain their functional integrity, they undergo fewer mutations than functionally less important amino acids. When a functional difference is observed between evolutionarily related proteins, it is assumed to arise from mutations at or near residues performing that function. These mutations define new branches of the protein family, and are under selective pressure not to mutate unless their critical roles are compromised. Thus, a protein should conserve its functional site, which should have a distinctly lower mutation rate, and be punctuated by events that causes its divergence.

Another example of an evolutionary approach makes use of phylogenetic trees of full-length sequences (Johnson and Church, 2000), and then uses phylogenetic trees of sequences lining the binding cleft of the ligands. It is then determined whether sub-branches of the tree correlate with ligand binding preference. In theory, this can be used to predict ligand-binding function for many uncharacterized database sequences—to identify specific ligand contacts in proteins without solved structures.

A final example of an evolutionary approach to protein–protein interaction prediction is that of Marcotte et al. (1999). This study inferred protein interactions from genome sequence based on the observation that some pairs of interacting proteins have homologs in other organisms fused into a single protein chain. This single protein chain was termed the "Rosetta Stone" sequence, and the method was termed "domain fusion analysis." Nearly 7000 pairs of nonhomologous sequences were found in which both sequences of the pair showed considerable similarity to a single protein in another genome (the Rosetta Stone sequence). The theory behind the presence of

Rosetta Stone sequences is that the affinity between two proteins is greatly enhanced when they are fused. Thus, some interacting pairs of proteins may have evolved from evolutionarily distant proteins, where the two proteins of the pair were fused in a single polypeptide. In further support of the presence of Rosetta Stone sequences is the observation that protein–protein interfaces have a strong similarity to interdomain interfaces within a single protein molecule. This method has the potential to predict protein pairs with related biological function in addition to the prediction of potential protein–protein interactions.

2.15. Surface Complementarity

Complementarity between the interacting surfaces of a protein complex is a fundamentally important parameter in most protein–protein recognition systems, and is used in most docking algorithms. Various methods have been implemented to estimate the complementarity between the two surfaces in contact in a complex. Geometric complementarity involves optimising the van der Waals contacts, and is a major element in the recognition process between two molecules. Lawrence and Colman (1993) evaluated a correlation using a shape correlation index, having described the molecular surface as a set of closely spaced points and then generated a function for each point that reflects the complementarity between the points of a docked conformation. Jones and Thornton (1996) used a gap index to assess the compactness of the interface. This gap index is found by dividing the gap volume by the interface area. Laskowski (1995) used a program called SURFNET to estimate the gap volume between two protein surfaces, which is based on summing the volumes of a set of spheres. Atomic packing analysis is an alternative approach to assess complementarity. Packing density is estimated by using Veronoi volumes for each atom, in which a Veronoi polyhedron is drawn around each atom according to its surrounding atoms. Calculations show that the interface is as closely packed as the protein interior ($V/V_0 \approx 1$), where V_0 is the sum of the reference volumes. However, water, which is not present in the core, makes an important contribution to this packing at the protein interface.

2.16. Conformational Change

Conformational change may be a necessary step leading to the required shape complementarity between interacting proteins. These conformational changes could occur at the point of binding or prior to binding. Issues to consider include: what causes the conformational change (such as a third binding factor), the degree of the conformational change, and what the degree of the conformation change is related to. These conformational changes are perhaps one of the greatest challenges to the prediction of protein–protein interactions such as docking, and at present they are difficult to predict. Methods exist to incorporate "softness" into prediction studies such that these conformational changes can somehow be accounted for (Rosenfeld et al., 1995; Betts and Sternberg, 1999).

Betts and Sternberg (1999) performed a study of 39 heterocomplexes, which showed that half underwent a large conformational change upon complex formation. The study also showed that conformational change at the interface is greater than for the noninterface surface.

Lo Conte et al. (1999) performed a study on 75 heterocomplexes, and showed that the size of the recognition site was related to conformational change on complex formation. This study also showed that of the atoms that lose accessibility on complex formation, half make contacts across the interface and a third become fully inaccessible to solvent. In addition, it was found that the conformational changes allowed the buried atoms at the interface to be as tightly packed as the interior.

Ma et al. (1999) describe the concept of the binding mechanism in terms of a funnel model. The model explains that all proteins exist as a set of different conformers (conformational isomers), and that binding occurs as a result of conformer selection. They propose that rigid body binding indicates the presence of fewer conformational isomers. Molecular flexibility is portrayed as a rugged energy surface round the bottom of a folding funnel. They describe that the larger the flexibility, the greater is the population of diverse conformers, and the lower are the barriers between them. On the other hand, they say rigid molecules have fewer minima and higher energy barriers between them. Nonspecific binding fits the former model, and rigid binding fits the latter. This does not imply that specific binding always requires rigidity. Extreme rigidity may interfere with biological function and therefore be unfavorable to binding.

In a minireview, Sundberg and Mariuzza (2000) describe how plasticity of the interface allows for accommodation of mutations as interacting proteins coevolve. Second, it is pointed out that plasticity allows for a protein to bind multiple partners. They state that proteins that bind multiple ligands at a single site have high conformational flexibility compared to other interactions.

3. METHODS USED TO PREDICT PROTEIN–PROTEIN INTERFACES AND CONTACTS

3.1. Docking Methods

Many docking algorithms have now been developed to predict protein–protein interfaces and contacts. An overview of these methods is given by Smith and Sternberg (2002) and Halperin et al. (2002). Docking is the major method currently used as a means to predict a model of the best near native bound complex from undocked proteins. There are basically two stages to docking. The first stage involves developing a search method that will be able to find a near-correctly docked orientation with reasonable likelihood. The Fast Fourier Transform (FFT) method is used for this first stage of docking, and has been used in docking programs such as FTDock (Gabb et al. 1997), GRAMM (Vakser et al., 1999), DOT (Mandell et al., 2001), and ZDOCK (Chen and Weng, 2002). The FFT method involved discretising molecules onto a voxel grid, and then an exhaustive search of the three-dimensional space of relative orientations is performed. The second stage involves developing a scoring function that is able to discriminate correct or nearly correctly docked orientations. This score function is based on measures such as surface complementarity, electrostatic complementarity, and hydrophobic potentials. Some softening of the electrostatic energy function in this stage is needed otherwise, even in near-native dockings, these overwhelm the complementarity that remains. However, this softening necessarily reduces the ability to discriminate correctly docked orientations, such that many false positives are generated. These false positives are a major issue of the docking problem.

The next step is to introduce flexibility to allow for conformational changes. Further filtering steps may then be used, in which biological information is available concerning residues known to be present in the interface. This may help to choose the correctly docked solution among a number of false positives.

There is less value in redocking proteins of a complex in which the proteins have been simply separated from the solved complex. This is because these proteins have already been locked into place and undergone rearrangements to produce a perfect fit. Instead, it is essential to incorporate "softness" into the docking algorithm that allows for side chain rotations and larger main chain rotations that occur on complex formation.

GRAMM is one example of a docking program, designed to study low-resolution recognition (Vakser et al., 1999). The idea of this algorithm was to smooth out atomic sized molecular detail, and then systematically dock resulting molecular images. The results showed that 52% of the dataset (of 475 protein complexes) showed existence of some degree of recognition.

3.2. Machine Learning Techniques

An artificial neural network is a system loosely modeled on the human brain. It is a machine learning technique that uses multiple layers of simple processing elements called neurons. These neurons attempt to simulate real biological neurons. Each neuron is linked to certain of its neighbors with varying coefficients of connectivity (weights) that represent the strength of these connections. Learning is accomplished by adjusting these weights to cause the overall network to output appropriate results.

Neural networks have been used to predict the interface residues of protein–protein interactions (Zhou and Shan, 2001; Fariselli et al., 2002) with a degree of success. Using evolutionary conservation and surface disposition, a cross-validation test determined the correct detection of 73% of the residues involved in protein interactions in a selected database of 226 heterodimers (Fariselli et al., 2002).

Neural nets have also been used by Fariselli et al. (2001) in order to predict inter-residue contacts. A residue contact is defined here as $C\beta$ atoms of two proteins within a distance threshold of 8 Å. The prediction used information from sequence profiles (evolutionary information), sequence conservation, correlated mutations, and predicted secondary structures. They were able to assign protein contacts with an accuracy of 0.21 (21% of correctly predicted interactions), which was an improvement over random by a factor greater than six. However, it can be seen that the prediction of these interacting residue pairs is much less successful than just the prediction of the interaction residues alone.

REFERENCES

Bader, G.D., Betel, D., and Hogue, C.W. (2003). BIND: the biomolecular interaction network database. *Nucleic Acids Res.* 31:248–250.

Bernstein, F.C., Koetzle, T.F., Williams, G.J., Meyer, E.F., Brice, M.D., Rogers, J.R., Kennard, O., Shimanouchi, T., and Tasumi, M. (1977). The Protein Data Bank: a computer-based archival file for macromolecular structures. *J. Mol. Biol.* 112:535.

Betts, M.J., and Sternberg, M.J. (1999). An analysis of conformational changes on protein–protein association: implications for predictive docking. *Protein Eng.* 12:271–283.

Blatch, G.L., and Lassle, M. (1999). The tetratricopeptide repeat: a structural motif mediating protein-protein interactions. *Bioessays* 21:932–939.

Bogan, A.A., and Thorn, K.S. (1998). Atonomy of hot spots in protein interfaces. *J. Mol. Biol.* 280:1–9.

Camacho, C.J., Weng, Z., Vadja, S., and DeLisi, C. (1999). Free energy landscapes of encounter complexes in protein-protein association. *Biophys. J.* 76:1166–1178.

Chakrabarti, P., Janin, J. (2002). Dissecting protein–protein recognition sites. *Proteins.* 47:334–43.

Chen, R., and Weng, Z. (2002). Docking unbound proteins using shape complementarity, desolvation and electrostatics. *Proteins* 47:281–294.

Chothia, C., and Janin, J. (1975). Principles of protein-protein recognition. *Nature* 256:705–708.

Clackson, T., and Wells, J.A. (1995). A hot spot of binding energy in a hormone-receptor interface. *Science* 267:383–386.

Cohen, F.E., and Goh, C.S. (2002). Co-evolutionary analysis reveals insights into protein–protein interactions. *J. Mol. Biol.* 324: 177–192.

Connolly, M.L. (1983). Analytical molecular surface calculation. *J. Appl. Crystallogr.* 16:548–558.

Covell, D.G., and Wallqvist, A. (1997). Analysis of protein-protein interactions and the effects of amino acid mutations on their energetics. The importance of water molecules in the binding epitope. *J. Mol. Biol.* 269:281–297.

Decanniere, K., Transue, T.R., Desmyter, A., Maes, D., Muyldermans, S., and Wyns, L. (2001). Degenerate interfaces in antigen-antibody complexes. *J. Mol. Biol.* 313:473–478.

DeLano, W.L. (2002). Unraveling hot spots in binding interfaces: progress and challenges. *Curr Opin Struct Biol.* 12:14–20.

Elcock, A.H. (2001). Prediction of functionally important residues based solely on the computed energetics of protein structure. *J. Mol. Biol.* 312:885–896.

Elcock, A.H., and McCammon, A. (2001). Identification of oligomerisation states by analysis of interface conservation. *Proc. Natl. Acad. Sci.* 98:2990–2994.

Fanning, A.S., and Anderson, J.M. (1996). Protein–protein interactions: PDZ domain networks. *Curr. Biol.* 6:1385–1388.

Fariselli, P., Olmea, O., Valencia, A., and Casadio, R. (2001). Prediction of contact maps with neural networks and correlated mutations. *Protein Eng.* 14:835–845.

Fariselli, P., Pasos, F., Valencia, A., and Casadio, R. (2002). Prediction of protein–protein interaction sites in heterocomplexes with neural networks. *Eur. J. Biochem.* 269:1356–1361.

Fischer, S., and Verma, C.S. (1999). Binding of buried structural water increases the flexibility of proteins. *Proc. Natl. Acad. Sci.* 96:9613–9615.

Gabb, H.A., Jackson, R.M., and Sternberg, M.J.E. (1997). Modelling protein-protein docking using shape complementarity, electrodtatics and biochemical information. *J. Mol. Biol.* 272:106–120.

Gee, S.H., Quenneville, S., Lombardo, C.R., and Chabot, J. (2000). Single-amino acid substitutions alter the specificity and affinity of PDZ domains for their ligands. *Biochemistry* 39:14638–14646.

Glaser, F., Steinberg, D.M., Vasker I.A., and Ben-Tal N. (2001). Residue frequencies and pairing preferences at protein–protein interfaces. *Proteins* 43:89–102.

Cohen, F.E. and Goh, C. (2002). Co-evolutionary analysis reveals insights into protein-protein interactions. *J. Mol. Biol.* 324:177–192.

Halperin, I., Ma, B., Wolfson, H., and Nussinov, R. (2002). Principles of docking: an overview of search algorithms and a guide to scoring functions. *Proteins* 47: 409–443.

Hendsch, Z.S., and Tidor, B. (1994). Do salt bridges stabilize proteins? A continuum electrostatic analysis. *Protein Sci.* 3:211–226.

Hermjakob, H., Montecchi-Palazzi, L., Lewington, C., Mudali, S., Kerrien, S., Orchard, S., Vingron, M., Roechert, B., Roepstorff, P., Valencia, A., Margalit, H., Armstrong, J., Bairoch, A., Cesareni, G., Sherman, D., Apweiler, R. (2004). Int Act: an open source molecular interaction database. *Nucleic Acids Res*: 32:D452–455.

Hood, L., and Galas, D. (2003). The digital code of DNA. *Nature* 421:444–448.

Hu Z., Ma, B., Wolfson, H., and Nussinov, R. (2000). Conservation of polar residues as hot spots at protein interfaces. *Proteins* 39:331–342.

Huang, K., Lu, W., Anderson, S., Laskowski, M. Jr., and James, M.N. (1995). Water molecules participate in protease-inhibitor interactions. *Protein Sci.* 4:1985–1997.

Hubbard, S.J., and Argos, P. (1994). Cavities and packing at protein interfaces. *Protein Sci.* 3:2194–2206.

Jackson, R.M. (1999). Comparison of protein-protein interactions in serine-protease and antibody-antigen complexes: implications for the protein docking problem. *Protein Sci.* 8:603–613.

Janin, J. (1995). Elusive affinities. *Proteins* 21:30–39.

Janin, J. (1997). The kinetics of protein-protein recognition. *Proteins* 28:153–161.

Janin, J. (1999). Wet and dry interfaces: the role of solvent in protein–protein and protein–DNA interactions. *Struct. Fold Des.* 7:R277–279.

Janin, J., and Chothia, C. (1990). The structure of protein-protein recognition sites. *J. Biol. Chem.* 265:16027–16030.

Janin and Rodier (1995) Protein-protein interaction at crystal contacts. Proteins 4: 580–587.

Jiang, L., and Lai, L. (2002). CH-O hydrogen bonds at protein-protein interfaces. *J. Biol. Chem.* 277:37732–37740.

Johnson, J.M., and Church, G.M. (2000). Predicting ligand-binding function in families of bacterial receptors. *Proc. Natl. Acad. Sci.* 97:3965–3970.

Jones, S., and Thornton, J.M. (1996). Principles of protein-protein interactions. *Proc. Natl. Acad. Sci. USA* 93:13–20.

Jones, S., Martin, A., and Thornton, J.M. (2000). Protein domain interfaces: characterization and comparison with oligomeric protein interfaces. *Protein Eng.* 13:77–82.

Kortemme, T., and Baker, D. (2002). A simple physical model for binding energy hot spots in protein-protein complexes. *Proc. Natl. Acad. Sci. USA* 99:14116–14121.

Larsen, T.A., Olson, A.J., and Goodsell, D.S. (1998). Morphology of protein–protein interfaces. *Structure* 6:421–427.

Laskowski, R.A. (1995). SURFNET: a program for visualizing molecular surfaces, cavities, and intermolecular surfaces. *J. Mol. Graph.* 13:323–330.

Lawrence, M.C., and Colman, P.M. (1993). Shape complementarity at protein–protein interfaces. *J. Mol. Biol.* 234:946–950.

Lee, L.P., and Tidor, B. (2001). Barstar is electrostatically optomoised for tight binding to barstar. *Nat. Struct. Biol.* 8:73–76.

Lichtarge, O., Bourne, H.R., and Cohen, F.E. (1996). An evolutionary trace method defines binding surfaces common to protein families. *J. Mol. Biol.* 257:342–358.

Noskov, S.Y. and Lim, C., (2001). Free energy decomposition of protein–protein interactions. *Biophys. J.* 81:737–750.

Lo Conte, L., Chothia, C., and Janin, J. (1999). The atomic structure of protein–protein recognition sites. *J. Mol. Biol.* 285:2177–2198.

Ma, X.H., Wang, C.X. and Li, C.H. (2002). A fast empirical approach to binding free energy calculations based on protein interface information. *Prot Eng.* 15:677–681.

Mandell, J.G., Roberts, V.A., Pique, M.E., Kotlovyi, V., Mitchell, J.C., Nelson, E., Tsigelny, I., and Ten Eyck, L.F. (2001). Protein electrostatics using continuum electrostatics and geometric fit. *Protein Eng.* 14:105–113.

Marcotte, E.M., Pellegrini, M., Ng, H.L., Rice, D.W., Yeates, T.O., and Eisenberg, D. (1999). Detecting protein function and protein–protein interactions from genome sequences. *Science* 285:751–753.

Matthews, J.M., Kowalski, K., Liew, C.K., Sharpe, B.K., Fox, A.H., Crossley, M., and Mackay J.P. (2000). A class of zinc fingers involved in protein–protein interactions. *Eur J Biochemistry.* 267:1030–1038.

McCoy, A.J., Chandana Epa, V., and Colman, P.M. (1997). Electrostatic complementarity at protein–protein interfaces. *J. Mol. Biol.* 268:570–584.

McDonald, I.K., and Thornton, J.M. (1994). Satisfying hydrogen bonding potential in proteins. *J. Mol. Biol.* 238:777–793.

Nooren, I.M., and Thornton, J.M. (2003). Structural classification and functional significance of transient protein–protein interactions. *J. Mol. Biol.* 325:991–1081.

Norel, R., Lin, S.L., Wolfson, H.J., and Nussinov, R. (1994). Shape complementarity at protein-protein interfaces. *Biopolymers* 34:933–940.

Norel, R., Lin, S.L., Wolfson, H.J., and Nussinov, R. (1995). Molecular surface complementarity at protein-protein interfaces. The critical role played by surface normals at well placed, sparse, points in docking. *J. Mol. Biol.* 252:263–273.

Norel, R., Petrey, D., Wolfson, H.J., and Nussinov, R. (1999). Examination of shape complementarity in docking of unbound proteins. *Proteins* 36:307–317.

Noskov, S.Y., and Lim, C. (2001). Free energy decomposition of protein-protein interactions. *Biophys. J.* 81:737–750.

Pazos, F., and Valencia, A. (2001). Similiarity of phylogenetic trees as indicator off protein-protein interaction. *Protein Eng.* 14:609–614.

Pazos, F., Helmer-Citterich, M., Ausiello, G., and Valencia, A. (1997). Correlated mutations contain information about protein–protein interaction. *J. Mol. Biol.* 271:511–523.

Pellegrini, M., Marcotte, E.M., Thompson, M.J., Eisenberg, D., and Yeates, T.O. (1999). Assigning protein functions by comparative genome analysis: protein phylogenetic profiles. *Proc. Natl. Acad. Sci. USA* 96:4285–4288.

Ponstingl, H., Henrick, K., Thornton, J.M. (2000). Discriminating between homodimeric and monomeric proteins in the crystalline state. 41:47–57.

Preißner, R., Goede, A., and Frömmel, C. (1999). Homonyms and synonyms in the Dictionary of Interfaces in Proteins (DIP). *Bioinformatics* 15:832–836.

Robert, C.H., and Janin, J. (1998). A soft, mean-field potential derived from crystal contacts for predicting protein-protein interactions. *J. Mol. Biol.* 283:1037–1047.

Rosenfeld, R., Vajda, S., and DeLisi, C. (1995). Flexible docking and design. *Annu. Rev. Biophys.* 24:677–700.

Sheinerman, F., and Honig, B. (2002). On the role of electrostatic interactions in the design of protein–protein interfaces. *J. Mol. Biol.* 318:161–177.

Sheinerman, F.B., Norel, R., and Honig, B. (2000). Electrostatic aspects of protein-protein interactions. *Curr. Opini. Struct. Biol.* 10:153–159.

Smith, G.R., and Sternberg, M.J., (2002). Prediction of protein–protein interactions by docking. *Methods* 12:28–35.

Sprinzak, E., and Margalit, H. (2001). Correlated sequence-signatures as markers of protein-protein interaction. *J. Mol. Biol.* 311:681–692.

Sundberg, E.J., and Mariuzza, R.A. (2000). Luxury accommodations: the expanding of structural plasticity in protein-protein interactions. *Structure* 8:137–142.

Teichmann, S.A. (2002). The constraints protein–protein interactions place on sequence divergence. *J. Mol. Biol.* 324:399–407.

Tsai, C.J., and Nussinov, R. (1997). Hydrophobic folding units at protein-protein interfaces. Implications to protein folding and to protein–protein association. *Protein Sci.* 6:1426–1437.

Tsai, C.J., Kumar, S., Ma, B. and Nussinov, R. (1999). Folding funnels, binding funnels, and protein function. *Protein Science.* 8:1181–1190.

Tsai, C.J., Lin, S.L., Wolfson, H.J., and Nussinov, R. (1997a). Studies of protein-protein interfaces: a statistical analysis of the hydrophobic effect. *Protein Sci.* 6:53–64.

Tsai, C.J., Xu, D., and Nussinof, R. (1997b). Structural motifs at protein-protein interfaces: protein cores verses two-state and three-state model complexes. *Protein Sci.* 6:1793–1805.

Uetz, P, Giot, L, Cagney, G, Mansfield, T.A., Judson, R.S., Knight, J.R., Lockshon, D., Narayan V., Srinivasan, M., Pochart, P., Qureshi-Emili, A., Li, Y., Godwin, B., Conover, D., Kalbfleisch, T., Vijayadamodar, G., Yang, M., Johnston, M., Fields, S., Rothberg, J.M. (2000). A comprehensive analysis of protein–protein interactions in *Saccharomyces cerevisiae. Nature* 403:623–627.

Vakser, I.A., Matar, O.G., and Lam, C.F. (1999). A systematic study of low-resolution recognition in protein-protein complexes. *Proc. Natl. Acad. Sci. USA* 96:8777–8482.

Valdar, W.S. (2002). Scoring residue conservation. *Proteins* 48:227–241.

Valdar, W.S., and Thornton, J.M. (2001a). Conservation helps to identify biologically relevant crystal contacts. *J. Mol. Biol.* 313:399–416.

Valdar, W.S., and Thornton, J.M. (2001b). Protein-protein interfaces: analysis of amino acid conservation in homodimers. *Proteins* 42:108–124.

Vaughan, C.K., Buckle, A.M., and Fersht, A.R. (1999). Structural response to mutation at a protein interface. *J. Mol. Biol.* 286:1487–1506.

Venter, J.C., et al. (2001). The sequence of the human genome. *Science* 291:1304–1351.

Villar, H.O. and Kauvar, L.M. (1994). Amino acid preferences at protein binding sites. *FEBS Lett.* 349:125–30.

Wang, J. (2002). Protein recognition by cell surface receptors: physiological receptors versus virus interactions. *Trends Biochem Sci.* 27:122–126.

Wodak, S.J., and Janin, J. (2002). Structural basis of macromolecular recognition. *Adv. Protein Chem.* 61:9–73.

Wright, P.E., and Dyson, H.J. (1999). Intrinsically unstructured proteins: re-assessing the protein structure-function paradigm. *J. Mol. Biol.* 293:321–331.

Xenarios, I., Rice, D.W., Salwinski, L., Baron, M.K., Marcotte, E.M., and Eisenberg, D. (2000). DIP: the database of interacting proteins. *NAR* 28:289–291.

Xu, D., Lin, S.L., and Nussinov, R. (1997a). Protein binding versus protein folding: the role of hydrophilic bridges in protein associations. *J. Mol. Biol.* 265:68–84.

Xu, D., Tsai, C.J., and Nussinov, R. (1997b). Hydrogen bonds and salt bridges across protein-protein interfaces. *Protein Eng.* 10:999–1012.

Zanzoni, A., Montecchi-Palazzi, L., Quondam, M., Ausiello, G., Helmer-Citterich, M., and Cesareni, G. (2002). MINT: a Molecular INTeraction database. *FEBS Lett.* 513:135–140.

Zhou, H., and Shan, Y. (2001). Prediction of protein interaction sites from sequence profile and residue neighbour list. *Proteins* 44:336–343.

6

Protein–Protein Docking Methods

Garland R. Marshall and Ilya A. Vakser

ABSTRACT

A critical review of published methodology used in docking proteins and of current understanding of the problems associated with the inherent flexibility of proteins is presented. The underlying assumption made in the past of docking two rigid bodies (six degrees of freedom) is clearly not applicable to most protein–protein interactions as induced fit is the rule rather than the exception. Nevertheless, significant progress is being made as investigators increase flexibility of the docking partners with the availability of increased computational power. In the extreme case, however, docking of two proteins is equivalent to predicting the structure of the complex from the two sequences alone.

1. INTRODUCTION

Determination of protein–protein interactions is one focal point of activity in computational structural biology. The three-dimensional (3D) structure of a protein–protein complex is, generally, more difficult to determine experimentally than the structure of an individual protein. Adequate computational techniques to model protein interactions are important because of the growing number of known protein 3D

GARLAND R. MARSHALL • Center for Computational Biology and Departments of Biochemistry and Molecular Biophysics and of Biomedical Engineering, Washington University, St. Louis, MO 63110, USA.
ILYA A. VAKSER • Center for Bioinformatics, The University of Kansas, Lawrence, KS 66045

Proteomics and Protein–Protein Interactions: Biology, Chemistry, Bioinformatics, and Drug Design, edited by Waksman. Springer, New York, 2005.

structures, particularly in the context of structural genomics (Skolnick et al., 2000; Goldsmith-Fischman and Honig, 2003; Sali et al., 2003). Recent estimates of the average number of intermolecular associations that any expressed protein has in yeast is eight (Tong et al., 2004). Protein-docking techniques offer tools to elucidate these interactions, for fundamental studies of protein interactions, and to provide a structural basis for drug design to modulate complex formation (Veselovsky et al., 2002).

Computational structural approaches to molecular recognition were introduced in early 1970s by Scheraga and co-workers (Platzer et al., 1972) for small ligand interactions with proteins. At the same time, Marshall and colleagues were developing molecular modeling tools and approaches to conformational analysis in addressing the conformational parameter in structure–activity studies of ligands (Marshall et al., 1979). Structure-based drug design began with the seminal experimental work of Goodford and his colleagues (Goodford, 1984) designing inhibitors of the DPG-binding site using the crystal structure of hemoglobin for the treatment of sickle-cell anemia. Specifically, protein–protein docking techniques were pioneered in 1978 by Wodak and Janin (Wodak and Janin, 1978) and Greer and Bush (Greer and Bush, 1978). Since then, the field has grown substantially, especially starting from early 1990s, through the development of powerful docking algorithms, rapid progress in computer hardware, and significant expansion of available experimental data on structures of protein–protein complexes.

Nevertheless, our understanding of the principles of protein recognition and of adequate protein-docking techniques is still limited. With the rapid advances in experimental and computational determination of structures of individual proteins, the importance of modeling of protein 3D interactions increases. We now face the enormous challenge of structural modeling of protein-interaction networks on the genome scale, requiring much more powerful docking methodologies.

In this chapter, we describe the structural and physicochemical foundations of modern protein-docking techniques, the fundamental elements of docking, and provide an overview of existing docking approaches and future directions in methodological developments, especially in light of the challenges of genomics/proteomics.

2. PROTEIN RECOGNITION AND CHARACTERIZATION OF INTERFACES

The increasing availability of crystal structures of protein complexes has allowed characterization of the interfaces between the proteins in the complex with the goal of understanding the interactions that stabilize such complexes and determine the specificity of interaction. Protein–protein complex formation can be viewed from either a more physical perspective as a minimization of the free energy of the system, or from a more empirical point of view as a match between various phenomenological structural and/or physicochemical motifs (so-called "recognition factors"). In living organisms, proteins recognize their partners among many other proteins and bind in a specific way in short physiological timeframes. Given the complexity of the system, from either the physical or empirical points of view, the formation of a protein–protein complex is a remarkable event, based on the nature's super-efficient "energy-minimization protocol" and guided by strong long-range and short-range recognition factors. Modern methods of protein docking are based on our efforts to simulate and navigate the intermolecular energy landscape, and on our current understanding of the recognition factors governing complex formation.

2.1. Parallels Between Protein Recognition and Protein Folding

Principles of protein folding and protein recognition are the basis for understanding life processes at the molecular level. They also provide the foundation for the algorithms of predicting protein structure and interactions. The underlying physics that determines the structure of individual proteins and the structure of protein complexes are identical, and the derived principles are, therefore, quite similar. The underlying physicochemical and structural principles of protein interactions are discussed in greater detail later in this chapter. Here we mention them briefly within the context of binding and folding similarity.

Databases of cocrystallized protein–protein complexes are used to study the interface properties and derive relevant principles. Statistically derived residue–residue and atom–atom preferences for protein–protein interfaces were found similar to those in protein cores (Tsai et al., 1996; Vajda et al., 1997; Keskin et al., 1998; Glaser et al., 2001). A major role of hydrophobicity in protein folding is well established (Richards, 1977; Dill, 1990). Studies of protein–protein interfaces confirm the importance of hydrophobicity in complex formation as well (Korn and Burnett, 1991; Vakser and Aflalo, 1994; Young, et al., 1994; Tsai, et al., 1996). The importance of the concept of the energy funnel, first demonstrated for protein folding (Bryngelson et al., 1995; Dill, 1999), has been expanded to the intermolecular energy landscape in protein–protein interactions (Tsai et al., 1999; Shoemaker et al., 2000; Tovchigrechko and Vakser, 2001; Baker and Lim, 2002).

Tight packing of structural elements inside proteins is one of the fundamental concepts in our understanding of protein structures (e.g., Ponder and Richards, 1987; Hubbard and Argos, 1994; Jiang et al., 2003). Galaktionov et al. (Galaktionov et al., 2001) have developed an effective approach based on the contact matrix that restricts sampling of candidate folds based on sequence to those consistent with the experimentally determined density of proteins. A variety of computational approaches to compactness have been suggested; the average inter-$C\alpha$ distance corrected for protein size was used in a scoring function derived for ab initio protein prediction by Berglund et al. (Berglund et al., 2004) and found to have the highest weight among the parameters examined. The same concept of compactness applies to protein–protein interfaces as well (See Section 2.2).

2.2. Geometric and Physicochemical Complementarity

A tight geometric complementarity between interacting protein surfaces has been a cornerstone of protein–protein docking methodology since its inception in 1978 (Wodak and Janin, 1978). Systematic database analysis of the rapidly growing number of cocrystallized protein–protein complexes provides an increasing amount of evidence supporting this concept (Tsai et al., 1996; Larsen et al., 1998). A number of investigations of packing and buried surface area at protein–protein interfaces (Lawrence and Colman, 1993; Hubbard and Argos, 1994; Janin, 1995) supported the general conclusion that the interacting proteins have a high degree of surface complementarity, but indicated that there is a significant variation in this regard between different complexes. For example, packing at the antigen–antibody interface is relatively loose (Lawrence and Colman, 1993; Mariuzza and Poljak, 1993). The contact surface area in protein–protein complexes generally varies 10-fold from 500 to 5000 $Å^2$ with many complexes having even larger contact areas (Lo Conte et al., 1999; Veselovsky et al., 2002).

Most protein–protein interfaces are found to be more hydrophobic than exposed areas (Korn and Burnett, 1991; Vakser and Aflalo, 1994; Young et al., 1994; Tsai et al., 1997). Hydrophobic amino acid residues tend to be enriched in the interface in hydrophobic patches of 200 to 400 $Å^2$ (Johns and Thornton, 1996; Tsai et al., 1996; Lijnzaad and Argos, 1997). A high degree of electrostatic and hydrogen-bonding complementarity is also observed for protein–protein interfaces (Janin, 1995; McCoy et al., 1997; Tsai et al., 1997; Larsen et al., 1998).

2.3. Structural-Recognition Motifs and Hot Spots

It has become apparent that some side chains within the interface play a more significant role ("hot spots") than others in the energetics of binding and in the precise relative orientation of the two proteins in the complex. Experimentally, this information has often been obtained by systematic mutagenesis of side chains to alanine within the interface and determination of the changes in binding affinity. Bogan and Thorn (Bogan and Thorn, 1998) collected a database of 2325 alanine mutants for which the change in free energy of binding on mutation to alanine had been measured (an updated database ASEdb is available at http://www.asedb.org). Analysis of the database by Bogan and Thorn (Bogan and Thorn, 1998) generated several observations; amino-acid side chains in hot spots are located near the center of protein–protein interfaces, are generally solvent inaccessible, and are self-complementary across protein–protein interfaces, that is, they align and pack against one another. Out of 31 contact residues involved in the interaction of growth hormone with its receptor, for example, two tryptophan residues of the receptor account for more than 75% of the free energy of binding as determined by mutation to alanine (Cunningham and Wells, 1993). In the case of growth hormone itself, 8 of the 31 side chains involved in the interface accounted for approximately 85% of the binding energy, thus the genesis of the "hot-spot" theory (Clackson and Wells, 1995) as a basis for drug discovery targeting protein–protein interfaces.

In an earlier analysis of experimental data on peptide recognition by proteins, Marshall (Marshall, 1992) concluded that "Two extremes of motifs have emerged: one linear, with the peptide backbone providing many of the recognition elements; the other dominated by interactions with the side chains, often held in β-turns."

The linear motif is usually associated with proteolytic enzymes where orientation of the peptide backbone within the active site gives precise orientation to the functional groups responsible for hydrolysis. Furthermore, in the case where side-chain recognition is dominant, "Aromatic residues are found consistently to play a special role in the recognition and activation of receptors . . . The rigid arrangement of atoms and the resulting fixed large-surface area of aromatic side chains, such as tyrosine, tryptophan, histidine and phenylalanine, combine to maximize the potential free energy of interaction as the entropic cost of assuming a particular geometry has already been paid." Charged groups, particularly the planar guanidium and carboxyl groups of Arg, Glu, and Asp, are also often essential recognition "hot" spots in peptide messages. Both observations are also valid within the interfaces of protein–protein interactions (Hu et al., 2000) as formation of salt bridges across the intermolecular interface is highly favorable (Xu et al., 1997; Drozdov-Tikhomirov et al., 2001). From the analysis of Bogan and Thorn of amino acid preference in "hot spots," similar conclusions support those of Marshall derived from structure–activity studies on peptide ligands; tryptophan is most highly enriched (almost fourfold having the largest planar surface), followed by Arg, Tyr, Ile, Asp, and His, respectively to a 50% enrichment. One significant difference between peptide–ligand binding and protein–protein interfaces, of course, is that interfaces between proteins are most often composed of amino acid sequences that are not contiguous, while the small size of peptides usually means that one member has an interacting surface composed of adjacent amino acids, such as occur in a reverse turn, or on the surface of an α-helix. One additional difference is that the peptide has little if any intrinsic structure in solution, and the entropic cost of binding is, therefore, much greater.

This difference has implications in the effort to development of small molecules inhibiting protein–protein interactions as has been recently reviewed (Chrunyk et al., 2000; Veselovsky et al., 2002; Berg, 2003). In the case of HIV protease, the enzyme is a homodimer with each of two active site aspartyl residues contributed from each monomer. A four-stranded β-sheet stabilizes the dimer and is responsible for more than 80% of the stabilization energy (Todd et al., 1998). Several attempts have been made to inhibit dimer formation with β-sheet mimetics (Zutshi et al., 1998; Shultz and Chmielewski, 1999; Bowman and Chmielewski, 2002).

The other common secondary structure element common in molecular recognition is the α-helix. Besides the obvious significant role of α-helices in regulation of expression by binding to nucleic acids, α-helices have been shown to play a major role in protein/protein recognition. The prototype for experimental study was ribonuclease S (Fig. 6.1) in which the amide bond between residues 20 and 21 of ribonuclease A had been cleaved by subtilisin (Finn and Hofmann, 1973). The cleaved enzyme remained enzymatically active and the 20-residue S-peptide could be reversibly dissociated, and the energetics of recognition studied (Varadarajan et al., 1992) by calorimetry where a mutation of Met-13 to glycine eliminates half the binding energy (some of which must be due to increased entropy of the glycine mutant in the free peptide). The S-peptide has no discernible solution structure in isolation. Another system of more current biological interest in which the α-helix plays a dominant role in complex

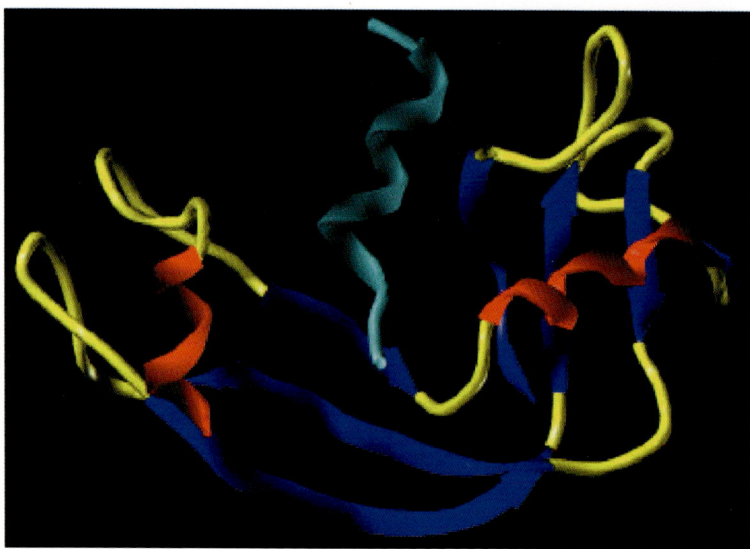

Figure 6.1. Ribonuclease S system (helical S-peptide, *azure*; β-sheet structure of S-protein in *blue*, helices in *red*, and loops in *yellow*).

formation is the binding of a helix of p53 to a hydrophobic cleft on the surface of Hdm2 (Fig. 6.2). Recently, Vassilev et al. (Vassilev et al. 2004) have described a small molecular weight drug candidate (Fig. 6.2) for the treatment of cancer that inhibits this interaction by mimicking the interaction of the three amino acid residues (Leu, Phe, Trp) crucial for binding of the helix (Kussie et al., 1996). A short helical octapeptide had previously been described (Chene et al., 2000; Garcia-Echeverria et al., 2000) as a nanomolar inhibitor that contained two α, α-dialkylamino acids to stabilize the helical conformation (Hodgkin et al., 1990) and orient the three side chains of Leu, Phe, and Trp. Developing inhibitors of protein–protein complexes in which the interacting surfaces arise from discontinuities in the peptide backbone is naturally more problematic.

2.4. Large-Scale Recognition Factors

An important insight into the basic rules of protein recognition is provided by the studies of large-scale structural recognition factors, such as correlation of the antigenicity of surface areas with their accessibility to large probes (Novotny et al., 1986), role of the surface clefts (Laskowski et al., 1996), automatic binding-site identification based on geometric criteria (Ho and Marshall, 1990; Peters et al., 1996), study of the "low-frequency" surface properties (Duncan and Olson, 1993), recognition of proteins deprived of atom-size structural features (Vakser, 1995, 1996; Vakser and Nikiforovich, 1995; Vakser et al., 1999), and backbone complementarity in protein recognition (Vakser, 1996). The practical importance of the large-scale

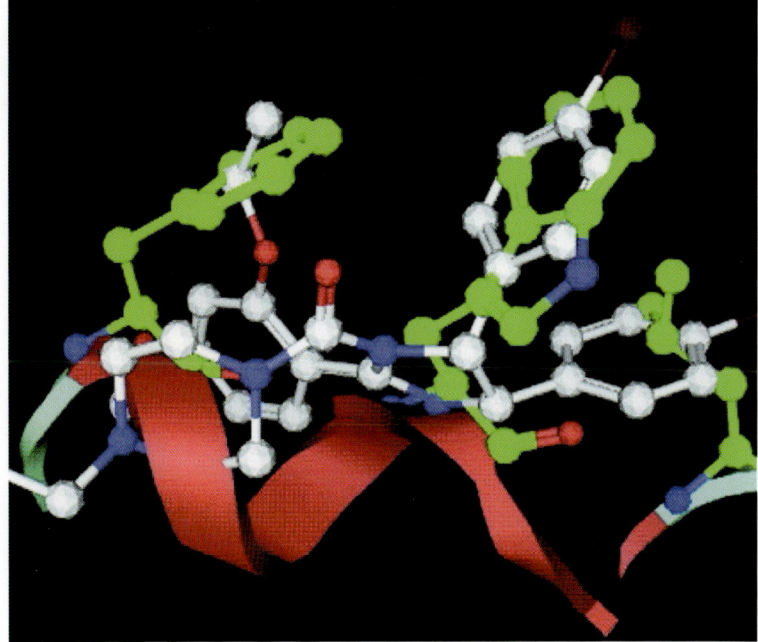

Figure 6.2. Bound conformation of Nutlin-X (*colored stick figure*) to Hdm2 compared with bound conformation of helical segment of P53 (*green*) (Used with permission from Vassilev et al.).

recognition factors for docking methodologies is that they often allow one to ignore local structural inaccuracies (e.g., those caused by conformational changes of the partners upon complex formation).

2.5. Intermolecular-Energy Landscape

The existence of the large-scale structural recognition factors in protein association has to do with the funnel-like intermolecular energy landscape. The concept of the funnel-like energy landscapes has had a significant impact on the understanding of protein folding (Dill, 1999). The kinetics of the amino-acid chain folding into a unique 3D structure is impossible to explain using "flat" energy landscapes, in which minima are located on the energy "surface" that do not favor the native structure (so-called "golf-course" landscapes). The general slope of the energy landscape toward the native structure ("the funnel") explains the kinetics of protein folding. It also provides the basis for protein-structure prediction. The basic physicochemical and structural principles of protein binding are similar, if not identical, to those of protein folding. Thus, the funnel concept can be naturally extended to intermolecular energy (Tsai et al., 1999; Tovchigrechko and Vakser, 2001). As in protein folding, this concept is necessary to explain the kinetics data for protein–protein association. The existence of a funnel in protein–protein interactions is supported by considerations regarding long-range electrostatic and/or hydrophobic "steering forces" and the geometry of proteins (Berg and von Hippel, 1985; McCammon, 1998), energy estimates for near-native complex structures (Camacho et al., 2000), and binding mechanism that involves protein folding (Shoemaker et al., 2000). Important evidence for the funnel in energy landscapes is low-resolution protein–protein recognition (Vakser et al., 1999; Tovchigrechko and Vakser, 2001). The transition to low-resolution images corresponds to the extension of the potential range and thus to averaging of contribution of neighboring atoms to the intermolecular energy. This averaging leads to the smoothing of the energy landscape and reveals the funnel less obscured by the local landscape fluctuations. The process is similar to the use of smoothed potential functions for force-field minimization (Piela et al., 1989; Pappu et al., 1999).

3. PRINCIPLES OF DOCKING

3.1. Docking versus Binding

The only computational approaches that directly model physical interactions between proteins are docking and binding simulations (McCammon, 1998). Docking approaches, as opposed to binding simulations, are not concerned with modeling of real binding pathways, but rather focus only on the final configuration(s) of the complex, that is, equilibrium versus kinetics. This makes docking computationally efficient and allows scanning of a wide variety of potential matches. Binding simulations potentially provide deeper insight into the mechanism of complex formation, assuming a correct force field and adequate sampling. Docking and binding simulations address modeling of protein–protein interactions from different perspectives and are highly complementary to each other. In this chapter, we focus exclusively on docking methodologies.

3.2. Protein–Protein versus Protein–Ligand, Protein–DNA Docking, and so Forth

Proteins naturally interact not only with other proteins, but also with small ligands, as well as with DNA and other biopolymers. These interactions share the same physical and empirical principles of molecular recognition. For practical purposes, however, the relative importance of major recognition factors, as well as the docking strategy, is different.

In small ligand–protein interactions, the small ligand size reduces the relative importance of surface complementarity, especially in the lower-resolution, first-approximation aspect. Correspondingly, from the start, it elevates the relative importance of other physicochemical factors (hydrophobicity, electrostatics, and hydrogen bonding), structural-recognition motifs on the receptor site, as well as the ligand's flexibility. These aspects are reflected in the docking strategy. Another major difference with the protein–protein case in docking strategy is that for the small ligand–protein case, the binding site on the receptor is often known or presumed. The main, and often the only, goal remaining is to determine the intricate details of the ligand–receptor interactions. Protein–protein docking usually requires prediction of the general docking mode for the two proteins, which combined with the large interaction interfaces makes determination of the intricate interatomic details impossible in existing docking approaches. A major distinction of small ligand–protein docking is that the design requirements often involve large libraries of candidate ligands, particularly with virtual screening in drug-design (Lamb et al., 2001; Bajorath, 2002).

In the protein–DNA case, the two major differences from the protein–protein case are the ultimate flexibility of nucleic acids except in multimeric helices, and related to it, the general absence of structural recognition factors beyond the local sequence of the nucleic acids. On the other hand, the binding site on the protein may be derived from known cocrystallized protein–DNA complexes. Also protein–DNA docking may involve a reduced dimensionality of the docking space, when the protein molecule basically slides along the DNA helix until a matching combination of the nucleic acids is found. The situation with RNA is more problematic owing to much less structural information being available.

The practice of protein–DNA docking, although very important, is still currently limited (Aloy et al., 1998; Sternberg et al., 1998). Small ligand–protein docking (Kuntz et al., 1982; Goodsell et al., 1993; Ewing and Kuntz, 1997; Morris et al., 1998; Sun et al., 1998; Knegtel and Wagener, 1999; Makino et al., 1999; Shoichet et al., 1999; Abagyan and Totrov, 2001; Lamb et al., 2001; Glen and Allen, 2003) enjoys huge popularity in both industrial and academic communities, and its practice preceded protein–protein docking. Realistically, although the small ligand–protein and protein–protein docking share many principles, algorithms, and procedures, they have diverged into two distinct fields. In this chapter, we focus primarily on protein–protein docking.

3.3. Homology, Threading, and Ab Initio Docking

Template-based modeling (homology and threading) has become the major driving force in modeling of individual protein structures in the absence of direct experimental data. A growing pool of structural templates has emerged because of the rapid progress in the experimental techniques of protein-structure determination, primarily X-ray crystallography and nuclear magnetic resonance (NMR). Compared with ab initio protein-structure predictions, the template-based techniques provide a significantly higher accuracy in predictions (Moult et al., 2003). The current situation in docking, however, is different because of two major factors. First, the structure of the complex is generally more difficult to obtain by experimental techniques (e.g., X-ray crystallography or NMR) than the structure of individual proteins. The second factor (related to the first) is that it is widely believed that the majority of functional protein–protein interactions are transient, and thus do not form complexes stable enough for crystallization. Thus, the pool of protein–protein structural templates is limited and heavily biased toward multisubunit proteins. The significance of template-based modeling of protein–protein complexes is growing, especially in such important applications as predictions of the existence of an interaction (Lu et al., 2002). For the prediction of protein–protein docking modes, however, docking techniques are virtually exclusively ab initio ones (Janin et al., 2003).

4. DOCKING METHODOLOGY

Energy minimization of structures in molecular biology is, in general, quite difficult to solve; the large molecular sizes and the rugged nature of the potential energy surface with many local minima result in a very challenging global-optimization problem. The objective is to find the conformation of lowest free energy, which should correspond to the native structure, that is, the potential well where the macromolecular ensemble spends most of its time. Thus, the task is split in two: development of an objective function (force field) as a representative of the potential energy surface that has its global minimum at the native conformation, or close to it, and the procedure used to locate that global minimum. The typical tradeoff is between the quality/complexity of the force field, and the realistic possibility of locating the global minimum of a more complex objective function. This dichotomy of function—search procedure (engine) is often complemented by a third stage—post-processing, aimed at improving the signal (correct prediction)-to-noise (false-positive prediction) ratio of the results using "scoring" functions that are too expensive computationally to be included in the main search procedure.

4.1. Docking Force Fields

Force fields in existing protein–protein docking procedures widely range between the most trivial ones (e.g., an empirical step-function approximation of the Lennard-Jones potential, which exclusively employs digitized steric complementarity [Vakser, 1996]) to full-scale "standard" force fields (e.g., OPLS [Jorgensen et al., 1996], AMBER [Weiner et al., 1984], CHARMM [MacKerell et al., 1998], ECEPP [Nemethy et al., 1983], etc.), and to the next generation force fields such as AMOEBA (Grossfield et al., 2003; Ponder and Case, 2003) that include multipole electrostatics and polarizability. For an objective overview of force fields used in protein modeling, see the recent review by Ponder and Case (Ponder and Case, 2003). The difference between the complex and the simple force fields often reflects two different paradigms of docking strategy—generating a large number of candidate matches based on accurate force fields and finding the correct match at the post-processing (refinement and scoring) stage, as opposed to locating the approximate position of the proteins within the complex using a simple force field and bringing in the details at the subsequent refinement stage using a more complex force field. These two paradigms are conceptually similar in the second stage (post-processing/refinement), but differ in the first stage of generating the candidate matches.

An important development in docking force fields has been introduction of statistically and evolutionary derived potentials, following similar developments in modeling of individual protein structures by Jernigan and Sippl independently (Jernigan and Bahar, 1996; Sippl et al., 1996; Miyazawa and Jernigan, 1999). The statistically derived residue–residue and atom–atom potentials (Keskin et al., 1998; Moont et al., 1999; Glaser et al., 2001; Gray et al., 2003; Lu et al., 2003) provide powerful tools for the detection of protein–protein matches with some built-in tolerance to structural mismatches originated from structural flexibility. Evolutionary-derived potentials are based on observations that the surface residues are evolutionary more conserved in the binding regions than in the nonbinding areas (Lichtarge and Sowa, 2002). Thus, evolution-based components are added to force fields to increase the score of matches within potential binding sites. Zhang et al. have simplified a statistically based potential using only three atoms per residue (C_α, C_β, and side-chain center of mass). This potential was tested against 96 decoy sets for protein-structure recognition, 21 docking-decoy sets, and with two sets of loop predictions, where it performed better than many of the current scoring functions in use (Zhang et al., 2004a,b).

4.2. Docking Engines

Sampling of the docking degrees of freedom is a crucial step in docking methodology. The docking degrees of freedom involve six external degrees of the rigid body movement (e.g., three translation coordinates and three angles of rotation) and any internal degrees of freedom, which determine the conformation of the proteins. To make the number of the internal degrees of freedom manageable, approximations are essential. The "rigid-body" approximation leaves only the external degrees of freedom and approximates internal degrees of freedom by making the protein "body" soft and thus tolerant to local structural mismatches. The rigid-body approximation is adequate for docking separated proteins from cocrystallized complexes (no structural changes on complex formation), low-resolution docking of the unbound/inaccurate structures, as well as in some cases of high-resolution docking of unbound proteins. For an atomic resolution docking of unbound structures, however, some form of explicit processing of internal degrees of freedom is required in general. It may be assumed (although not proven by systematic studies) that for most complexes of unbound proteins atomic resolution accuracy in docking may be achieved by properly designed limited conformational search of the surface side chains.

The existing search procedures used in the sampling of the docking degrees of freedom vary dramatically, from exhaustive search on a grid to Monte Carlo, molecular dynamics, and genetic algorithms. An important direction in protein docking is based on the correlation technique by the fast Fourier transformation (FFT), introduced in 1992 by Katchalski-Katzir et al. (Katchalski-Katzir et al., 1992). It predicts the structure of a complex by maximizing surface overlap between the two molecular images. The images are digital 3D representations of molecular shape that distinguish between the surface and the interior of a molecule (Fig. 6.3). The algorithm is based on the correlation between these images, which is calculated rapidly using FFT.

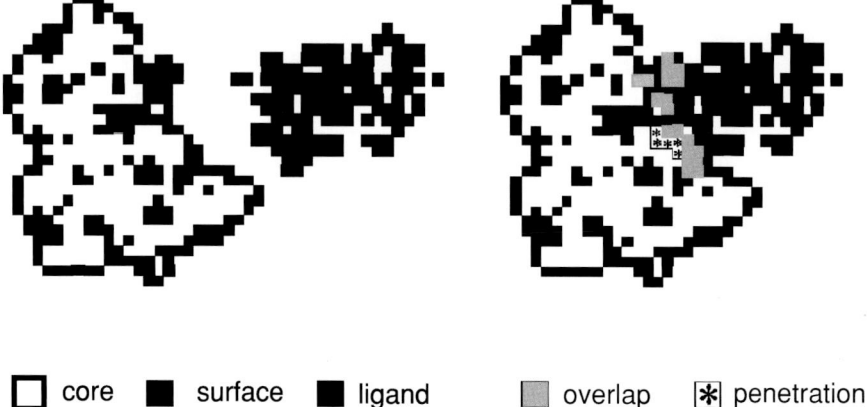

☐ core   ■ surface   ■ ligand   ▨ overlap   ✱ penetration

Figure 6.3. Cross section through the 3D representation of the molecules.

Correlation peaks found by the procedure indicate geometric match and thus represent a potential complex. The procedure is equivalent to the full six-dimensional search (three translations and three rotations of the ligand) but much faster by design. The approach has been implemented in a number of algorithms (Harrison et al., 1994; Vakser and Aflalo, 1994; Ackermann et al., 1995; Ten Eyck et al., 1995; Vakser, 1995; Meyer et al., 1996; Blom and Sygusch, 1997; Friedman, 1997; Gabb et al., 1997; Bliznyuk and Gready, 1999; Ritchie and Kemp, 2000; Chen and Weng, 2002; Heifetz et al., 2002). These algorithms share the same approach to sampling of three or more external docking coordinates by integral operations. They differ in representation of molecules, structural tolerance, and physicochemical characterization of surfaces. Other grid-based algorithms perform the search explicitly, but contain empirical structure-based filters that reduce the search space (Jiang and Kim, 1991; Palma et al., 2000; Jiang et al., 2002).

A docking algorithm developed by Nussinov, Wolfson, and co-workers (Nussinov and Wolfson, 1991; Fischer et al., 1995; Schneidman-Duhovny, et al., 2003) represents protein surfaces by a reduced set of critical points that contains roughly the same information about the surface as a more dense surface representation. The triplets or pairs of the critical points with associated normals representing the proteins are matched using an efficient computational algorithm based on computer vision techniques. An alternative approach to matching surface points is based on the genetic algorithm (Gardiner et al., 2001).

Abagyan and co-workers (Totrov and Abagyan, 1994; Fernandez-Recio et al., 2002) and Scheraga and co-workers (Trosset and Scheraga, 1999) developed flexible docking algorithms that use Monte Carlo–based energy-minimization protocols, which include both external and internal coordinates. The method developed by Baker and colleagues (Gray et al., 2003) docks two proteins using a Monte Carlo protocol with low-resolution residue-scale statistical potentials, followed by a high-resolution refinement. A simple long-range electrostatic guidance method based on the balance of intermolecular atom–atom forces was developed (Fitzjohn and Bates, 2003).

One approach to flexibility of proteins that should be combined with docking is the prediction of low-frequency vibrational modes of protein dynamics by elastic models that are primarily determined by shape. This approach has been pioneered by Jernigan, whose group has shown a strong correlation between simulations of elastic models and the experimentally observed b-factors from crystallography (Kim et al., 2002a,b), suggesting that such models capture the larger scale motions to a first approximation. Each protein could be represented by a limited ensemble of conformers reflecting normal-mode motions, and complex formation between the ensembles tested for the most stable complex.

In case of objective functions designed to be sufficiently realistic, the global minimum is often never found owing to time constraints and the complexity of the energy landscape, and the minimization determines only local minima which can be far away from the native conformations seen in the complex. Thus, global-optimization methodologies for complex objective functions in protein docking remain of high priority despite the multitude of existing techniques. There is a consensus among researchers that progress in solving structure-prediction problems depends on the presence of certain properties of the native energy landscape (which must be reflected in the objective-function landscape). Most importantly, equilibrium energy landscapes in folding and docking often reveal a funnel-like character of the energy landscape near the native structures (Miller and Dill, 1997; Bouzida et al., 1999; Camacho et al., 1999; Tsai et al., 1999; Vakser et al., 1999; Zhang et al., 1999; Shoemaker et al., 2000; Tovchigrechko and Vakser, 2001).

Several protein-recognition techniques use smoothed potential functions (Piela et al., 1989; Ma and Straub, 1994; Vakser, 1996; Trosset and Scheraga, 1998; Pappu et al., 1999; Hart et al., 2000). These methods transform the original objective function in such a way that the number of local minima is significantly decreased, the energetic barriers between minima are reduced, and global optimization is correspondingly much easier. The underlying idea is to transform minimum found on the smoothed function gradually back to the original function in a way that allows one to track the location of the global minima through successive applications of smoothing. In particular, this approach aims to follow the path starting from the global minimum of the smoothed function to the global minimum of the original function. It should be noted that one is essentially removing higher frequency perturbations from the underlying funnel; thus, a narrow deep minimum can be eliminated.

Ponder and co-workers (Hart et al., 2000) revealed deep analogies between potential smoothing techniques and simulated annealing. The authors argue that the former is expected to be a more efficient method for conformational sampling because the search process is deterministically focused on specific regions, whereas simulated annealing is done in a probabilistic way. They further note that while potential smoothing scales reasonably well with the size of the system, simulated annealing requires Markov chain or time-series trajectory; thus the extent of search scales exponentially with the size of the system.

4.3. Post-Processing—Scoring and Refinement

A variety of approaches to scoring functions have been advocated in the literature in protein/protein docking. These vary from smoothed potentials (even as crude as smoothing the Leonard-Jones potential in GRAMM [Vakser, 1996]) to use of molecular dynamics simulations with current force fields and explicit solvation for very limited applications. Considering the magnitude of the problem of docking two reasonably sized proteins, especially if one tries to accommodate conformational changes in the side chains, or even small conformational changes in the backbone, one is, by necessity, reduced to a set of successive filters of increasing resolution as the most plausible approach.

Elimination of those candidate complexes from consideration that do not satisfy any of the criteria, such as compactness that are clearly properties of protein–protein interfaces, is a first step. Some docking approaches such as GRAMM provide a low-resolution filtering and focus on potential complexes for further, more detailed, analysis. The detection of near-native conformations (<5 Å RMSD) and their further refinement is a challenging task that requires multiple stage protocols of physicochemical filtering and structure minimization (Camacho et al., 2000). In reviews (Oprea and Marshall, 1998; Marshall et al., 2000) of the approaches to prediction of affinity in ligand/protein interactions, Marshall and colleagues have emphasized the difficulty of predicting the change in entropy (ΔS) on complex formation because of conformational restrictions on the two binding partners and changes in solvation. Unfortunately, binding free energies (ΔG) are on the order of 5 to 20 kcal/mol, but represent only a small difference ($\Delta G = \Delta H - T\Delta S$) between two much larger energetic components (ΔH and ΔS), often of a magnitude of hundreds of kilocalories per mole. Thus, small errors are estimates of binding entropies and enthalpies can lead to significant deviations in predicted affinities from those observed experimentally.

The change in enthalpy of binding can be reasonably approximated by molecular modeling recognizing, however, that the monopole approximation of electrostatics used in current force fields does not accurately reproduce the electrostatic potential of the two partners (Williams, 1988, 1991). With an error in the electrostatic potential of each molecule approaching 10% from the monopole approximation and no consideration of polarizability on proximity of charge, it is not surprising that accurate reproduction of experimental results by molecular mechanics simulations is difficult. Fortunately, a next-generation force field AMOEBA that includes multipole electrostatics and polarizability has recently been parameterized for proteins by Ponder and co-workers (Grossfield et al., 2003; Ponder and Case, 2003). Use of force fields of this quality should vastly improve our ability to refine the atomic details of protein–protein complexes, and more accurately predict affinities.

What is desired at the low-to-medium level of resolution are objective scoring functions that can be evaluated rapidly so that candidate complexes can be selected for refinement. Certainly, the statistical knowledge-based potentials (so-called "inverse Boltzmann" potentials) derived from experimental observation of residue–residue frequencies, as pioneered by Jernigan and Sippl (Jernigan and Bahar, 1996; Sippl et al., 1996; Keskin et al., 1998; Miyazawa and Jernigan, 1999) for proteins, have a useful role as they presumably represent the free energy (ΔG) of interaction. These potentials have been applied in prediction of protein–protein complexes (Keskin et al., 1998; Moont et al., 1999; Glaser et al., 2001; Gray et al., 2003; Lu et al., 2003).

Che and Marshall (Che and Marshall, 2003) have derived a statistical-based atom-atom potential from the analysis of 179 high-resolution nonhomologous protein complexes with data sampled only from the complex interface and interactions weighted by the degree of atom burial to reflect "hot-spot" residues at the center of the interface surrounded by hydrophobic interactions. In support of this approach, the distance-dependent pair preferences of polar atoms had a minimum in the 2.6 to 3 Å range corresponding to hydrogen bonds and salt bridges. The strongest interacting atom type for nonpolar aromatic carbons was the positively charged nitrogen representing a π-cation interaction and a minimum strength of approximately one fifth that seen for either the salt bridge or hydrogen bond. While this relative energetics is dramatically different from that seen in the calculated comparison of π-cation and salt-bridge interactions by Gallivan and Dougherty (Gallivan and Dougherty, 2000), it is excellent agreement with relative strengths observed in model systems when the counter-ion in the π-cation interaction is included (Bartoli and Roelens, 2002). Although this test of the potential function is encouraging, its ability to correctly score possible binding modes has yet to be explored in CAPRI.

An alternative approach that has proven successful in the prediction of affinities is to combine easily calculated parameters that have some correlation with ΔG and ΔS and utilize modern statistical techniques (partial least squares of latent variables [PLS] and cross-validation) to develop a predictive model of the ΔG of binding by training with a set of experimental data. Head et al. used such an approach for the prediction of binding affinities of a wide variety of ligands binding to proteins and DNA in the program VALIDATE (Head et al., 1996). This approach has recently been extended by Berglund et al. (Berglund et al., 2004) to score candidate folds in the ab initio prediction of protein structure from sequence data alone with the program ProVal. The scoring function was able to correctly find the crystal structure when embedded in a decoy set of alternative folds in most cases, and scored the crystal structure in the top 10% in all 28 decoy sets examined. What is of particular interest is that this low-resolution model considers only the protein backbone and side chains are represented by multipoles centered at the β-carbons. Extension of this approach to scoring protein/protein interfaces is being considered.

5. LARGE-SCALE DOCKING

Protein interactions form a major basis for life processes at the molecular level. Most protein interactions are with other proteins, although interactions with other macromolecules/ligands play essential roles. Thus, efforts to recreate the network of protein–protein interactions are important for the interpretation of the information encoded in genomes. The number of protein-protein interactions is significantly larger than the number of individual proteins. For example, the average number of protein/ protein interaction estimated is approximately eight for each protein expressed in yeast (Tong et al., 2004). Thus, high-throughput methods are needed for studies of these interactions on a genome scale. The existing methodologies, both experimental and computational, may be roughly separated into methods detecting direct physical interactions between proteins (e.g., two-hybrid analysis, mass spectrometry, etc.) and the function-assigning methods (e.g., correlation of mRNA levels, method of phylo-genetic profiles, fusion pattern method, sequence alignment, and fold comparison). The outline of "the post-genomic" methods is presented in several reviews (Eisenberg et al., 2000; Oliver, 2000; Skolnick et al., 2000; Vukmirovic and Tilghman, 2000; Sali et al., 2003).

Since proteins are 3D objects, the importance of the direct 3D analysis of protein–protein interactions is obvious. Such analysis is necessary for the prediction of these interactions, their adequate study, and for further applications (e.g., structure-based drug design). Direct experimental approaches (primarily, X-ray crystallography, and NMR) are developing fast. However, they are capable of determining only a fraction of all protein structures. Thus, the structures of most individual proteins in genomes have to be modeled by high-throughput modeling approaches (Burley, 2000; Sali et al., 2003). The growing availability of the experimentally determined structures of representative protein folds makes the template-based modeling of the majority of proteins in genomes quite realistic. The limitations of the direct experimental techniques are even more evident in the case of the structures of protein–protein complexes, which are, in general, more difficult to determine than the structures of individual proteins. The fact that most individual protein structures from the genome will be models—makes the computational docking approaches indispensable for the direct 3D analysis of probable protein–protein interactions in genomes.

The number of potential protein–protein interactions and the nature of protein structures to be docked impose strong requirements on the docking techniques. Be-cause of the large number of proteins to dock, docking has to be fast. At the same time, since the majority of individual protein structures in a genome will be mod-els, the docking has to be capable of predicting complexes of modeled proteins. The major difference between an experimental (X-ray) protein 3D structure and a model, in general, is the substantially lower accuracy of the latter (Moult et al., 2003). The accuracy of the protein models may vary significantly, based on the availability of an appropriate structural template and the degree of target-template similarity, from approximately 1 Å RMSD (high-sequence similarity to templates) to >6 Å RMSD (low-sequence similarity to templates, or no templates). Thus, the docking procedure has to be capable of tolerating very significant structural inaccuracies.

Obviously, docking cannot yield greater precision than the precision of the par-ticipating protein structures. Even the low precision of approximately 10 Å relative displacement of the two proteins, however, results in meaningful predictions of the binding interfaces and the gross structural features of the complex (Vakser et al., 1999). GRAMM was shown to adequately address the variable resolution docking of pro-tein structures, by performing fast, approximate docking of low-resolution molecular images and slower, precision docking of more accurate molecular representations

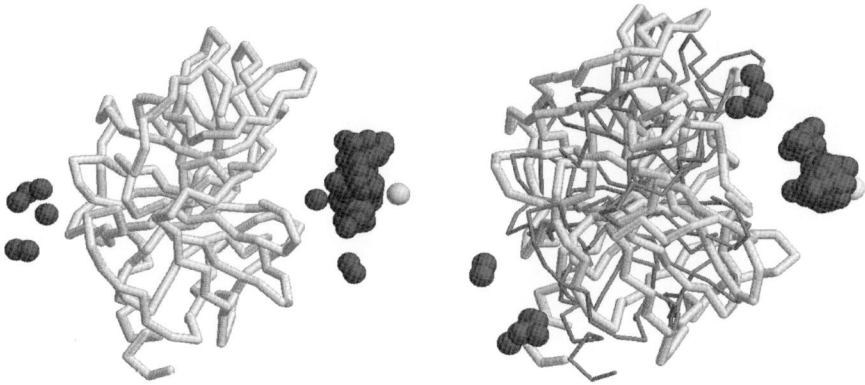

Figure 6.4. Results of the low-resolution docking of trypsin and BPTI. The experimental structures are on the **left** and the low-resolution models (RMS = 6 Å, both trypsin and BPTI) are on the **right**. The dark spheres are the BPTI center of mass in 100 lowest energy positions. The light gray sphere (indicated by an *arrow*) is the BPTI center of mass in the cocrystallized complex. For comparison, the experimental structure of trypsin (*thin, dark backbone*) is overlapped with the model. The docking of the models clearly preserves the cluster of correct predictions in the area of the binding site.

(Katchalski-Katzir et al., 1992; Vakser and Aflalo, 1994; Vakser and Nikiforovich, 1995). These studies suggested the possibility of docking inaccurate protein models.

Vakser et al. reported (Vakser et al. 1999) the application of GRAMM at low resolution to X-ray protein structures from a nonredundant database of 475 cocrystallized protein–protein complexes. The results of the study were analyzed further in a subsequent report (Tovchigrechko and Vakser, 2001) using various statistical models. The same techniques have been applied to docking of protein models of different accuracies (Tovchigrechko et al., 2002). To simulate the precision of protein models, all proteins in the protein/protein database were structurally modified in the range of 1 to 10 Å RMSD, with 1 Å intervals. A sophisticated procedure was specifically designed and implemented for that purpose. All resulting models of the proteins were docked. The statistical significance of the docking was analyzed, and the results were correlated with the precision of the models. The data showed that even highly imprecise protein models (>6 Å RMSD) still yield structurally meaningful docking results, that are accurate enough to predict binding interfaces and to serve as starting points for further structural analysis. An example of docking protein models of low accuracy is shown in Fig. 6.4. The study demonstrated the applicability of existing docking techniques to genome-wide modeling of protein–protein interactions.

6. DOCKING IN TERTIARY-STRUCTURE PREDICTION

The interaction of secondary structure elements in protein structures may be formulated in terms of docking, even though docking is traditionally considered to be a problem of matching two separate molecules. The main difference in matching secondary structure elements and matching separate molecules is in the constraints imposed by the environment. A number of studies explored the applicability of docking to secondary structure packing (Ausiello et al., 1997; Nikiforovich et al., 1998; Yue and Dill, 2000; Vakser and Jiang, 2002; Inbar et al., 2003; Jiang et al., 2003). A multiplicity of physicochemical factors obviously plays a role in the packing of secondary structure elements in proteins and in the formation of protein complexes. However, the well-known tight packing of structural elements suggests the importance of the geometric fit. Steric complementarity in protein interactions has been studied extensively (for a review see Halperin et al., 2002). Obviously, the role of steric complementarity in the interaction of secondary structure elements deserves similar attention.

Earlier studies of this subject were primarily focused on helix–helix packing (e.g., Richmond and Richards, 1978; Cohen et al., 1979; Chothia et al., 1981; Murzin and Finkelstein, 1988; Reddy and Blundell, 1993; Walther et al., 1996). One reason was the limited number of high-quality crystal structures, mostly containing helices. A traditional biochemical view on interactions of secondary structure elements largely neglected geometric complementarity as an important factor (with the exception of helix–helix interactions). A docking algorithm based on geometric complementarity was applied to a comprehensive database of secondary structure elements derived from the Protein Data Bank (PDB) (Jiang, Tovchigrechko et al., 2003). The results show that the steric fit plays an important role in the interaction of all secondary structure elements. Docking procedures have started to be utilized in protein-structure prediction (Yue and Dill, 2000; Haspel et al., 2002; Inbar et al., 2003). In such cases, the secondary structure elements are docked by rigid-body procedures followed by structural refinement.

Docking approaches are popular in modeling the structure of transmembrane (TM) helix bundles in G-protein–coupled receptors (GPCR) and other integral membrane proteins. The few existing crystal structures of integral membrane proteins provide useful information on the TM bundle configurations. The TM helices are roughly parallel to each other; they are of similar length (determined by the thickness of the membrane), and are well packed. Thus, it is reasonable to assume that the structure of the bundle is determined primarily by the helix–helix interactions, rather than by the interhelical loops (which, of course, still determine the general topology of the bundle). Most helix–helix interfaces in TM bundles are predominantly binary–if two interfaces overlap, one of them is usually dominant. In that regard TM bundles are ideal objects for docking predictions. At the same time helices are simple enough to provide validation ground for new docking concepts (e.g., see Pappu et al., 1999). It has been noted that the side chains at the helix–helix interfaces, on average, are shorter than those at the noninterface helix areas (Jiang and Vakser, 2000, 2004). This structural characteristic creates a low-resolution recognition factor that allows one to model the TM bundle at low resolution (Vakser and Jiang, 2002). However, a high-resolution model of the TM bundle requires explicit conformational search of the helix internal coordinates (primarily side chains) (Nikiforovich et al., 1998, 2001). Thus, from the practical point, the high-resolution modeling of TM bundles is currently useful only if accompanied by an ample set of experimentally derived structural constraints. Nikiforovich and Marshall (Nikiforovich and Marshall, 2003) have recently generated a model of photoactivated rhodopsin consistent with experimental measurements by de novo helix packing with all-*trans*-retinal attached to TM7. A detailed review of the TM modeling, including application of docking techniques, extends beyond the scope of this chapter.

7. COMMUNITY-WIDE EVALUATIONS OF DOCKING TECHNIQUES

Following recent dramatic progress in genomics, accompanied by advances in structural and computational biology, the importance of modeling protein–protein interactions has grown significantly. Accordingly, the visibility of protein–protein docking field has increased and the protein-docking community has begun to organize and actively develop community-wide activities. At the First Conference on Modeling of Protein Interactions in Genomes at Charleston, SC, 2001 (Vajda et al., 2002), a number of such activities were discussed and decided on, including CAPRI and other benchmarking community-wide experiments. These activities were further developed at the Second Conference at Stony Brook, NY, 2003; the CAPRI meeting at La Londe-des-Maures, France, 2002 (Janin et al., 2003); and other meetings.

7.1. Protein–Protein Decoy Sets

Protein-structure decoy sets are extremely popular in the protein-modeling community. Decoys are protein structures artificially put in a "wrong" (nonnative) conformation, which otherwise look like native structures, at least in terms of structural packing. They are used by the developers of modeling techniques to gauge potentials and scoring functions in their ability to discriminate false-positive predictions. Recently, in addition to decoy sets of individual protein structures (Samudrala and Levitt, 2000), several groups put together decoy sets of protein–protein complexes. These sets contain false-positive matches of proteins and serve the same purpose as the decoys of individual structures—development of better potentials and scoring functions for the discrimination of false-positive predictions. The first limited set of protein/protein docking decoys was suggested by Vajda (personal communication) and compiled by Vakser's group (http://www.bioinformatics.ku.edu). Currently, other decoy sets are available from Sternberg's group (http://www.sbg.bio.ic.ac.uk/docking/all_decoys.html), Baker's group (http://depts.washington.edu/bakerpg), and the group of Weng (http://zlab.bu.edu/~rong/dock/software.shtml).

7.2. Benchmarking

A nonredundant benchmark set of protein–protein complexes was developed in Weng's group (Chen et al. 2003). The set currently contains 59 complexes of cocrystallized protein complexes and their components crystallized separately (unbound structures). The idea behind the benchmark is to have pairs of unbound structures with the correct match known (cocrystallized complex) for the development of algorithms for the prediction of complexes of unbound proteins. The set (http://zlab.bu.edu/~rong/dock/benchmark.shtml) has been used in the docking community for the development of docking methodology.

7.3. CASP/CAPRI

An important activity in the field of protein–protein docking is the community-wide experiments on Critical Assessment of Structure Prediction (CASP; http://predictioncenter.llnl.gov) and Critical Assessment of Predicted Interactions (CAPRI; http://capri.ebi.ac.uk). These experiments allow a comparison of different computational methods on a set of prediction targets (experimentally determined structures unknown to the predictors). The protein–protein docking category was introduced at CASP2 (Dixon, 1997; Vakser, 1997) and has been successfully continued in CAPRI (Janin et al., 2003). The CAPRI paradigm solicits yet unpublished structures of cocrystallized protein–protein complexes from experimentalists (primarily, X-ray crystallographers) and distributes the separately crystallized structures of the components, when available, to the community of predictors. The CAPRI experiment is conducted on a continued basis, updated with the availability of new prediction targets. Currently, approximately 3 years from its inception, six rounds of CAPRI have been completed. CAPRI generated great interest in the protein-docking and protein-modeling communities and has already led to visible progress in docking methodologies.

8. FUTURE OF DOCKING

8.1. Accuracy Through Structural Flexibility

It has become obvious that static docking is feasible with current improvement in sampling and force fields that are being introduced. It has also become obvious that docking of two rigid proteins, in general, cannot provide atomic-resolution details for real protein–protein complexes. Dynamics and flexibility are inherent in protein structures and ignoring this inherent complication renders most current approaches problematic. As an example of the problem, consider the protein CheY that complexes with a variety of modulatory proteins in bacterial two-component systems involved in chemotaxis. The two-component system is composed of a histidine kinase that is activated by extracellular ligands and a response regulator that further transmits the signal. The response regulator is usually composed of two domains: an N-terminal CheY domain that receives the phosphoryl group and activates target proteins that is conserved (Volz, 1993). The C-terminal $\alpha_4 - \beta_5 - \alpha_5$ surface is CheY interface for

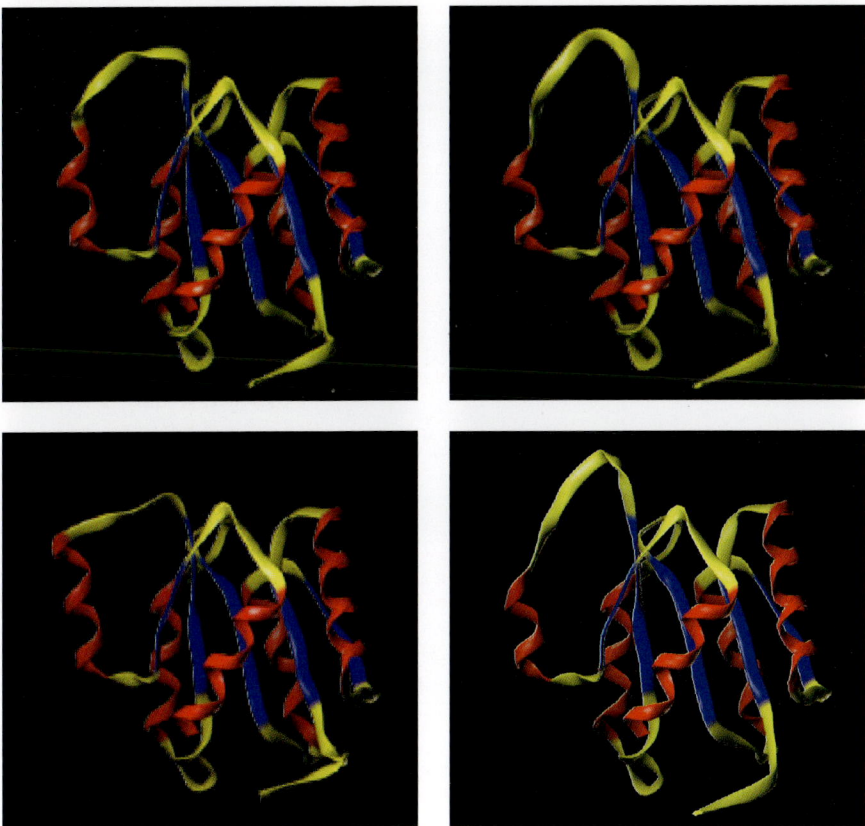

Figure 6.5. Crystal structures of CheY from four complexes with different proteins superimposed to show variation in loop conformation (*yellow ribbon*, **top left**.)

protein–protein interactions. CheA/CheY is the best biochemically and structurally characterized two-component system, playing a key role in chemotaxis. Convergent binding sites on CheY, with three other proteins—CheA, FliM and CheZ—have been structurally determined (Zhu et al., 1997; Welch et al., 1998; Lee et al., 2001; Zhao et al., 2002). What is different in the complexes is the detailed conformation of the surface loops of CheY that interacts with its protein partners (Fig. 6.5). Certainly, the concept of induced fit is applicable, and numerous other examples of dramatic changes of conformation on complex formation can be cited; two immediate cases are the change in β-hairpin flap position (Fig. 6.6) in HIV protease on inhibitor binding (Miller et al., 1989) and the dramatic helix distortion (Fig. 6.7) of the calmodulin dumbbell (Babu et al., 1988) on the binding of helical peptides to calmodulin (Meador et al., 1992). Until docking algorithms accommodate flexibility in both partners, there

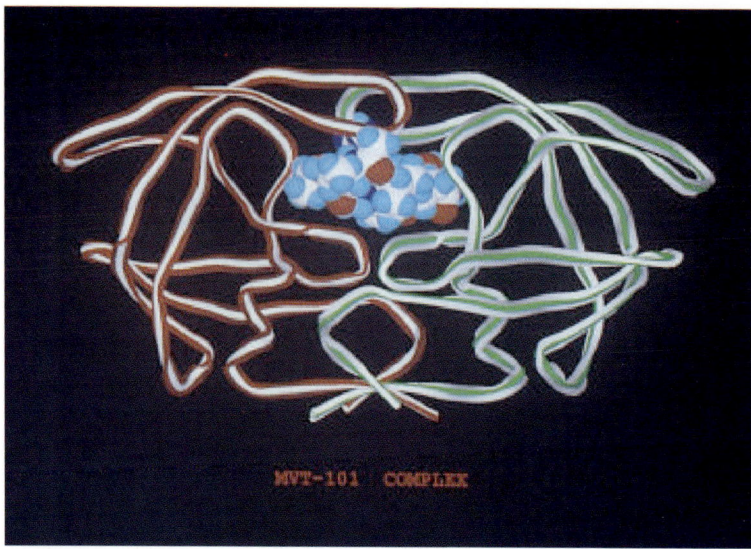

Figure 6.6. Ribbon diagram (candy-cane ribbons for each monomer) of HIV protease with inhibitor (MVT-101, space filling; 4HPV, PDB) bound. The two β-hairpin flaps (*on top*) have closed down to complete binding site for inhibitor; this conformation change is seen with essentially every inhibitor bound to HIV protease.

is limited chance that the predicted complex will reflect reality at the atomic level even if the binding surfaces at the interface may be predicted correctly.

8.2. Genome-Wide Modeling, High-Throughput Docking of Models

With rapid progress in the experimental structural determination of proteins, currently, about one third of individual protein structures can be modeled by relatively accurate template-based techniques (Sali et al., 2003). This percentage is expected to grow significantly in the near future as more fold templates are determined. At the same time, new experimental and computational techniques yield genome-wide maps of protein–protein interactions with increasingly greater precision (Uetz et al., 2000; Aebersold and Mann, 2003; Salwinski and Eisenberg, 2003). Combination of these two factors paves the way for future genome-wide structural modeling of protein–protein interactions. Such modeling will reveal deep insights into the complexity of life at the molecular level. It will also lead to structural characterization of drug targets and facilitate structure-based drug design. Currently, a working prototype of a genome-wide docking database has been created by the Vakser group (unpublished results) as a proof of principle in a general sense. To become practical, however, such a database will require development of advanced high-throughput docking/modeling approaches and more accurate methods of building genome-wide maps of protein–protein interactions.

A

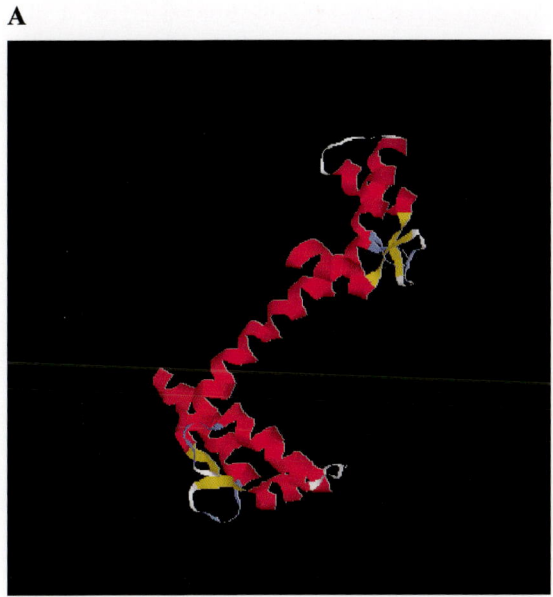

B

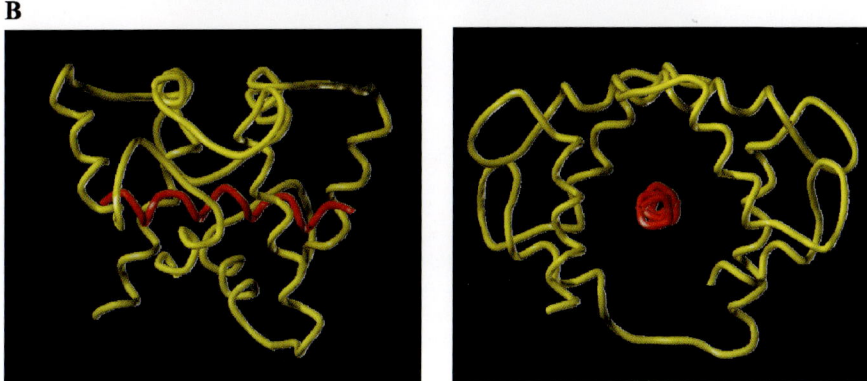

Figure 6.7. (**A**, *left*). crystal structure of calmodulin (ribbon diagram, 2CLN, PDB). Note long magenta helix connecting two calcium-binding domains. (**B**, below). Orthogonal views of calmodulin (*yellow ribbon*, 1CDL, PDB) bound to helical peptide (red ribbon).

8.3. Automated Servers

In current CAPRI, performance of docking procedures is complemented by expert human intervention. Such intervention often plays a crucial role in the quality of the submitted predictions. This greatly diminishes the usefulness of docking software for the broad community of biomedical scientists, who do not necessarily have extensive expertise in protein recognition. An important development expected in docking evaluations is the routine use of automated docking web servers. Currently, several protein-docking servers are available http://capri.ebi.ac.uk). Progress in developing docking servers in the near future will be facilitated by advances in docking methodology and computer hardware. Reliable prediction of the structure of a protein complex without human intervention is an important goal in the rapidly developing protein-docking field.

Acknowledgments

We acknowledge the input of our students, collaborators, colleagues, and mentors whose interactions have shaped and stimulated our research. In particular, G.M. acknowledges the efforts of a recent graduate Ye Che to force him to seriously consider protein–protein recognition as exemplified in the Che Y system, and the efforts of Stan Galaktionov, Gregory Nikiforovich, Anders Berglund, Richard Head, and Eric Welsh to devise better sampling and scoring approaches to protein-structure prediction. I.V. is deeply indebted to Ephraim Katchalski-Katzir, who introduced him to the protein-docking field; Sandor Vajda, with whom he enjoys frequent discussion of the protein-docking field; and all his co-workers, especially Andrei Tovchigrechko and Sulin Jiang. The sine qua non, of course, is funding from the National Institutes of Health.

REFERENCES

Abagyan, R., and Totrov, M. (2001). High-throughput docking for lead generation. *Curr. Opin. Chem. Biol.* 5:375–382.

Ackermann, F., G. Herrmann, F. Kummert, S. Posch, G. Sagerer, and D. Schromburg. (1995). Protein docking combining symbolic descriptions of molecular surfaces and grid-based scoring functions. In: Rawlings, C., Clark, D., Altmanet, R. (eds), *Intelligent Systems for Molecular Biology*. Menlo Park, CA, AAAI Press, pp. 3–11.

Aebersold, R., and Mann, M. (2003). Mass spectrometry-based proteomics. *Nature* 422:198–207.

Aloy, P., G. Moont, H.A. Gabb, E. Querol, F.X. Aviles, and M.J.E. Sternberg. (1998). Modelling repressor proteins docking to DNA. *Proteins* 33:535–549.

Ausiello, G., G. Cesareni, and M. Helmer-Citterich. (1997). ESCHER: a new docking procedure applied to the reconstruction of protein tertiary structure. *Proteins* 28:556–567.

Babu, Y.S., C.E. Bugg, and W.J. Cook. (1988). Structure of calmodulin refined at 2.2 A resolution. *J. Mol. Biol.* 204:191–204.

Bajorath, J. (2002). Integration of virtual and high-throughput screening. *Nat. Rev. Drug Discov.* 1:882–894.

Baker, D., and Lim,W.A. (2002). Folding and binding. From folding towards function. *Curr. Opin. Struct. Biol.* 12:11–13.

Bartoli, S., and Roelens, S. (2002). Binding of acetylcholine and tetramethylammonium to a cyclophane receptor: anion's contribution to the cation-pi interaction. *J. Am. Chem. Soc.* 124:8307–8315.

Berg, O.G., and von Hippel, P.H. (1985). Diffusion-controlled macromolecular interactions. *Annu. Rev. Biophys. Biophys. Chem.* 14:131–160.

Berg, T. (2003). Modulation of protein-protein interactions with small organic molecules. *Angew Chem. Int. Ed. Engl.* 42:2462–2481.

Berglund, A., R.D. Head, E. Welsh, and G.R. Marshall. (2004). ProVal: a protein scoring function for the selection of native and near-native folds. *Proteins Struct. Funct. Bioinform.* 54:289–302.

Bliznyuk, A.A., and Gready, J.E. (1999). Simple method for locating possible ligand binding sites on protein surfaces. *J. Comput. Chem.* 20:983–988.

Blom, N.S., and Sygusch, J. (1997). High resolution fast quantitative docking using fourier domain correlation techniques. *Proteins* 27:493–506.

Bogan, A.A., and Thorn, K.S. (1998). Anatomy of hot spots in protein interfaces. *J. Mol. Biol.* 280:1–9.

Bouzida, D., P.A. Rejto, and G.M. Verkhivker. (1999). Monte Carlo simulations of ligand-protein binding energy landscapes with the weighted histogram analysis method. *Int. J. Quantum Chem.* 73:113–121.

Bowman, M.J., and Chmielewski, J. (2002). Novel strategies for targeting the dimerization interface of HIV protease with cross-linked interfacial peptides. *Biopolymers* 66:126–133.

Bryngelson, J.D., J.N. Onuchic, N.D. Socci, and P.G. Wolynes. (1995). Funnels, pathways, and the energy landscape of protein folding: a synthesis. *Proteins* 21:167–195.

Burley, S.K. (2000). An overview of structural genomics. *Nat. Struct. Biol.* 7:932–934.

Camacho, C.J., Z. Weng, S. Vajda, and C. DeLisi. (1999). Free energy landscapes of encounter complexes in protein-protein association. *Biophys. J.* 76:1166–1178.

Camacho, C.J., D.W. Gatchell, S.R. Kimura, and S. Vajda. (2000). Scoring docked conformations generated by rigid-body protein-protein docking. *Proteins* 40:525–537.

Che, Y., and Marshall, G.R. (2003). A statistical-based atom-atom based potential for protein/protein complex evaluation. *Ph.D. Thesis, Protein-Protein Recognition: Structure, Energetics and Drug Design, Washington University St. Louis, August, 2003.*

Chen, R., and Weng, Z. (2002). Docking unbound proteins using shape complementarity, desolvation, and electrostatistics. *Proteins* 47:281–294.

Chen, R., J. Mintseris, J. Janin, and Z. Weng. (2003). A protein-protein docking benchmark. *Proteins* 52:88–91.

Chene, P., J. Fuchs, J. Bohn, C. Garcia-Echeverria, P. Furet, and D. Fabbro. (2000). A small synthetic peptide, which inhibits the p53-hdm2 interaction, stimulates the p53 pathway in tumour cell lines. *J. Mol. Biol.* 299:245–253.

Chothia, C., M. Levitt, and D. Richardson. (1981). Helix to helix packing in proteins. *J. Mol. Biol.* 145:215–250.

Chrunyk, B.A., M.H. Rosner, Y. Cong, A.S. McColl, I.G. Otterness, and G.O. Daumy. (2000). Inhibiting protein-protein interactions: a model for antagonist design. *Biochemistry* 39:7092–7099.

Clackson, T., and Wells, J.A. (1995). A hot spot of binding energy in a hormone-receptor interface. *Science* 267:383–386.

Cohen, F.E., T.J. Richmond, and F.M. Richards. (1979). Protein folding: evaluation of some simple rules for the assembly of helices into tertiary structures with myoglobin as an example. *J. Mol. Biol.* 132:275–288.

Cunningham, B.C., and Wells, J.A. (1993). Comparison of a structural and a functional epitope. *J. Mol. Biol.* 234:554–563.

Dill, K.A. (1990). Dominant forces in protein folding. *Biochemistry* 29:7133–7155.

Dill, K.A. (1999). Polymer principles and protein folding. *Protein Sci.* 8:1166–1180.

Dixon, J.S. (1997). Evaluation of the CASP2 docking section. *Proteins* (Suppl. 1):198–204.

Drozdov-Tikhomirov, L.N., D.M. Linde, V.V. Poroikov, A.A. Alexandrov, and G.I. Skurida. (2001). Molecular mechanisms of protein-protein recognition: whether the surface placed charged residues determine the recognition process? *J. Biomol. Struct. Dyn.* 19:279–284.

Duncan, B.S., and Olson, A.J. (1993). Approximation and characterization of molecular surfaces. *Biopolymers* 33:219–229.

Eisenberg, D., E.M. Marcotte, I. Xenarios, and T.O. Yeates. (2000). Protein function in the post-genomic era. *Nature* 405:823–826.

Ewing, T.J.A., and Kuntz, I.D. (1997). Critical evaluation of search algorithms for automated molecular docking and database screening. *J. Comput. Chem.* 18:1175–1189.

Fernandez-Recio, J., M. Totrov, and R. Abagyan. (2002). Soft protein-protein docking in internal coordinates. *Protein Sci.* 11:280–291.

Finn, F.M., and Hofmann, K. (1973). The S-peptide S-protein system: a model for hormone-receptor interaction. *Acc. Chem. Res.* 6:169–176.

Fischer, D., S.L. Lin, H.L. Wolfson, and R. Nussinov. (1995). A geometry-based suite of molecular docking processes. *J. Mol. Biol.* 248:459–477.

Fitzjohn, P.W., and Bates, P.A. (2003). Guided docking: first step to locate potential binding sites. *Proteins* 52:28–32.

Friedman, J.M. (1997). Fourier-filtered van der Waals contact surfaces: accurate ligand shaped from protein structures. *Protein Eng.* 10:851–863.

Gabb, H.A., R.M. Jackson, and M.J.E. Sternberg. (1997). Modelling protein docking using shape complementarity, electrostatics and biochemical information. *J. Mol. Biol.* 272:106–120.

Galaktionov, S., G.V. Nikiforovich, and G.R. Marshall. (2001). Ab initio modeling of small, medium, and large loops in proteins. *Biopolymers* 60:153–168.

Gallivan, J.P., and Dougherty, D.A. (2000). A computational study of cation-pi interactions vs. salt bridges in aqueous media: implications for protein engineering. *J. Am. Chem. Soc.* 122:870–874.

Garcia-Echeverria, C., P. Chene, M.J. Blommers, and P. Furet. (2000). Discovery of potent antagonists of the interaction between human double minute 2 and tumor suppressor p53. *J. Med. Chem.* 43:3205–3208.

Gardiner, E.J., P. Willett, and P.J. Artymiuk. (2001). Protein docking using a genetic algorithm. *Proteins* 44:44–56.

Glaser, F., D. Steinberg, I.A. Vakser, and N. Ben-Tal. (2001). Residue frequencies and pairing preferences at protein-protein interfaces. *Proteins* 43:89–102.

Glen, R.C., and Allen, S.C. (2003). Ligand-protein docking: cancer research at the interface between biology and chemistry. *Curr. Med. Chem.* 10:763–767.

Goldsmith-Fischman, S., and Honig, B. (2003). Structural genomics: computational methods for structure analysis. *Protein Sci.* 12:1813–1821.

Goodford, P.J. (1984). Drug design by the method of receptor fit. *J. Med. Chem.* 27:557–564.

Goodsell, D.S., H. Lauble, C.D. Stout, and A.J. Olson. (1993). Automated docking in crystallography: analysis of the substrates of aconitase. *Proteins Struct. Funct. Genet.* 17:1–10.

Gray, J.J., S. Moughon, C. Wang, O. Schueler-Furman, B. Kuhlman, C.A. Rohl, and D. Baker. (2003). Protein–protein docking with simultaneous optimization of rigid body displacement and side-chain conformations. *J. Mol. Biol.* 331:281–299.

Greer, J., and Bush, B.L. (1978). Macromolecular shape and surface maps by solvent exclusion. *PNAS* 75:303–307.

Grossfield, A., P. Ren, and J.W. Ponder. (2003). Ion solvation thermodynamics from simulation with a polarizable force field. *J Am Chem Soc* 125:15671–15682.

Halperin, I., B. Ma, H. Wolfson, and R. Nussinov. (2002). Principles of docking: An overview of search algorithms and a guide to scoring functions. *Proteins* 47:409–443.

Harrison, R.W., I.V. Kourinov, and L.C. Andrews. (1994). The Fourier-Green's function and the rapid evaluation of molecular potentials. *Protein Eng.* 7:359–369.

Hart, R.K., R.V. Pappu, and J.W. Ponder. (2000). Exploring the similarities between potential smoothing and simulated annealing. *J. Comput. Chem.* 21:531–552.

Haspel, N., C.J. Tsai, H. Wolfson, and R. Nussinov. (2002). Reducing the computational complexity of protein folding via fragment folding and assembly. *Protein Sci.* 12:1177–1187.

Head, R.D., M.L. Smythe, T.I. Oprea, C.L. Waller, S.M. Green, and G.R. Marshall. (1996). Validate—a new method for the receptor-based prediction of binding affinities of novel ligands. *J. Am. Chem. Soc.* 118:3959–3969.

Heifetz, A., E. Katchalski-Katzir, and M. Eisenstein. (2002). Electrostatistics in protein-protein docking. *Protein Sci.* 11:571–587.

Ho, C.M.W., and Marshall, G.R. (1990). Cavity search: an algorithm for the isolation and display of cavity-like binding regions. *J. Comput. Aided Mol. Des.* 4:337–354.

Hodgkin, E.E., J.D. Clark, K.R. Miller, and G.R. Marshall. (1990). Conformational analysis and helical preferences of normal and α,α-dialkyl amino acids. *Biopolymers* 30:533–546.

Hu, Z., B. Ma, H. Wolfson, and R. Nussinov. (2000). Conservation of polar residues as hot spots at protein interfaces. *Proteins* 39:331–342.

Hubbard, S.J., and Argos, P. (1994). Cavities and packing at protein interfaces. *Protein Sci.* 3:2194–2206.

Inbar, Y., H. Benyamini, R. Nussinov, and H.J. Wolfson. (2003). Protein structure prediction via combinatorial assembly of substructural units. *Bioinformatics* 19:i158–i168.

Janin, J. (1995). Principles of protein-protein recognition from structure to thermodynamics. *Biochimie* 77:497–505.

Janin, J., K. Henrick, J. Moult, L. Ten Eyck, M.J.E. Sternberg, S. Vajda, I. Vakser, and S.J. Wodak. (2003). CAPRI: A critical assessment of predicted Interactions. *Proteins* 52:2–9.

Jernigan, R.L., and Bahar, I. (1996). Structure-derived potentials and protein simulations. *Curr. Opin. Struct. Biol.* 6:195–209.

Jiang, S., and Vakser, I.A. (2000). Side chains in transmembrane helices are shorter at helix-helix interfaces. *Proteins* 40:429–435.

Jiang, S., and Vakser, I.A. (2004). Shorter side chains optimize helix-helix packing. *Protein Sci.* 13:1426-1429.

Jiang, F., and Kim, S.-H. (1991). "Soft Docking": matching of molecular surface cubes. *J. Mol. Biol.* 219:79–102.

Jiang, F., W. Lin, and Z. Rao. (2002). SOFTDOCK: understanding of molecular recognition through a systematic docking study. *Protein Eng.* 15:257–263.

Jiang, S., A. Tovchigrechko, and I.A. Vakser. (2003). The role of geometric complementarity in secondary structure packing: a systematic docking study. *Protein Sci.* 12:1646–1651.

Jones, S., and Thornton, J.M. (1996). Principles of protein-protein interactions. *Proc. Natl. Acad. Sci. USA* 93:13–20.

Jorgensen, W.L., D.S. Maxwell, and J. Tirado-Rives. (1996). Development and testing of the OPLS all-atom force field on conformational energetics and properties of organic liquids. *J. Am. Chem. Soc.* 118:11225–11236.

Katchalski-Katzir, E., I. Shariv, M. Eisenstein, A.A. Friesem, C. Aflalo, and I.A. Vakser. (1992). Molecular surface recognition: determination of geometric fit between proteins and their ligands by correlation techniques. *Proc. Natl. Acad. Sci. USA* 89:2195–2199.

Keskin, O., I. Bahar, A.Y. Badretdinov, O.B. Ptitsyn, and R.L. Jernigan. (1998). Empirical solvent-mediated potentials hold for both intra-molecular and inter-molecular inter-residue interactions. *Protein Sci.* 7:2578–2586.

Kim, M.K., G.S. Chirikjian, and R.L. Jernigan. (2002a). Elastic models of conformational transitions in macromolecules. *J Mol Graph Model* 21:151–160.

Kim, M.K., R.L. Jernigan, and G.S. Chirikjian. (2002b). Efficient generation of feasible pathways for protein conformational transitions. *Biophys J* 83:1620–1630.

Knegtel, R.M., and Wagener, M., (1999). Efficacy and selectivity in flexible database docking. *Proteins* 37:334–345.

Korn, A.P., and Burnett, R.M. (1991). Distribution and complementarity of hydropathy in multisubunit proteins. *Proteins* 9:37–55.

Kuntz, I.D., J.M. Blaney, S.J. Oatley, R. Langridge, and T.E. Ferrin. (1982). A geometric approach to macromolecule-ligand interactions. *J. Mol. Biol.* 161:269.

Kussie, P.H., S. Gorina, V. Marechal, B. Elenbaas, J. Moreau, A.J. Levine, and N.P. Pavletich. (1996). Structure of the MDM2 oncoprotein bound to the p53 tumor suppressor transactivation domain. *Science* 274:948–953.

Lamb, M.L., K.W. Burdick, S. Toba, M.M. Young, K.G. Skillman, X.Q. Zou, J.R. Arnold, and I.D. Kuntz. (2001). Design, docking, and evaluation of multiple libraries against multiple targets. *Proteins Struct. Funct. Genet.* 42:296–318.

Larsen, T.A., A.J. Olson, and D.S. Goodsell. (1998). Morphology of protein-protein interfaces. *Structure* 6:421–427.

Laskowski, R.A., N.M. Luscombe, M.B. Swindells, and J.M. Thornton. (1996). Protein clefts in molecular recognition and function. *Protein Sci.* 5:2438–2452.

Lawrence, M.C., and Colman, P.M. (1993). Shape complementarity at protein/protein interfaces. *J. Mol. Biol.* 234:946–950.

Lee, S.Y., H.S. Cho, J.G. Pelton, D. Yan, R.K. Henderson, D.S. King, L. Huang, S. Kustu, E.A. Berry, and D.E. Wemmer. (2001). Crystal structure of an activated response regulator bound to its target. *Nat Struct Biol* 8:52–56.

Lichtarge, O., and Sowa, M.E. (2002). Evolutionary predictions of binding surfaces and interactions. *Curr. Opin. Struct. Biol.* 12:21–27.

Lijnzaad, P., and Argos, P. (1997). Hydrophobic patches on protein subunit interfaces: charactersitics and prediction. *Proteins* 28:333–343.

Lo Conte, L., C. Chothia, and J. Janin. (1999). The atomic structure of protein-protein recognition sites. *J Mol Biol* 285:2177–2198.

Lu, L., H. Lu, and J. Skolnick. (2003). Development of unified statistical potentials describing protein-protein interactions. *Biophys. J.* 84:1895–1901.

Lu, L., H. Lu, and J. Skolnick. (2002). MULTIPROSPECTOR: an algorithm for the prediction of protein-protein interactions by multimeric threading. *Proteins* 49:350–364.

Ma, J., and Straub, J.E. (1994). Simulated annealing using the classical density distribution. *J. Chem. Phys.* 101:533–541.

MacKerell, A.D., D. Bashford, M. Bellott, R.L. Dunbrack, J.D. Evanseck, M.J. Field, S. Fischer, J. Gao, H. Guo, S. Ha, D. Joseph-McCarthy, L. Kuchnir, K. Kuczera, F.T.K. Lau, C. Mattos, S. Michnick, T. Ngo, D.T. Nguyen, B. Prodhom, W.E. Reiher, B. Roux, M. Schlenkrich, J.C. Smith, R. Stote, J. Straub, M. Watanabe, J. Wiorkiewicz-Kuczera, D. Yin, and M. Karplus. (1998). All-atom empirical potential for molecular modeling and dynamics studies of proteins. *J. Phys. Chem. B* 102:3586–3616.

Makino, S., T.J.A. Ewing, and I.D. Kuntz. (1999). DREAM++: Flexible docking program for virtual combinatorial libraries. *J. Comput. Aided Mol. Des.* 13:513–532.

Mariuzza, R.A., and Poljak, R.J. (1993). The basics of binding: mechanisms of antigen recognition and mimicry by antibodies. *Curr. Opin. Immunol.* 5:50–55.

Marshall, G.R. (1992). Three-dimensional structure of peptide-protein complexes: implications for recognition. *Curr. Opin. Struct. Biol.* 2:904–919.

Marshall, G.R., C.D. Barry, H.E. Bosshard, R.A. Dammkoehler, and D.A. Dunn. (1979). The conformational parameter in drug design: the active analog approach. In: E.C. Olson, and Christoffersen, R.E. (eds), *Computer-Assisted Drug Design*. Washington, D.C., American Chemical Society. ACS Symposium ll2:205–226.

Marshall, G.R., R.H. Head, and R. Ragno. (2000). Affinity prediction: the sina qua non. In: Di Cera, E. (eds), *Thermodynamics in Biology*. Oxford University Press, New York. pp.87–111.

McCammon, J.A. (1998). Theory of biomolecular recognition. *Curr. Opin. Struct. Biol.* 8:245–249.

McCoy, A.J., V.C. Epa, and P.M. Colman. (1997). Electrostatic complementarity at protein/protein interfaces. *J. Mol. Biol.* 268:570–584.

Meador, W.E., A.R. Means, and F.A. Quiocho. (1992). Target enzyme recognition by calmodulin: 2,4:° A structure of a calmodulin-peptide complex. *Science* 257:1251–1255.

Meyer, M., P. Wilson, and D. Schomburg. (1996). Hydrogen bonding and molecular surface shape complementarity as a basis for protein docking. *J. Mol. Biol.* 264:199–210.

Miller, D.W., and Dill, K.A. (1997). Ligand binding to proteins: the binding landscape model. *Protein Sci.* 6:2166–2179.

Miller, M., J. Schneider, B.K. Sathyanarayana, M.V. Toth, G.R. Marshall, L. Clawson, L. Selk, S.B. Kent, and A. Wlodawer. (1989). Structure of complex of synthetic HIV-1 protease with a substrate-based inhibitor at 2.3 A resolution. *Science* 246:1149–1152.

Miyazawa, S., and Jernigan, R.L. (1999). Self-consistent estimation of inter-residue protein contact energies based on an equilibrium mixture approximation of residues. *Proteins* 34:49–68.

Moont, G., H.A. Gabb, and M.J.E. Sternberg. (1999). Use of pair potential across protein interfaces in screening predicted docked complexes. *Proteins* 35:364–373.

Morris, G.M., D.S. Goodsell, R.S. Halliday, R. Huey, W.E. Hart, R.K. Belew, and A.J. Olson. (1998). Automated docking using a Lamarckian genetic algorithm and an empirical binding free energy function. *J. Comput. Chem.* 19:1639–1662.

Moult, J., K. Fidelis, A. Zemla, and T. Hubbard. (2003). Critical assessment of methods of protein structure prediction (CASP)-round V. *Proteins* 53:334–339.

Murzin, A.G., and Finkelstein, A.V. (1988). General architecture of the alpha-helical globule. *J. Mol. Biol.* 204:749–769.

Nemethy, G., M.S. Pottle, and H.A. Scheraga. (1983). Energy parameters in polypeptides. 9. Updating of geometrical parameters, nonbonded interactions, and hydrogen bond interactions for the naturally occuring amino acids. *J. Phys. Chem.* 87:1883–1887.

Nikiforovich, G.V., and Marshall, G.R. (2003). 3D Model for meta-II rhodopsin, an activated G-protein coupled receptor. *Biochemistry* 42:9110–9120.

Nikiforovich, G.V., S.G. Galaktionov, V.M. Tseitin, D.R. Lowis, M.D. Shenderovich, and G.R. Marshall. (1998). 3D Modeling for TM receptors: algorithms and validations. *Lett. Pept. Sci.* 5:413–415.

Nikiforovich, G.V., S. Galaktionov, J. Balodis, and G.R. Marshall. (2001). Novel approach to computer modeling of seven-helical transmambrane proteins: current progress in test case of bacteriorhodopsin. *Acta Biochim. Polon.* 48:53–64.

Novotny, J., M. Handschumacher, E. Haber, R.E. Bruccoleri, W.B. Carlson, D.W. Fanning, J.A. Smith, and G.D. Rose. (1986). Antigenic determinants in proteins coincide with surface regions accessible to large probes (antibody domains). *Proc. Natl. Acad. Sci. USA* 83:226–230.

Nussinov, R., and Wolfson, H.J. (1991). Efficient detection of three-dimensional structural motifs in biological macromolecules by computer vision techniques. *PNAS* 88:10495–10499.

Oliver, S. (2000). Guilt-by-association goes global. *Nature* 403:601–603.

Oprea, T.I., and Marshall, G.R. (1998). Receptor-based prediction of binding affinities. *Perspect. Drug Discov. Des.* 9–11:35–61.

Palma, P.N., L. Krippahl, J.E. Wampler, and J.J.G. Moura. (2000). BiGGER: A new (soft) docking algorithm for predicting protein interactions. *Proteins* 39:372–384.

Pappu, R.V., G.R. Marshall, and J.W. Ponder. (1999). A potential smoothing algorithm accurately predicts transmembrane helix packing. *Nat. Struct. Biol.* 6:50–55.

Peters, K.P., J. Fauck, and C. Frommel. (1996). The automatic search for ligand binding sites in proteins of known three-dimensional structure using only geometric criteria. *J. Mol. Biol.* 256:201–213.

Piela, L., J. Kostrowicki, and H.A. Scheraga. (1989). The multiple-minima problem in the conformational analysis of molecules. Deformation of the potential energy hypersurface by the diffusion equation method. *J. Phys. Chem.* 93:3339–3346.

Platzer, K.E.B., F.A. Momany, and H.A. Scheraga. (1972). Conformational energy calculations of enzyme-substrate interactions. I. Computation of preferred conformations of some substrates of chymotrypsin. *Int. J. Pept. Protein Res.* 4:187–200.

Ponder, J.W., and Case, D.A. (2003). Force fields for protein simulations. *Adv. Protein Chem.* 66:27–85.

Ponder, J.W., and Richards, F.M. (1987). Internal packing and protein structural classes. *Cold Spring Harbor Symp. Quant. Biol.* LII:421–428.

Reddy, B.V.B., and Blundell, T.L.(1993). Packing of secondary structure elements in proteins. Analysis and prediction of inter-helix distances. *J. Mol. Biol.* 233:464–479.

Richards, F.M. (1977). Areas, volumes, packing, and protein structure. *Annu. Rev. Biophys. Bioeng.* 6:151–176.

Richmond, T.J., and Richards, F.M. (1978). Packing of alpha-helices: geometrical constraints and contact areas. *J. Mol. Biol.* 119:537–555.

Ritchie, D.W., and Kemp, G.J.L. (2000). Protein docking using spherical polar Fourier correlations. *Proteins* 39:178–194.

Sali, A., R. Glaeser, T. Earnest, and W. Baumeister. (2003). From words to literature in structural proteomics. *Nature* 422:216–225.

Salwinski, L., and Eisenberg, D. (2003). Computational methods of analysis of protein–protein interactions. *Curr. Opin. Struct. Biol.* 13:377–382.

Samudrala, R., and Levitt, M. (2000). Decoys 'R' Us: a database of incorrect conformations to improve protein structure prediction. *Protein Sci.* 9:1399–1401.

Schneidman-Duhovny, D., Y. Inbar, V. Polak, M. Shatsky, I. Halperin, H. Benyamini, A. Barzilai, O. Dror, N. Haspel, R. Nussinov, and H.J. Wolfson. (2003). Taking geometry to its edge: fast unbound rigid (and hinge-bent) docking. *Proteins* 52:107–112.

Shoemaker, B.A., J.J. Portman, and P.G. Wolynes. (2000). Speeding molecular recognition by using the folding funnel: the fly-casting mechanism. *Proc. Natl. Acad. Sci. USA* 97:8868–8873.

Shoichet, B.K., A.R. Leach, and I.D. Kuntz. (1999). Ligand solvation in molecular docking. *Proteins* 34:4–16.

Shultz, M.D., and Chmielewski, J. (1999). Probing the role of interfacial residues in a dimerization inhibitor of HIV-1 protease. *Bioorg. Med. Chem. Lett.* 9:2431–2436.

Sippl, M.J., M. Ortner, M. Jaritz, P. Lackner, and H. Flockner. (1996). Helmholtz free energies of atom pair interactions in proteins. *Fold. Des.* 1:289–298.

Skolnick, J., J.S. Fetrow, and A. Kolinski. (2000). Structural genomics and its importance for gene function analysis. *Nat. Biotech.* 18:283–287.

Sternberg, M.J.E., H.A. Gabb, and R.M. Jackson. (1998). Predictive docking of protein-protein and protein-DNA complexes. *Curr. Opin. Struct. Biol.* 8:250–256.

Sternberg, M.J.E., H.A. Gabb, and R.M. Jackson. (1998). CombiDOCK: Structure-based combinatorial docking and library design. *J. Comput. Aided Mol. Des.* 12:597–604.

Ten Eyck, L.F., J. Mandell, V.A. Roberts, and M.E. Pique. (1995). Surveying molecular interactions with DOT. *ACM/IEEE Supercomputing Conference*, San Diego, CA.

Todd, M.J., N. Semo, and E. Freire. (1998). The structural stability of the HIV-1 protease. *J. Mol. Biol.* 283:475–488.

Tong, A.H.Y., G. Lesage, G.D. Bader, H. Ding, H. Xu, X. Xin, J. Young, G.F. Berriz, R.L. Brost, M. Chang, Y. Chen, X. Cheng, G. Chua, H. Friesen, D.S. Goldberg, J. Haynes, C. Humphries, G. He, S. Hussein, L. Ke, N. Krogan, Z. Li, J.N. Levinson, H. Lu, P. Menard, C. Munyana, A.B. Parsons, O. Ryan, R. Tonikian, T. Roberts, A.-M. Sdicu, J. Shapiro, B. Sheikh, B. Suter, S.L. Wong, L.V. Zhang, H. Zhu, C.G. Burd, S. Munro, C. Sander, J. Rine, J. Greenblatt, M. Peter, A. Bretscher, G. Bell, F.P. Roth, G.W. Brown, B. Andrews, H. Bussey, and C. Boone. (2004). Global mapping of the yeast genetic interaction network. *Science* 303:808–813.

Totrov, M., and Abagyan, R. (1994). Detailed ab initio prediction of lysozyme-antibody complex with 1.6A accuracy. *Nat. Struct. Biol.* 1:259–263.

Tovchigrechko, A., and Vakser, I.A. (2001). How common is the funnel-like energy landscape in protein-protein interactions? *Protein Sci.* 10:1572–1583.

Tovchigrechko, A., C.A. Wells, and I.A. Vakser. (2002). Docking of protein models. *Protein Sci.* 11:1888–1896.

Trosset, J.-Y., and Scheraga, H.A. (1998). Reaching the global minimum in docking simulations: a Monte Carlo energy minimization approach using Bezier splines. *Proc. Nat. Acad. Sci. USA* 95:8011–8015.

Trosset, J.Y., and Scheraga, H.A. (1999). PRODOCK: software package for protein modeling and docking. *J. Comput. Chem.* 20:412–427.

Tsai, C.-J., S.L. Lin, H.J. Wolfson, and R. Nussinov. (1996). Protein-protein interfaces: architectures and interactions in protein-protein interfaces and in protein cores. Their similarities and differences. *Crit. Rev. Biochem. Mol. Biol.* 31:127–152.

Tsai, C.-J., S.L. Lin, H. Wolfson, and R. Nussinov. (1997). Studies of protein-protein interfaces: a statistical analysis of the hydrophobic effect. *Protein Sci.* 6:53–64.

Tsai, C.-J., S. Kumar, B. Ma, and R. Nussinov. (1999). Folding funnels, binding funnels, and protein function. *Protein Sci.* 8:1181–1190.

Uetz, P., L. Giot, G. Cagney, T.A. Mansfield, R.S. Judson, J.R. Knight, D. Lockshon, V. Narayan, M. Srinivasan, P. Pochart, A. Qureshi-Emili, Y. Li, B. Godwin, D. Conover, T. Kalbfleisch, G. Vijayadamodar, M. Yang, M. Johnson, S. Fields, and J.M. Rothberg. (2000). A comprehensive analysis of protein-protein interactions in *Saccaromyces cerevisiae*. *Nature* 403:623–627.

Vajda, S., M. Sippl, and J. Novotny. (1997). Empirical potentials and functions for protein folding and binding. *Curr. Opin. Struct. Biol.* 7:222–228.

Vajda, S., I.A. Vakser, M.J.E. Sternberg, and J. Janin. (2002). Meeting report: modeling of protein interactions in genomes. *Proteins* 47:444–446.

Vakser, I.A. (1995). Protein docking for low-resolution structures. *Protein Eng.* 8:371–377.

Vakser, I.A. (1996a). Long-distance potentials: an approach to the multiple-minima problem in ligand-receptor interaction. *Protein Eng.* 9:37–41.

Vakser, I.A. (1996b). Low-resolution docking: prediction of complexes for underdetermined structures. *Biopolymers* 39:455–464.

Vakser, I.A. (1996c). Main-chain complementarity in protein-protein recognition. *Protein Eng.* 9:741–744.

Vakser, I.A. (1997). Evaluation of GRAMM low-resolution docking methodology on the hemagglutinin-antibody complex. *Proteins* (**Suppl.1**):226–230.

Vakser, I.A., and Aflalo, C. (1994). Hydrophobic docking: a proposed enhancement to molecular recognition techniques. *Proteins* 20:320–329.

Vakser, I.A., and Jiang, S. (2002). Strategies for modeling the interactions of the transmembrane helices of G-protein coupled receptors by geometric complementarity using the GRAMM computer algorithm. *Methods Enzymol.* 343:313–328.

Vakser, I.A., and Nikiforovich, G.V. (1995). Protein docking in the absence of detailed molecular structures. In: Atassi, M.Z and Appella, E. (eds.), *Methods in Protein Structure Analysis*. New York, Plenum Press, pp. 505–514.

Vakser, I.A., O.G. Matar, and C.F. Lam. (1999). A systematic study of low-resolution recognition in protein-protein complexes. *Proc. Natl. Acad. Sci. USA* 96:8477–8482.

Varadarajan, R., P.R. Connelly, J.M. Sturtevant, and F.M. Richards. (1992). Heat capacity changes for protein–peptide interactions in the ribonuclease S system. *Biochemistry* 31:1421–1426.

Vassilev, L.T., B.T. Vu, B. Graves, D. Carvajal, F. Podlaski, Z. Filipovic, N. Kong, U. Kammlott, C. Lukacs, C. Klein, N. Fotouhi, and E.A. Liu. (2004). In vivo activation of the p53 pathway by small-molecule antagonists of MDM2. *Science.* 303:844–848.

Veselovsky, A.V., Y.D. Ivanov, A.S. Ivanov, A.I. Archakov, P. Lewi, and P. Janssen. (2002). Protein–protein interactions: mechanisms and modification by drugs. *J. Mol. Recognit.* 15:405–422.

Volz, K. (1993). Structural conservation in the CheY superfamily. *Biochemistry* 32:11741–11753.

Vukmirovic, O.G., and Tilghman, S.M. (2000). Exploring genome space. *Nature* 405:820–822.

Walther, D., F. Eisenhaber, and P. Argos. (1996). Principles of helix-helix packing in proteins: the helical lattice superposition model. *J. Mol. Biol.* 255:536–553.

Weiner, S.J., P.A. Kollman, D.A. Case, U.C. Singh, C. Ghio, G. Alagona, J. Salvatore Profeta, and P. Weiner. (1984). A new force field for molecular mechanical simulation of nucleic acids and proteins. *J. Am. Chem. Soc.* 106:765–784.

Welch, M., N. Chinardet, L. Mourey, C. Birck, and J.P. Samama. (1998). Structure of the CheY-binding domain of histidine kinase CheA in complex with CheY. *Nat. Struct. Biol.* 5:25–29.

Williams, D.E. (1988). Representation of the molecular electrostatic potential by atomic multipole and bond dipole models. *J. Comput. Chem.* 9:745–763.

Williams, D.E. (1991). Net atomic charge and multipole models for the *ab initio* molecular electric potential. *Rev. Comput. Chem.* 2:219–271.

Wodak, S.J., and Janin, J. (1978). Computer analysis of protein-protein interactions. *J. Mol. Biol.* 124:323–342.

Xu, D., C.-J. Tsai, and R. Nussinov. (1997). Hydrogen bonds and bridges across protein–protein interfaces. *Protein Eng.* 10:999–1012.

Young, L., R.L. Jernigan, and D.G. Covell. (1994). A role for surface hydrophobicity in protein–protein recognition. *Protein Sci.* 3:717–729.

Yue, K., and Dill, K.A. (2000). Constraint-based assembly of tertiary protein structures from secondary structure elements. *Protein Sci.* 9:1935–1946.

Zhang, C., J. Chen, and C. DeLisi. (1999). Protein-protein recognition: exploring the energy funnels near the binding sites. *Proteins* 34:255–267.

Zhang, C., S. Liu, H. Zhou, and Y. Zhou. (2004a). An accurate, residue-level, pair potential of mean force for folding and binding based on the distance-scaled, ideal-gas reference state. *Protein Sci.* 13:400–411.

Zhang, C., S. Liu, and Y. Zhou. (2004b). Accurate and efficient loop selections by the DFIRE-based all-atom statistical potential. *Protein Sci.* 13:391–399.

Zhao, R., E.J. Collins, R.B. Bourret, and R.E. Silversmith. (2002). Structure and catalytic mechanism of the E. coli chemotaxis phosphatase CheZ. *Nat. Struct. Biol.* 9:570–575.

Zhu, X., K. Volz, and P. Matsumura. (1997). The CheZ-binding surface of CheY overlaps the CheA- and FliM-binding surfaces. *J. Biol. Chem.* 272:23758–23764.

Zutshi, R., M. Brickner, and J. Chmielewski. (1998). Inhibiting the assembly of protein-protein interfaces. *Curr. Opin. Chem. Biol.* 2:62–66.

7

Thermochemistry of Binary and Ternary Protein Interactions Measured by Titration Calorimetry: Complex Formation of CD4, HIV gp120, and Anti-gp120

Michael L. Doyle and Preston Hensley

ABSTRACT

This chapter describes the use of solution biophysical technologies for quantitative analysis of protein–protein interactions. Biophysical technologies have the potential for revealing detailed, molecular information about the binding mechanism of protein interactions, such as stoichiometry, binding thermodynamics, coupled protonation events, and accurate determination of affinity. However, binding interactions between proteins can be more complex than often assumed. An accurate interpretation of biophysical data in terms of molecular binding mechanism normally requires the use of multiple biophysical methods to evaluate the roles of various processes that may be coupled to binding such as oligomerization, protein folding, protonation, and changes in hydration. This chapter presents two applications. The first

---

MICHAEL L. DOYLE • Gene Expression and Protein Biochemistry Group, Bristol-Myers Squibb Pharmaceutical Research Institute, Route 206 and Province Line Road, Princeton, NJ 08543-4000, USA. PRESTON HENSLEY • Protein and Cell Sciences, Pfizer Global Research and Development Eastern Point Road, Groton, CT 06340, USA.

Proteomics and Protein–Protein Interactions: Biology, Chemistry, Bioinformatics, and Drug Design, edited by Waksman. Springer, New York, 2005.

is the analysis of the binary interaction between the human receptor soluble CD4 and the human immunodeficiency vius envelop protein gp 120. A strategy is outlined for characterizing binary protein interactions, and evaluating the roles of various commonly occuring side reactions such as oligomerization, protonation, protein folding, and conformational change. The sCD4-gp 120 reaction has one of the largest binding enthalpy energies of any protein interaction. The origin of this highly unusual feature is assignes, after considering several possible molecular origins, to a large conformational change within gp 120 that is coupled to binding of sCD4. The second application builds on the first, and includes the quantitative analysis of ternary protein complex formation of sCD4, gp 120, and the anti-gp 120 monoclonal antibody 48d. The importance of characterizing ternary interactions within the energetic constrainst set by a thermodynamic cycle analysis for equilibrium binding reactions is described.

1. INTRODUCTION

Cellular infection by the human immunodeficiency virus (HIV) is initiated by binding of the viral exterior envelope glycoprotein, gp120, with the human T-cell core-ceptor CD4. Binding to CD4 induces conformational change in gp120 that stimulates binding of gp120 to one of the host chemokine receptors, CCR5 or CXCR4. Binding to a chemokine receptor in turn elicits further conformational change in gp120 that triggers membrane fusion and cellular infection.

Most of the evidence for CD4-induced conformational changes in gp120 has come from indirect methods, such as CD4-dependent enhancement of protease sensitivity, shedding of gp120 from virus, and stimulation of chemokine receptor and neutralizing antibody binding (literature reviewed in Myszka et al., 2000). These types of studies demonstrated the existence of structural changes but did not provide a quantitative measure of the extent of structural reorganization. One of the goals of this chapter is to describe how analysis of the CD4 binding thermodynamics of gp120 has been used to quantify the extent of the conformational reorganization in gp120.

Structural insight into the CD4-induced conformational change in gp120 comes from the X-ray crystallographic structure of the core domain of gp120, which has been solved as a ternary complex containing domains D1D2 of CD4, and an Fab fragment of the neutralizing anti-gp120 monoclonal (mAb) 17b (Kwong et al., 1998). In the ternary complex, gp120 is organized into inner and outer domains that are connected by a four-stranded bridging sheet. As pointed out by Kwong et al. (1998), elements of this structure appear to be dependent on association with CD4 for their stabilization. In particular, the extended conformation of the bridging sheet is stabilized by direct contacts with CD4. The bridging sheet also constitutes key residues that are involved in chemokine receptor binding, but the residues are on the opposite side of the sheet as those that bind CD4. Thus, CD4-induced formation of the bridging sheet appears to play an important role in stimulating chemokine receptor binding. In contrast to the substantial structural rearrangements that occur in gp120, structural change in CD4 is minimal. Inspection of the apo and gp120-complexed forms of CD4 D1D2 show that it undergoes very little structural rearrangement on binding gp120 (Ryu et al., 1990; Wang et al., 1990).

Structural rearrangement in gp120 serves not only as an allosteric trigger for cellular invasion, but also as a mechanism for evading the host immune system. HIV has evolved several molecular strategies for evading the immune system. First, gp120 contains a set of hypervariable regions. The frequent mutational changes that occur in these regions render host antibodies ineffective. Second, a large surface area of gp120 is covered by an "umbrella" of carbohydrate, and is relatively silent toward antibody formation. Third, the conserved regions of gp120, which are necessary for CD4 and chemokine receptor binding, are in theory large enough that they should be good targets for antibody-based neutralization. However, they are believed to be hidden from the immune system, and are exposed only on CD4-induced structural changes. These molecular properties of gp120 have greatly hindered vaccine development.

The goals of this chapter are to describe the thermochemistry of (1) the binary binding interaction of HIV gp120 with human CD4 and (2) the ternary interaction between gp120, CD4, and a neutralizing anti-gp120 antibody 48d. We also outline an experimental approach for quantitative analysis of protein–protein interactions that includes measurements of oligomeric states and thermal stabilities of the proteins. The results provide a quantitative measure of gp120 structural changes coupled to CD4 and antibody binding, and have ramifications for cellular infection, immune evasion, and vaccine design.

2. METHODOLOGIES

The primary technology used to determine the binding thermodynamics for CD4, gp120, and anti-gp120 is the isothermal titration calorimeter (ITC). ITC is, in principle, capable of providing a high-resolution description of the binding mechanism for protein–protein interactions. However, a rigorous interpretation of the data requires information from auxiliary methods that examine the thermal stability and oligomeric behavior of the proteins, as well as information on other possible side reactions such as protonation or ion binding. In this chapter we outline an experimental approach for discerning whether these types of side reactions contribute to the overall binding thermodynamics measured by ITC. Here we have used circular dichroism spectroscopy for evaluating thermal stabilities, and analytical ultracentrifugation for evaluating potential oligomerization reactions, but other methods such as differential scanning calorimetry or light scattering can also be used.

2.1. Isothermal Titration Calorimetry

The ITC measurements in this chapter were carried out with Microcal Inc. MCS and VP-ITC instruments (Wiseman et al., 1989). Because gp120 was the reagent in most limited supply, it was chosen to be the reactant placed inside the calorimeter cell. The syringe reactants were thus either CD4 or antibody. The concentration of gp120 in the calorimeter cell was typically about 4 μM. The concentrations of CD4, full-length gp120 WD61, and core gp120 were determined by absorbance at 280 nm from theoretical extinction coefficients of 1.4, 0.74, and 1.2 mL cm^{-1} mg^{-1}, respectively. Data were analyzed with Microcal Origin software according to the single-site binding model that is implicated not only from goodness of the fit from the data analysis, but also from evidence of 1:1 binding in the X-ray structure (Kwong et al., 1998). Titrations were carried out in phosphate buffer to minimize artifactual ionization heats that would arise if proton uptake or release were coupled to binding (see Baker and Murphy, 1996 for review of proton ionization effects on ITC data). A general protocol for ITC analysis of protein–protein interactions has been reported by Doyle (1999).

2.2. Circular Dichroism Spectroscopy

The goal of the circular dichroism experiments was to evaluate the thermal stability of CD4, gp120, and antibody, by monitoring protein secondary structure as a function of temperature (Eftink, 1995). The circular dichroism measurements were conducted with a JASCO J-710 spectropolarimeter in a water-jacketed 0.1-cm pathlength cuvette. Buffer conditions were 100 mM NaCl, 3 mM Na$_2$HPO$_4$, and various temperatures from 12° to 42°C. Several scans were made at a rate of 20 nm/min, a bandwidth of 1 nm, and a response time of 2 s. All spectra were buffer corrected and normalized to difference molar extinction coefficient units as $\Delta \varepsilon = I/(32.98\,l\,C)$, where l is pathlength in cm, and C is the molarity of peptide bonds in the sample.

2.3. Analytical Ultracentrifugation

The purpose of the analytical ultracentrifugation experiments was to examine whether the reactants (CD4, gp120, or antigp120) or products (complexes involving CD4, gp120, and/or anti-gp120) are monomeric or oligomeric. Any changes in oligomeric status of the individual proteins that may occur during binding would generate a corresponding enthalpy change that would need to be accounted for in the interpretation of the overall binding thermodynamics. Sedimentation equilibrium experiments were conducted with a Beckman XL-A analytical ultracentrifuge as described previously (Hensley, 1996). Conditions were 20 mM NaPO$_4$, 200 mM NaCl, pH 7.0, and studies were conducted over a range of temperature from 15° to 40°C in order to explore possible temperature-induced oligomerization side reactions.

The absorbance at 280 nm at equilibrium is related to the distribution of a homogeneous species as:

$$A_{280,r} = A_{280,m} \exp \left[\frac{M\left(1 - \bar{v}\rho\right)\omega^2 \left(r^2 - r_m^2\right)}{2RT} \right] + \textit{offset} \tag{1}$$

$A_{280,m}$ is absorbance at the meniscus, M is protein molecular mass, ω is angular velocity, r is distance in cm from the center of rotation, r_m is radial position of the reference position (the meniscus) in cm, $\bar{v}$ is partial specific volume, ρ is solvent density, *offset* is an offset absorbance that is constant along radial position, R is the gas constant, and T is temperature. Equilibrium was defined by no change in protein distributions in scans acquired 4 h apart. Partial specific volumes, calculated from amino acid and carbohydrate compositions (Laue et al., 1992) for CD4, full-length glycosylated gp120, and core gp120 were 0.738, 0.695, and 0.716 mL/g, respectively.

2.4. Interpretation of Binding Thermodynamics

Interpretation of binding thermodynamics for protein–protein interactions can be a tricky undertaking and requires a thorough understanding of the binding reaction being analyzed. The interpretation will almost certainly require information about the binding mechanism from orthogonal biophysical methods that probe structure and function, such as X-ray crystallography, nuclear magnetic resonance (NMR), fluorescence, and the methods for assessing thermal stability and oligomeric state described above. In spite of the challenges for interpreting binding thermodynamics, there are examples where sufficient biophysical information has been obtained on protein–protein interactions that were sufficiently well behaved to enable rigorous interpretation. The interaction of gp120 with CD4 described in this chapter is one such example.

Several approaches have been reported for interpreting protein binding thermodynamics. Here we utilize two that are based on very different theoretical underpinnings. The first one derives from an extensive data base of protein folding thermodynamic measurements and correlates thermodynamics to amount of surface area buried in a protein reaction (Murphy and Freire, 1992; Xie and Freire, 1994). The amount of apolar (ΔASA_{ap}) and polar (ΔASA_{pol}) surface area buried upon binding is calculated from the empirical relationships to the binding heat capacity (ΔC°) and enthalpy (ΔH°) changes as: $\Delta C^\circ = 0.45 \cdot \Delta ASA_{ap} - 0.26 \cdot \Delta ASA_{pol}$; ΔH° (60) $= -8.44 \cdot \Delta ASA_{apol} + 31.4 \cdot \Delta ASA_{pol}$, where ΔH° (60) is the binding enthalpy change at 60°C.

A second, orthogonal approach correlates thermodynamics to the number of residues that are known to fold based on inspection of high resolution X-ray and NMR structure data (Spolar and Record, 1994). The number of residues that become ordered during binding, R^{th}, is predicted from the experimental association entropy change ΔS°_{assoc} of the binding reaction as: $\Delta S^\circ_{assoc} = \Delta S^\circ_{HE} + \Delta S^\circ_{RT} + \Delta S^\circ_{other}$ where ΔS°_{HE} and ΔS°_{RT} are entropy change contributions due to the hydrophobic effect and loss of rotational and translational degrees of freedom, respectively. ΔS°_{HE} is estimated by $\Delta S^\circ_{HE} = 1.35 \Delta C^\circ \ln(T_s/386)$, where T_s is the characteristic temperature for which $\Delta S^\circ_{assoc} = 0$ for a given interaction. ΔS°_{rt} has been deduced empirically to be -50 e.u. for binary protein–protein interactions. The remaining term, ΔS°_{other}, has been shown, based on thermodynamic and high-resolution structural data, to relate to the number of residues $\mathfrak{R}^{th}$ that fold upon binding as $\mathfrak{R}^{th} = \Delta S^\circ_{other}/-5.6$ e.u. (Spolar and Record, 1994).

3. BINARY REACTION OF GP120 AND sCD4: THERMODYNAMIC SIGNATURE FOR A LARGE STRUCTURAL REARRANGEMENT IN THE CORE OF GP120

3.1. Isothermal Titration Calorimetry Data with Full-Length and Core Gp120

Two binary reactions of gp120 with sCD4 were studied by titration microcalorimetry: one with full-length glycosylated gp120 and one with an engineered core form of gp120. The core form of gp120 is the same as was used by Kwong et al. (1998) to solve the X-ray structure of gp120 in a complex with sCD4 and anti-gp120 Fab 17b. The core gp120 has deletions of 52 and 19 residues from the N- and C-termini, respectively. It also has Gly-Ala-Gly substitutions for 67 V1/V2 loop residues and 32 V3 loop residues, and has been deglycosylated down to the two core N-acetylglucosamine residues. It is therefore of interest to know whether these rather substantial modifications in gp120 alter the CD4 binding functional chemistry. Figure 7.1A,B shows the ITC data for CD4 binding to both the full-length

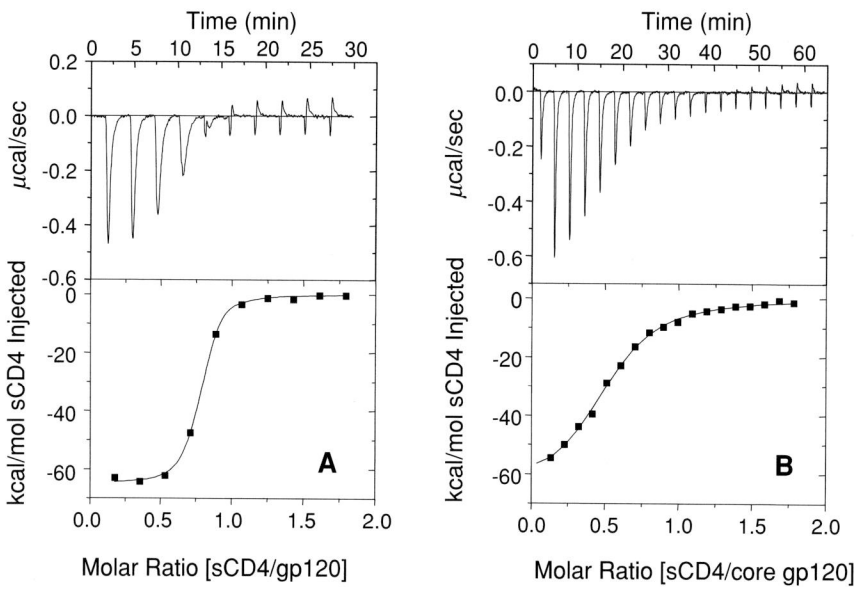

Figure 7.1. Isothermal titration calorimetry data for binding of human soluble CD4 to HIV gp120. Titration of WD61 full-length glycosylated gp120 is shown in (**A**) and titration of core deglycosylated gp120 is shown in (**B**). Conditions were: 10 mM NaPO$_4$, 200 mM NaCl, 0.5 mM EDTA, pH 7.4 and 37°C. *Top panel* shows raw data in power versus time. Area under each spike is proportional to heat produced at each injection. *Lower panel* shows integrated areas normalized to the number of moles of CD4 injected at each injection step. Best-fit curves represent binding enthalpy changes of −63 and −62 kcal/mol CD4 for full length and core gp120, respectively. Equilibrium binding K_D values were determined as 5 nM and 190 nM, respectively.

Table 7.1. Thermodynamics of the CD4-gp120 interaction at 37°C by ITC

	$\Delta G°$ (kcal/mol)	$\Delta H°$ (kcal/mol)	$-T\Delta S°$ (kcal/mol)	$\Delta C°$ (kcal/mol/K)	K_D (nM)
WD61 full length gp120	-11.8 ± 0.3	-63 ± 3	51.2 ± 3	-1.2 ± 0.2	5 ± 3
core gp120	-9.5 ± 0.1	-62 ± 3	52.5 ± 3	-1.8 ± 0.4	190 ± 30

and core gp120 proteins, and a summary of the CD4 binding thermodynamics are listed in Table 7.1. To a good measure, the CD4 binding thermochemistry of these two forms of gp120 are the same. A striking feature with both of these interactions is the enormously favorable binding enthalpy ($\Delta H° \simeq -60$ kcal/mol). This demonstrates that a very large number of favorable bonding interactions (e.g., hydrogen bonds and van der Waals interactions) occur during complex formation. However, the binding entropy change for both forms of gp120 is also very large ($-T\Delta S° \simeq 52$ kcal/mol) and unfavorable. This demonstrates that there is a substantial loss in the degrees of freedom upon binding. Interestingly, the net binding free energy changes of -11.8 and -9.5 kcal/mol for full-length and core gp120 are quite modest relative to the enormous potential energy available from the enthalpic term. The net binding free energies thus result from a balance between highly favorable bonding interactions and highly unfavorable molecular ordering. The similar thermodynamics for core and full-length gp120 proteins indicates they bind CD4 with a similar binding mechanism. The 40-fold difference in affinity between the gp120 proteins, although experimentally distinguishable, reflects a difference of only 2.2 kcal/mol in Gibbs free energy change and indicates only a minor difference in binding mechanism.

The thermochemistry for the CD4 binding reactions of full-length and core gp120 can be further explored by measuring the temperature dependence of their binding enthalpy changes. For a binary protein–protein interaction having a conserved binding mechanism over the range of temperature studied, the temperature dependence of the binding enthalpy change should be approximately linear. The slope of the line is equal to the observed binding heat capacity change ($\Delta Cp°$). Deviations from linearity would be indicative of temperature-dependent change in binding mechanism. For example, Thomson et al. (1994) showed that a pronounced nonlinearity for the enthalpy change versus temperature plot of S-peptide binding to S-protein was due to the onset of global unfolding of S-protein at the high temperature range of the analysis. Figure 7.2 shows the results for full-length and core gp120. The slopes from the lines in Figure 7.2 yield $\Delta Cp°$ values of -1.2 ± 0.2 and -1.8 ± 0.4 kcal/mol/deg, for full-length and core gp120, respectively. These values are significantly greater than the -0.2 to -0.7 kcal/mol/deg typically observed for protein–protein interactions (Stites, 1997), and suggest that extensive apolar surface area is buried during binding (Spolar et al., 1989). These results are also consistent, within error, for a common CD4 binding mechanism for full-length and core gp120, and indicate that the unusual CD4 binding mechanism is a property of the core of gp120.

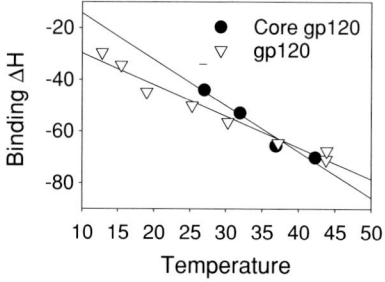

Figure 7.2. Temperature dependence of the CD4 binding enthalpy change for WD61 full-length gp120 (*open triangles*) and HXBc2 core gp120 (*solid circles*). Slopes of the plots, obtained from the best fit lines shown, yield the binding heat capacity change values of $\Delta C_p = -1.2 \pm 0.2$ and $\Delta C_p = -1.8 \pm 0.4$, respectively.

3.2. Possible Explanations for the Unusually Large Binding Enthalpy Change

The magnitudes of the $\Delta H°$ and $-T\Delta S°$ terms for the CD4–gp120 interaction are unusually large relative to other binary protein interactions (Stites, 1997; Wilcox et al., 1999). One may suspect that for the reaction of gp120 with CD4 the large enthalpy and entropy changes may be due to a large conformational change coupled to binding. But there are several types of side reactions that accompany many protein–protein interactions and may contribute to the CD4 binding thermodynamics of gp120 (see Fig. 7.3).

3.2.1. Protein Folding

The side reaction that can potentially generate the largest amount of artifactual enthalpy change is the coupling of binding to protein refolding. That is, the observed

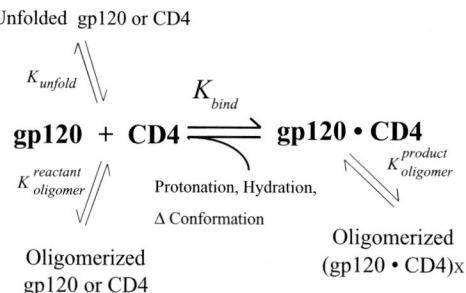

Figure 7.3. Hypothetical side reactions that could potentially be coupled to the gp120-CD4 binding reaction. The figure serves as a reminder of the various possible reactions that must be considered when interpreting binding thermodynamic data at the molecular level. Each of the side reactions (oligomerization, thermal unfolding, protonation, etc.) would contribute energetically to the overall thermodynamic parameters measured by any binding method. Equilibrium constants represent protein unfolding (K_{unfold}), oligomerization ($K_{oligomer}$), and the binding constant for the protein–protein interaction of interest (K_{bind}). The methodologies section describes an experimental protocol for evaluating the extent to which these side reactions contribute to a given protein–protein interaction.

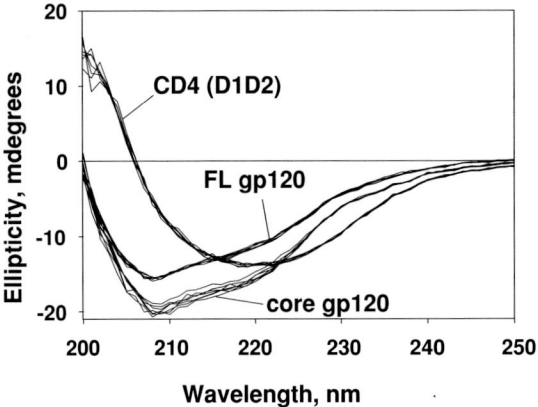

Figure 7.4. Secondary structural analysis of the thermal stabilities of CD4 (D1D2), full-length gp120, and core gp120 as indicated as monitored by circular dichroism spectroscopy wavelength scans made over a range in temperature from 12 to 42°C. The overlay of spectral scans for each protein indicates no loss in secondary structure over the temperature range.

binding enthalpy change will be artifactually much more exothermic if the binding reaction is studied at a temperature where one of the proteins is partially or wholly thermally unfolded, and refolds on binding. Of course, the unfolded protein must be capable of refolding on addition of the ligand. To put this in perspective, the average binding enthalpy changes for protein–protein and antibody–antigen interactions found in the literature has been reported to be about −5 and −20 kcal/mol, respectively (Stites, 1997). In contrast, the refolding enthalpy change for a globular protein the size of core gp120 (330 residues) can be approximated to be about −150 kcal/mol at 37°C from the data of Privalov (1979). This is about 10-fold larger than the typical binding enthalpy change for proteins/antibodies. Thus 20% to 30% global refolding of gp120 could in theory largely contribute to the CD4 binding thermodynamics.

To investigate the possible role of protein refolding in the observed CD4 binding thermodynamics, the thermal stabilities of the CD4, full length gp120 and core gp120 were examined by CD spectroscopy. Figure 7.4 shows far-UV CD scans for CD4, and both forms of gp120. The scans were measured at temperatures covering the range from 12° to 42°C. As can be seen, there is no evidence for thermally induced changes in structure. Similar experiments were conducted with the CD4-bound complexes of full length and core gp120, and again, there was no evidence of thermally induced changes in structure. The thermal stability of sCD4 has also been studied previously by CD spectroscopy and differential scanning calorimetry and is known to have a melting temperature near 62°C (Doyle and Hensley, 1998; Doyle et al., 2000).

Additional information on the possible role of protein refolding on the binding thermodynamics of CD4 and gp120 comes from inspection of the temperature dependence of the binding enthalpy change (Fig. 7.2). If either CD4 or gp120 were susceptible to temperature-induced unfolding over the temperature range studied, there would be a sizeable deviation from linearity. Precisely this type of behavior has been reported by Thomson et al. (1994) with S-peptide and S-protein. In the S-protein case, the unfolding/refolding reaction was reversible and contributed substantially to the observed binding thermodynamics at the higher temperature range of the study.

3.2.2. Oligomerization

The observed binding enthalpy for a protein–protein interaction may also have contributions from coupled oligomerization side reactions (Fig. 7.3). In principle, ligand binding can increase or decrease the oligomeric state of either the reactants or products. This phenomenon is widespread with biological macromolecules and has been described in theoretical terms by Wyman and colleagues (Colosimo et al. 1976), including explicit analysis of the contributions to the observed binding enthalpy change (Wyman and Gill (1990). The contribution to the binding enthalpy would depend, in a qualitative sense, on the enthalpy change for oligomerization and the directionality (i.e., whether binding increases or decreases oligomeric state). Generally, the enthalpy changes for oligomerization are expected to be quite modest when normalized to moles of monomer (assuming protein refolding is not also coupled to oligomerization). For proteins that oligomerize on ligand binding the observed ligand binding enthalpy change would be expected to be more exothermic, since oligomerization is analogous to hetero-protein–protein interactions. For proteins that decrease oligomeric status, the opposite would be expected and the observed binding enthalpy change would be less exothermic.

To investigate whether oligomerization equilibria are coupled to the binding of CD4 and gp120 we analyzed the reactants and products by analytical ultracentrifugation. Figure 7.5 depicts the results of a sedimentation equilibrium analysis of sCD4 at 25°C. The CD4 was found to be a homogeneous monomer over the concentration range of about 0.02 to 1.0 mg/mL shown in Figure 7.5, and there was no evidence of oligomerization. Importantly, the concentration range for the analytical ultracentrifugation analysis covers the concentrations at which the binding enthalpy changes were measured by ITC (typically 4 μM gp120 in the calorimeter cell). Similar centrifugation experiments were conducted for full-length gp120, core gp120, and mixtures of sCD4 with both of these gp120 proteins. Moreover, these experiments were conducted at various temperatures ranging from 12°C to 40°C. No evidence for oligomerization of any of the reactants or products was detected for any of the temperatures studied. The results for the 37°C experiments are shown in Table 7.2 and are compared to the monomeric masses as determined by mass spectrometry.

3.2.3. Coupling to Protonation Side Reactions

Most protein–protein interactions are coupled to changes in protonation state of one or both of the proteins. Near neutral pH the coupled protonation side reactions may be due to alterations in pK_a values of histidines, lysines, arginines, and N-termini that occur when their chemical environments change during binding. In fact, one should regard such linked reactions as being a natural part of the overall biochemical reaction. However, these coupled protonation reactions create artifactual heats in calorimetry

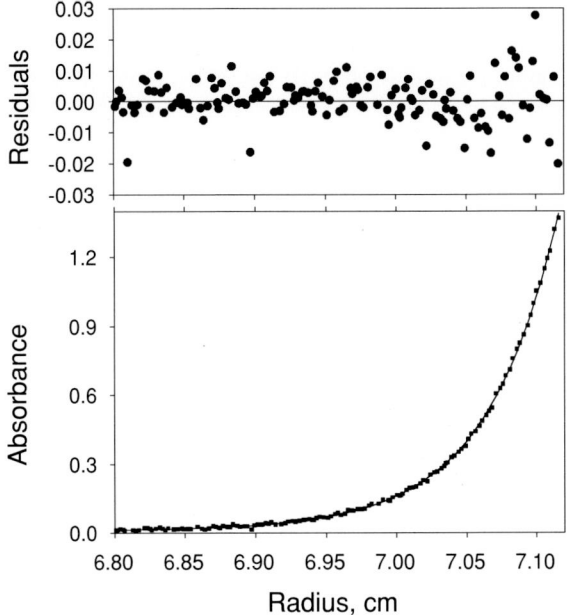

Figure 7.5. Analytical ultracentrifugation data for soluble human CD4. The data are shown in the lower panel as absorbance versus radial position of the sample cell. Best-fit curve is shown going through the data points and corresponds to a single mass species model. The residuals for the least-squares analysis are shown in the top panel. The best-fit mass value is determined as 44 ± 1 kDa. Conditions were 200 mM NaCl, pH 7.4, 0.5 mM EDTA, and 25°C. (Reprinted from Doyle et al. [2000] with permission from Elsevier, Copyright 2000.)

titrations. The coupled protons will either be added to or extracted from the pH buffer used in the experiment, and the buffer ionization side reactions are not part of the biochemical reaction of interest (see Baker and Murphy, 1996 for review of linked protonation effects on ITC). When protons are added or extracted from buffer, there is an enthalpy change equal to the ionization enthalpy of the buffer and the fractional change in protonation. To minimize such artifactual enthalpy effects it is wise to use buffers with small ionization enthalpies. Phosphate was used for the present analysis

Table 7.2. Assembly State of the CD4-gp120 complex by mass by MALDI and AUC

	Mass by MALDI	Mass by AUC
CD4	44,600	$45,000 \pm 300$
WD61 full length gp120	99,600	$97,000 \pm 300$
core gp120	39,000	$35,000 \pm 1,000$
CD4-WD61 full length gp120 complex	144,100*	$138,000 \pm 800$
CD4-core gp120 complex	83,600*	$83,000 \pm 200$

*Predicted mass for 1:1 complex.

and has a very low ionization enthalpy change of 0.4 kcal/mol of proton at 37°C (Christensen et al., 1976). The magnitudes of coupled protonation events for most protein–protein interactions at neutral pH are on the order of one proton or fraction thereof. Thus, it is highly unlikely that the very large CD4 binding enthalpy change for gp120 can be much affected by buffer ionization heats, especially in phosphate.

In addition to coupled protonation causing artifactual heats in ITC, it is worth mentioning that these effects can be exploited to measure the extent of coupled protonation for protein–ligand reactions. The extent of coupled protonation can be measured by conducting ITC experiments in different buffers having distinct ionization enthalpies. A plot of observed enthalpy change versus buffer ionization enthalpy change yields a straight line with slope equal to the number of protons coupled to binding (Baker and Murphy, 1996). In the present study we measured the CD4 binding enthalpy change in both phosphate and TRIS buffers, which have ionization enthalpy changes of 0.4 and −11.6 kcal/mol proton, respectively (Christensen et al., 1976). The resulting CD4 binding enthalpies were −65.7 ± 1.5 and −65 ± 12 kcal/mol, respectively. These values indicate that there is little net change in protonation of CD4 and gp120 on binding at pH 7.4 and 37°C. Within error, particularly with the measurement done in TRIS, there could be as much as about one proton taken up or released.

3.2.4. Coupling to Hydration Reactions

Protein–protein interactions are also usually accompanied by changes in water molecules bound to the surface of the protein interface. These coupled water molecules can contribute significantly to the observed binding thermodynamics (Ladbury, 1996) and should be considered when interpreting binding thermodynamics. Constriction of water molecules at interfaces can contribute favorably to the binding enthalpy change and unfavorably to the entropy change of protein interactions. In the case of the CD4–gp120 complex, 15 water molecules are within van der Waals radius + 1 Å of both molecules (Kwong et al., 1998). This value is close to the average of 18 (range 3 to 50) that was reported for analysis of 36 X-ray structures (Lo Conte et al., 1999). This suggests that the exceptionally large $\Delta H°$ and $-T\Delta S°$ values for the CD4–gp120 interaction do not arise predominantly from entrapment of water.

3.3. Interpretation of Binding Thermodynamics with Conformational Change Model

The extent of structural rearrangement that occurs during protein–protein interactions can be estimated (according to the caveats and methods described above) from relationships that correlate structure with thermodynamics. One approach derives from the correlation that exists between protein folding thermodynamics and the amount of water-accessible surface area buried during the reaction (Murphy and Freire, 1992; Xie and Freire, 1994). Using empirical relationships developed by Freire and colleagues (see Methods), the large $\Delta H°$ and $\Delta C°$ values observed for the CD4–gp120 interaction infer that 10,000 (± 2000) Å² of surface area are buried during complex formation. This value is much greater than the 1500 $\mathring{A}^2$ observed at the CD4–gp120 interface (Kwong et al., 1998), suggesting that complex formation includes burial of surface area well outside the observed binding interface.

In an independent approach, Spolar and Record (1994) have related the binding ΔC° and ΔS° of interactions to the number of residues that fold during binding by comparing thermodynamics to high-resolution structures (see Methodologies above). Application of this method to the CD4–gp120 binding thermodynamics suggests the folding of 94 ($\pm$15) and 126 ($\pm$30) residues on CD4 binding to full-length and core gp120, respectively. Within experimental error these values are the same and are among the largest reported for protein–protein interactions.

Interestingly, the extent of conformational "refolding" suggested by analysis of the binding thermodynamics is consistent with the model deduced by Kwong et al. (1998) from the three-dimensional structure of the CD4–gp120–Fab17b ternary complex. The authors pointed out the assembly of the bridging sheet, and inner and outer domains at the Phe43 nexus where CD4 binds. The structural rearrangements involving folding of the bridging sheet (54 residues) and burying the surface areas between inner and outer domains, the bridging sheet, and CD4 accounts well for the extent of refolding and burial predicted by the binding thermodynamics.

4. THERMOCHEMISTRY OF TERNARY COMPLEX FORMATION WITH CD4, GP120, AND ANTI-GP120 MAB 48d

The X-ray structure of gp120 shows that it can exist in a ternary complex with CD4 and a Fab fragment from a neutralizing anti-gp120 mAb 17b (Kwong et al., 1998). There are two mAbs, 17b and 48d, that are known to bind to CD4-induced discontinuous epitopes on gp120 (Thali et al., 1993) and are broadly but nonpotently neutralizing (reviewed in Kwong et al., 2002). Cross-competition analysis has shown mAbs 17b and 48d bind to overlapping epitopes (Moore and Sodroski, 1996), and bind to a region that is involved in chemokine receptor binding site (Trkola et al., 1996; Wu et al., 1996). The X-ray structure of gp120 reveals that this region includes the surface of the bridging sheet that is on the opposite side of that which is bound by CD4, and that binding of either CD4 or Fab 17b involves all four strands of the beta sheet (Kwong et al., 1998). Given the large structural reorganization that accompanies CD4 binding, and the proximity of the CD4 and Fab 48d binding sites, one would suspect that there would be considerable codependency on the binding thermodynamics of these ligand proteins.

4.1. Binding Synergy and Thermodynamic Signature for Structural Reorganization

To probe the mechanism of the CD4-induced epitope for mAb 48d, we conducted ITC experiments with all four binding reactions involving gp120, CD4, and mAb 148d. We measured binding thermodynamics for (1) the reaction of CD4 with gp120,

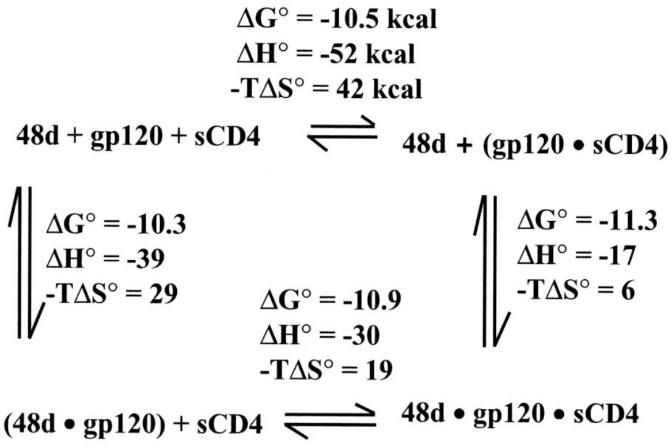

$$\Delta G^\circ = -10.5 \text{ kcal}$$
$$\Delta H^\circ = -52 \text{ kcal}$$
$$-T\Delta S^\circ = 42 \text{ kcal}$$

48d + gp120 + sCD4 $\rightleftharpoons$ 48d + (gp120 • sCD4)

$\Delta G^\circ = -10.3$ $\Delta G^\circ = -11.3$
$\Delta H^\circ = -39$ $\Delta H^\circ = -17$
$-T\Delta S^\circ = 29$ $\Delta G^\circ = -10.9$ $-T\Delta S^\circ = 6$
                    $\Delta H^\circ = -30$
                    $-T\Delta S^\circ = 19$

(48d • gp120) + sCD4 $\rightleftharpoons$ 48d • gp120 • sCD4

Figure 7.6. Thermodynamic cycle for ternary complex formation of gp120, CD4, and anti-gp120 mAb 48d. The four sides of the cycle were measured by ITC titrations in 10 mM NaPO$_4$, 200 mM NaCl, 0.5 mM EDTA, pH 7.4 and 37°C.

(2) the reaction of CD4 with the binary complex of gp120 and mAb 48d, (3) the reaction of mAb 48d with gp120, and (4) the reaction of mAb 48d with the binary complex of gp120 and CD4. The results are summarized in Figure 7.6. As can be seen from the binding free energy changes, there is a synergy between binding CD4 and mAb 48d to gp120, as should be the case for an antibody having a CD4-induced epitope. Both CD4 and mAb 48d increase the affinity of each other by an average of 0.7 kcal (threefold on the equilibrium binding constants).

Interestingly, the gp120 binding enthalpy change for mAb 48d is found to be quite large (−39 kcal/mol). In view of the structural juxtapositioning of the CD4 and 48d epitope on opposite sides of the bridging sheet, the large binding enthalpy for mAb 48d probably reflects a substantial structural reorganization in gp120 that is similar to that induced by CD4. Two predictions can be made if CD4 and mAb 48d are causing the same structural rearrangement in gp120. (1) CD4 and mAb 48d should bind synergistically. (2) Binding of either CD4 or mAb 48d to gp120 that is precomplexed with the other one should result in a substantial reduction (less exothermic) in the binding enthalpy change. The results shown in Figure 7.6 reveal that both of these predictions are borne out. The binding enthalpy change for CD4 is greatly reduced (less exothermic) by 22 kcal if gp120 is already complexed with mAb 48d. The binding enthalpy change for mAb 48d is also considerably reduced (by 22 kcal also) when gp120 is already complexed with CD4. These results, together with the X-ray structural data, suggest CD4 and mAb 48d induce similar structural reorganization of gp120.

4.2. Thermodynamic Cycle Analysis

Finally, it should be pointed out that the results in Figure 7.6 represent a thermodynamic cycle. One of the great powers of equilibrium thermodynamics is the pathway-independent nature of evaluating state function quantities, such as free energy, enthalpy and entropy changes (Nash, 1962). Thus, assuming all four of the binding reactions that we measured are equilibrium binding reactions, the sums of either ΔG, ΔH, or ΔS parameters for the individual reactions that lead to the ternary complex must be equal, regardless of path around the cycle. For example, the sum of the binding enthalpy changes for binding CD4 to gp120 (-52) plus that for mAb 48d binding the CD4–gp120 complex (-17) is equal to -69 kcal/mol gp120. Going around the cycle in the other direction, the value for mAb 48d binding to gp120 (-39) plus that for CD4 binding to the mAb 48d–gp120 complex (-30) is also equal to -69 kcal/mol gp120. One of the benefits of evaluating protein ternary complex formation as a thermodynamic cycle is that the equality of summing the thermodynamic terms in both directions lends additional validation to each of the individual measurements made. Alternatively, any disagreement in these sums would serve as a indication that an error exists in the measurements, or the assumption of equilibrium binding may not hold.

5. CONCLUDING REMARKS

Our investigation of the binding thermodynamics of gp120 binding to natural ligand CD4 reveals that this interaction is accompanied by an unusually large, favorable enthalpy change that is off set by an almost equally large, but unfavorable, entropy change. Previous studies have indicated that conformational change occurs in gp120 upon binding CD4, but were based on indirect methods that did not provide a measure of the magnitude of structure change. The unusually large thermodynamic parameters for CD4 binding gp120 led us (Myszka et al., 2000) to propose that they reflect large structural rearrangement in gp120. We therefore sought additional biophysical evidence to test this proposal, and adopted an experimental strategy for doing so that has been described in this chapter. Because the observed binding thermodynamics for protein–protein interactions are a composite of all molecular processes coupled to binding (including oligomerization, refolding, uptake, or release of small molecules such as waters, protons, etc.) we characterized CD4, gp120 and the CD4–gp120 complex by biophysical technologies that probe these potential side reactions (Fig. 7.3). In the end we found that the dominant contributor to the unusual binding thermodynamics of CD4 and gp120 was in fact the result of large structural rearrangement in gp120.

In this chapter we have also presented an example of how ITC can be used to obtain a detailed analysis for ternary protein interactions. We have characterized the binding thermodynamics for formation of the ternary complex of CD4, gp120, and anti-gp120 mAb 48d. The results revealed that (1) a threefold synergy exists in the gp120 binding affinity between CD4 and mAb 48d, (2) both ligands bind gp120 with very large enthalpy changes on their own, and (3) prior binding of one of these proteins to gp120 greatly diminishes (less exothermic by 22 kcal/mol gp120) the binding enthalpy of the other. The mechanism that is inferred from these results is that both CD4 and mAb 48d induce a similar, large conformational change in gp120. Inspection of the X-ray structure of CD4, gp120, and the related neutralizing Fab 17b suggests the conformational change involves organization of the bridging sheet and bringing together the inner and outer domains as pointed out by Kwong et al. (1998).

These results bear on the known difficulty in obtaining an effective vaccine. Recent thermodynamic studies of 20 gp120-reactive antibodies known to bind to a variety of defined regions on gp120 have shown that most of the broadly neutralizing mAbs bind with large enthalpy changes (Kwong et al., 2002). The implication of those studies is that antibodies that are capable of binding to the conserved regions seem to require induction of a large structural rearrangement in gp120 in order to achieve an epitope surface area of sufficient size. Thus, an energy barrier exists against binding of neutralizing mAbs, making it difficult to generate such mAbs and also difficult to achieve high enough affinity to be efficacious. It has been pointed out that it may be desirable to elicit and search for neutralizing antibodies that bind with low enthalpy changes such as in the case of mAb b12 (Jardetzky, 2002; Kwong et al. 2002), as this would reflect less structural rearrangement and a lower energy barrier for antibody binding.

Acknowledgments

We would like to thank Drs. Peter D. Kwong, Richard Wyatt, Wayne A. Hendrickson, and Joseph Sodroski for their insightful collaboration and support of the studies reviewed in this chapter.

REFERENCES

Baker, B.M., and Murphy, K.P. (1996). Evaluation of linked protonation effects in protein binding reactions using isothermal titration calorimetry. *Biophys. J.* 71:2049–2055.

Christensen, J.J., Hansen, L.D., and Izatt, R.M. (eds.) (1976). *Handbook of Proton Ionization Heats and Related Thermodynamic Quantities.* John Wiley & Sons, New York.

Colosimo, A., Brunori, M., and Wyman, J. (1976). Polysteric linkage. *J. Mol. Biol.* 100:47–57.

Doyle, M.L. (1999). Titration microcalorimetry. In J.E. Coligan, B.M. Dunn, H.L. Ploegh, D.W. Speicher, and P.T. Wingfield (eds.), *Current Protocols in Protein Science.* John Wiley & Sons, New York, Unit 20.4: 20.4.1–20.4.24.

Doyle, M.L., and Hensley, P. (1998). Tight binding affinities determined from van't Hoff analysis by titration calorimetry. *Methods Enzymol.* 295:88–99.

Doyle, M.L., Brigham-Burke, M., Brooks, I.S., Blackburn, M.N., Smith, T.M., Newman, R., Reff, M., Stafford, III, W.F., Sweet, R.W., Truneh, A., Hensley, P., and O'Shannessy, D.J. (2000). Measurement of protein interaction energetics: application to structural variants of an anti-CD4 antibody. *Methods Enzymol.* 323:207–230.

Eftink, M.R. (1995). Use of multiple spectroscopic methods to monitor equilibrium unfolding of proteins. *Methods Enzymol.* 259:487–512.

Hensley, P. (1996). Defining the structure and stability of macromolecular assemblies in solution: the re-emergence of analytical ultracentrifugation as a practical tool. *Structure* 4:367–373.

Jardetzky, T. (2002). Conformational camouflage. *Nature* 420:623–624.

Kwong, P.D., Wyatt, R., Robinson, J., Sweet, R.W., Sodroski, J., and Hendrickson, W.A. (1998). Structure of an HIV gp120 envelope glycoprotein in complex with the CD4 receptor and a neutralizing human antibody. *Nature* 393:648–659.

Kwong P.D., Doyle M.L., Casper D.J., Cicala C., Leavitt S.A., Majeed S., Steenbeke T.D., Venturi M., Chaiken I., Fung M., Katinger H., Parren P.W., Robinson J., Van Ryk D., Wang L., Burton D.R., Freire E., Wyatt R., Sodroski J., Hendrickson W.A., and Arthos J. (2002). HIV-1 evades antibody-mediated neutralization through conformational masking of receptor-binding sites. *Nature* 420:678–682.

Ladbury, J.E. (1996). Just add water! The effect of water on the specificity of protein-ligand binding sites and its potential application to drug design. *Chem. Biol.* 3:973–980.

Laue, T.M., Shah, B.D., Ridgeway, T.M., and Pelletier, S.L. (1992). Computer-aided interpretation of analytical sedimentation data, for proteins. In: S.E. Harding, A.J. Rowe & J.C. Horton (eds.), *Analytical Ultracentrifugation in Biochemistry and Polymer Science*. Royal Society of Chemistry, Cambridge, pp. 90–125.

Lo Conte, L., Chothia, C., and Janin, J. (1999). The atomic structures of protein–protein recognition sites. *J. Mol. Biol.* 285:2177–2198.

Moore, J.P., and Sodroski, J. (1996). Antibody cross-competition analysis of the human immunodeficiency virus type 1 gp120 exterior envelope glycoprotein. *J. Virol.* 70:1863–1872.

Murphy, K.P., and Freire, E. (1992). Thermodynamics of structural stability and cooperative folding behavior in proteins. *Adv. Protein Chem.* 43:313–361.

Myszka, D.G., Sweet, R.W., Hensley, P., Brigham-Burke, M., Kwong, P.D., Hendrickson, W.A., Wyatt, R., Sodroski, J., and Doyle, M.L. (2000). Energetics of the HIV gp120-CD4 binding reaction. *Proc. Natl. Acad. Sci. USA* 97:9026–9031.

Nash, L.K. (1962) *Elements of Chemical Thermodynamics*. Addison-Wesley, Reading, MA.

Privalov, P.L. (1979). Stability of proteins: small globular proteins. *Adv. Protein Chem.* 33:167–241.

Ryu, S.E., Kwong, P.D., Truneh, A., Porter, T.G., Arthos, J., Rosenberg, M., Dai, X.P., Xuong, N.H., Axel, R., Sweet, R.W., and Hendrickson W.A. (1990). Crystal structure of an HIV-binding recombinant fragment of human CD4. *Nature* 348:419–426.

Spolar, R.S., and Record, M.T. (1994). Coupling of local folding to site-specific binding of proteins to DNA. *Science* 263:777–784.

Spolar, R.S., Ha, J.-H., and Record, M.T. (1989). Hydrophobic effect in protein folding and other noncovalent processes involving proteins. *Proc. Natl. Acad. Sci. USA* 86:8382–8385

Stites, W.E. (1997). Protein–protein interactions: interface structure, binding thermodynamics, and mutational analysis. *Chem. Rev.* 97:1233–1250.

Thali, M., Moore J.P., Furman C., Charles M., Ho D.D., Robinson J., Sodroski J. (1993). Characterization of conserved human immunodeficiency virus type 1 gp120 neutralization epitopes exposed upon gp120-CD4 binding. *J. Virol.* 67:3978–3988.

Thomson, J., Ratnaparkhi, G.S., Varadarajan, R., Sturtevant, J., and Richards, F. M. (1994) Thermodynamic and structural consequences of changing a sulfur atom to a methylene group in the M13Nle mutation in ribonuclease-S. *Biochemistry* 33:8587–8593.

Trkola A., Dragic T., Arthos J., Binley J.M., Olson W.C., Allaway G.P., Cheng-Mayer C., Robinson J., Maddon P. J., Moore J. P. (1996). CD4-dependent, antibody-sensitive interactions between HIV-1 and its co-receptor CCR-5. *Nature* 384:184–187.

Wang, J.H., Yan, Y.W., Garrett, T.P., Liu, J.H., Rodgers, D.W., Garlick, R.L., Tarr, G.E., Husain, Y., Reinherz, E.L., and Harrison, S.C. (1990). *Nature* 348: 411–418.

Willcox, B.E., Gao, G.F., Wyer, J.R., Ladbury, J.E., Bell, J.I., Jakobsen, B.K., and van der Merwe, P.A. (1999). TCR Binding to Reptide-MHC Stabilizes a Flexible Recognition Interface. *Immunity* 10:357–365.

Wiseman, T., Williston, S., Brandts, J.F., and Lin, L.N. (1989). Rapid measurement of binding constants and heats of binding using a new titration calorimeter. *Anal. Biochem.* 179:131–137.

Wu, L., Gerard, N.P., Wyatt, R., Choe H, Parolin C, Ruffing N, Borsetti A, Cardoso AA, Desjardin E., Newman W., Gerard, C., and Sodroski J. (1996). CD4-induced interaction of primary HIV-1 gp120 glycoproteins with the chemokine receptor CCR-5. *Nature* 384:179–183.

Wyman, J., and Gill, S.J. (1990). Ligand control of Aggregation. In *Binding and Linkage: Functional Chemistry of Biological Macromolecules*. University Science Books, Mill Valley, CA, pp. 203–236.

Xie, D., and Freire, E. (1994). Molecular basis of cooperativity in protein folding. V. Thermodynamic and structural conditions for the stabilization of compact denatured states. *Proteins* 19:291–301.

8

Protein–Protein Recognition in Phosphotyrosine-Mediated Intracellular Signaling

John E. Ladbury

ABSTRACT

It is apparent that most fundamental cellular processes are transduced through tyrosine kinase–mediated pathways involving the modification of tyrosine to phosphotyrosine. Therefore, for transduction without corruption, the protein–protein interactions involved have to be mutually exclusive. Many of these proteins bind via homologous domains whose binding characteristics suggest that their innate specificity is not sufficiently high to account for the integrity of signal transduction. Here two such phosphotyrosine-binding domains (Src homology 2 (SH2) and phosphotyrosine binding (PTB)) are described and their capability to impose the required level of specificity in a signal transduction pathway is analyzed. The data available suggest that the domains are not highly specific, and indeed in the case of the SH2 domain a high level of promiscuity is observed. How then is mutual exclusivity in signaling achieved? It appears that models other than linear pathways need to be invoked to gain an understanding of phosphotyrosine-mediated signal transduction.

1. INTRODUCTION

Stimulation of a membrane-bound receptor on a cell surface results in a change to the intracellular region of the receptor. This results in recruitment of proteins that

JOHN E. LADBURY • Department of Biochemistry and Molecular Biology, University College London, Gower Street, London, WC1E 6BT, UK.

Proteomics and Protein–Protein Interactions: Biology, Chemistry, Bioinformatics, and Drug Design, edited by Waksman. Springer, New York, 2005.

propagate an intracellular signal transduction pathway. This activation of a multitude of proteins usually occurs over a proscribed time period, and involves a defined group of proteins that interact in a distinct order. In this way control is exerted over the ultimate cellular response. In some signaling pathways the signal appears to be transduced by protein–protein interactions that are based on recognition of the presence of posttranslational phosphorylation of tyrosine amino acids. Thus, signaling of this type depends on the stimulation or upregulation of tyrosine kinase activity. This can be derived from the membrane-bound receptor itself or from other proteins recruited to the signaling pathway. Recognition of the tyrosyl phosphate moiety (pTyr or pY) has been shown to occur via distinct domains (e.g., Src homology 2 [SH2] and phosphotyrosine binding [PTB] domains) of the interacting proteins. Interaction with these domains alone (or in combination with other commonly found domains) is thought to be largely responsible for discrete recognition of downstream targets and thus, at least in part, required for mutual exclusivity in the signaling process (Schlessinger, 2000; Hunter, 2000; Pawson and Nish, 2000; Pawson et al., 2001). It is still widely believed that these signals are based on a series of protein–protein interactions that result in a linear pathway. Thus, to ensure that the signal derived from a given type of receptor gives rise to the correct cellular response it is clearly absolutely necessary that the signal is highly specific. This chapter focuses on proteins involved in signal transduction pathways that are mediated by interactions that recognize the phosphotyrosine moiety with a view to shedding some light on the issue of integrity, or lack thereof, in this mode of signal transduction.

2. Src HOMOLOGY 2 (SH2) DOMAINS

2.1. Introduction

In most tyrosine kinase–mediated signaling pathways the stimulation of a receptor by a growth factor or other specific cytokine results in a change in conformation in the intracellular region (often in the context of a receptor dimer) that effects kinase activity. In many cases a tyrosine in the receptor itself is phosphorylated (often by *trans*-kinase activity between the dimer molecules). This increase in tyrosine phosphorylation above a basal level increases the net concentration of binding sites for the downstream effector in the early signaling process. Phosphotyrosine forms the basis of recognition for the interaction of SH2 domains. These domains whose sequences appear within in excess of 100 genes in the human genome are expressed as polypeptides consisting of approximately 100 amino acids. Aberrations in the interactions of SH2 domains are responsible numerous disease states including various cancers, osteoporosis, and immunodeficiency. These domains have highly homologous secondary and tertiary structures that can be described simplistically as a three- or four-stranded β-sheet sandwiched between two parallel α-helixes (Kuriyan and Cowburn; 1993, 1997).

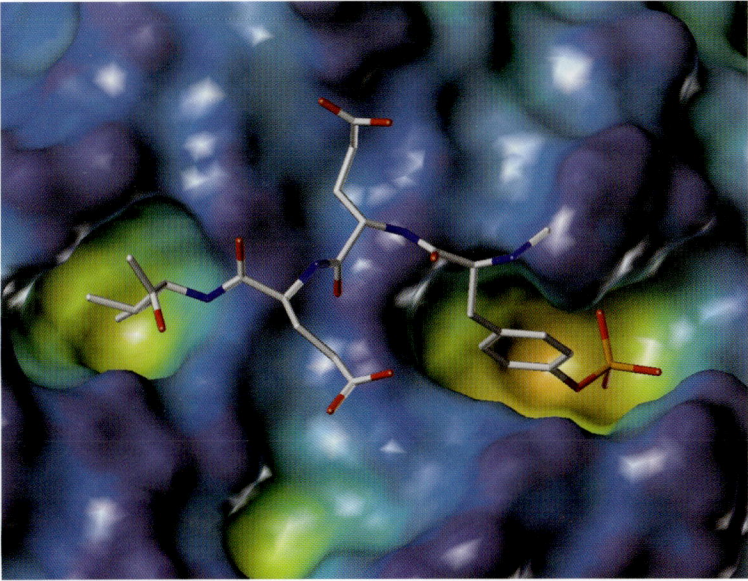

Figure 8.1. Space filling model of the binding surface of the Src SH2 domain in complex with the "specific" peptide (*stick model*) showing the "two-pronged plug" mode of recognition. The pTyr moiety binds in the deep pocket on the right, whilst the Ile interacts in the pocket on the left.

2.2. Specificity of SH2 Domains

Clearly, if any given cell can contain proteins expressing many SH2 domains that are involved in binding to cognate partners in different cellular signaling pathways, to ensure the integrity of a signal a high level of specificity must be imposed in their interactions. To understand this issue it is important to get a sense of what is meant by specificity. Specificity for a given binding site is best considered as the relative affinities of the appropriate binding partner versus those of another competing ligand. Thus, a competing ligand that has a K_D an order of magnitude more than the "specific" binding partner will compete for a binding site equally if that ligand has only a 10-fold higher local concentration. Since local signaling protein concentrations in cells are believed to fluctuate over several orders of magnitude, to ensure an interaction provides a mutually exclusive signal one would expect a specific interaction to be at least three to four orders of magnitude tighter than possible competing interactions. These are the ranges of affinities typically experienced between specific and nonspecific protein–DNA interactions (Ladbury et al. 1994). Bearing in mind that a cell can have in excess of 100 SH2 domain-containing proteins being expressed, the requirement for at least three orders of magnitude difference between specific and nonspecific would seem appropriate (Ladbury and Arold, 2000).

The first large-scale investigation of the issue of specificity in SH2 domain signaling was based on using a library of short tyrosyl-phosphopeptides, which were screened against a range of immobilized SH2 domains (Songyang et al., 1993). These studies demonstrated that a level of specificity could be largely derived from the residues proximal and C-terminal to the pY moiety. Thus, for example, the SH2 domain from the Src protein recognized the sequence pYEEI, whereas that from the N-terminus of the protein PI3-kinase would bind preferentially to the sequence pYMXM (where X is any amino acid). Within these studies there appeared to be a level of degeneracy such that, for example, the Src protein also seemed to show some affinity for sequences such as pYDNV/l or pYTDM (Songyang et al., 1993). Furthermore, these studies did not really provide an appreciation of the quantitative differences between specific and nonspecific binding.

2.3. Implications of Structure in SH2 Domain Interactions

The structural basis for ligand recognition became clearer at about the same time as the above peptide screening studies. As series of high-resolution structures revealed that although the binding sites of SH2 domains show some variation, the recognition of their cognate, peptide-based ligands is based on the pY moiety binding in a deep pocket (Kuriyan and Cowburn, 1993). The residues C-terminal of the pY were shown in the majority to comply with three broad structural motifs. In the first case the first two residues extend across a polar surface. The third residue then delves into a pocket in a manner forming a recognition motif resembling a "two-pronged plug" (see Fig. 8.1 and 8.2A). The second case involves extending the residues C-terminal of the pY across a furrow that is largely apolar (see Fig. 8.2B). The final case involves the second C-terminal residue being forced into a β-turn by the positioning of an amino acid bulky hydrophobic side chain on the SH2 domain binding site surface (see Fig. 8.2C).

The SH2 domains of Src family proteins recognize their cognate ligands by the former of these modes (Fig. 8.2A; Waksman et al., 1992, 1993; Xu et al., 1995; Tessari et al., 1997; Williams et al., 1997). The interaction with the "specific" sequence is based on the pY making an intimate series of hydrogen bonds in a deep pocket. The crystal structure of the peptide complex with the SH2 domain from Src shows that the pY can sustain of the order of 11 hydrogen bonds or charge–charge contacts. Numerous studies have revealed how important this basic interaction is for ligand binding affinity. Mutations in the pY binding site preclude binding and ligands without the pY moiety do not bind, or bind very weakly (see Table 8.1; Lemmon and Ladbury, 1994; Bradshaw et al., 1999). The first two residues C-terminal of the pY in the "specific" sequence are glutamic acid groups. These are seen to splay out across the domain surface, which is largely polar in this region. In the X-ray crystal structure of the Src SH2 domain–peptide complex the interactions made by these residues support a network of water molecules (Waksman et al., 1993). These water molecules play a key role in dictating both binding affinity and specificity (Chung et al., 1998). Removal of these water molecules in ligand binding appears to have a significantly

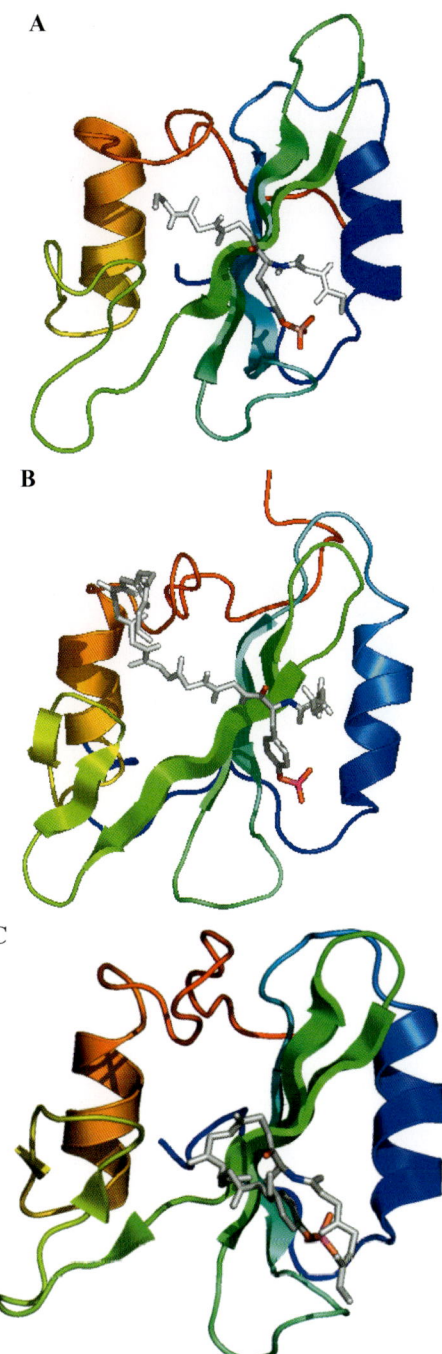

Figure 8.2. (**A**) The structure of the SH2 domain from Src bound to cognate tyrosyl-phosphopeptide. The secondary structural elements are discussed in the text and listed in Table 8.2 and are identified as follows: Starting at the N-terminus (*dark blue*) β-strand A (βA); loop AA, α-helix A (αA); loop AB (AB); β-strand B (βB); loop BC; βC, CD; βD; βD′; DE; βE; EF; βF; FB αB; BG; βG; C-terminus (*red*). (**B**) The structure of the SH2 domain from PLCγ bound to a cognate ligand. (**C**) The structure of the SH2 domain from Grb2 bound to a cognate ligand.

Table 8.1. Binding of peptides and peptide-mimetics to the Src SH2 domain[a]. Residues in bold type correspond to those interacting with the Src SH2 domain binding site.

Peptide/ ligand	T (°C)	K_D (μM)	ΔH (kJ · mol^{-1})	$T\Delta S$ (kJ · mol^{-1})	ΔG (kJ · mol^{-1})	Reference
Peptides based on "specific" sequence						
KGGQ**pYEEIPIP**	25	0.55	−35.4	0.5	−36.0	(1)
KGGQ**pYEEIPIP**[c]	25	0.77	−33.4	1.7	−35.1	(1)
EPQ**pYEEIPIYL**	25	0.09	−38.7	1.4	−40.1	(2)
PQ**pYEEIPI**	25	0.18	−32.3	6.3	−38.6	(3)
PQ**pYEEIPI**	25	0.25	−31.5	6.3	−37.8	(4)
PQ**pYEEIPI**	25	0.27	−31.5	6.3	−37.8	(5)
PQ**pYEEIPI**	25	0.2	−27.3	10.9	−38.2	(6)[g]
pYEEIE	25	0.05	−28.6	13.8	−42.3	(7)
pYEEIQ	25	0.09	−26.0	15.0	−40.1	(7)
pYEEI	25	0.5	−21.4	15.4	−36.8	(7)
Constrained-**pYEEI**	25	0.10	−18.8	21.3	−40.1	(8)
PQ**YEEIPI**	25	2222[c]	−16.4	−1.3	−15.1	(9)
pY	25	333[c]	−0.4	19.3	−19.7	(9)
PQ**pSEEIPI**	25	1818[b,c]	—	—	−16.4[b]	(9)
Peptides with substitution in pY +1 position of "specific" sequence						
PQ**pYQEIPI**	25	0.47	−31.9	4.2	−36.1	(9)
PQ**pYDEIPI**	25	0.18	−37.4	1.3	−38.6	(9)
PQ**pYAEIPI**	25	0.34	−32.3	4.2	−36.5	(9)
PQ**pYGEIPI**	25	6.25	−20.2	9.7	−29.8	(9)
PQ**pYAEIPI**	25	0.35	−32.3	4.2	−36.5	(5)
PQ**pYREIPI**	25	8.20	−21.8	7.56	−29.4	(5)
Peptides with substitution in pY +2 position of "specific" sequence						
PQ**pYEpYIPI**	25	0.07	−29.4	11.3	−40.7	(4)
PQ**pYEYIPI**	25	0.66	−26.8	8.4	−35.2	(4)
PQ**pYEQIPI**	25	0.53	−31.9	4.2	−36.1	(3)
PQ**pYEDIPI**	25	0.42	−26.0	10.5	−36.5	(3)
PQ**pYEAIPI**	25	1.04	−26.5	8.0	−34.4	(3)
PQ**pYEAIPI**	25	1.0	−26.46	7.98	−34.44	(5)
PQ**pYEGIPI**	25	1.96	−25.2	7.6	−32.8	(3)
Peptides with substitution in pY +3 position of "specific" sequence						
EPQ**pYEEVPIYL**	25	0.16	−28.6	10.2	−38.8	(2)
EPQ**pYEEEPIYL**	25	0.21	−32.7	5.4	−38.1	(2)
EPQ**pYEEWPIYL**	25	0.31	−32.2	4.9	−37.1	(10)
EPQ**pYEEDPIYL**	25	0.38	−27.5	9.1	−36.6	(10)
PQ**pYEELPI**	25	0.43	−23.5	13.0	−36.5	(3)
PQ**pYEEVPI**	25	0.46	−22.7	13.9	−36.5	(3)
PQ**pYEEAPI**	25	1.75	−21.4	11.3	−32.7	(3)
PQ**pYEEGPI**	25	0.39	−15.1	16.0	−31.1	(3)
Peptides with substitutions in pY +1 and +2 positions of "specific" sequence						
PQ**pYAAIPI**	25	2.4	−25.62	6.30	−31.92	(5)
PQ**pYKAIPI**	25	83.3	−18.48	4.62	−23.10	(5)
PQ**pYRAIPI**	25	59.8	−15.12	9.24	−24.36	(5)
Peptides with substitutions in pY +2 and +3 positions of "specific" sequence						
PQ**pYEAAPI**	25	4.76	−22.3	8.0	−30.2	(3)

Table 8.1. (Continued)

Peptide/ligand	T (°C)	K_D (μM)	ΔH (kJ · mol^{-1})	$T\Delta S$ (kJ · mol^{-1})	ΔG (kJ · mol^{-1})	Reference
Peptides with substitutions in all positions of "specific" sequence						
PQ**p**YAEAPI	25	7.14	−16.4	13.0	−29.4	(3)
PQ**p**YAAAPI	25	21.27	−16.4	10.5	−26.9	(3)
P**qp**Y**Ip**YVPI	25	0.40	−6.3	30.2	−36.5	(4)
TQ**p**YVMLEI	25	5.88	−14.3	15.5	−29.8	(6)
PQ**p**YQPGEN	25	29.4	−19.3	6.7	−26.0	(6)
EPQ**p**YQPGEN	25	14.3	−25.7	2.0	−27.7	(2)
Peptidomimetic ligands						
pYE-N-$(C_5H_{11})_2$	25	0.4	−18.1	19.1	−37.2	(7)
pYM-N-$(C_5H_{11})_2$	25	4.2	−16.4	14.6	−31.0	(7)
pYC-N-$(C_5H_{11})_2$	25	1.4	−18.1	15.9	−34.0	(7)
pYE-N-$C_3H_6 - C_5H_9$	25	0.4	−29.8	7.1	−37.0	(7)
pYE-N-$(C_5H_9)(C_4H_9)$	25	1.0	−22.7	12.5	−35.2	(7)
pYE-N-*hexanol*	25	3.4	−13.4	18.4	−31.8	(7)
pYE-N-*heptanol*	25	2.3	−19.3	13.4	−32.7	(7)

[a] All data derived from isothermal titration calorimetry.
[b] Data determined at 25°C.
[c] GST-fusion SH2.
[d] Not accurately determined.
[e] 50 mM MOPS, pH 6.8; 100 mM NaCl.
[f] 20 mM MES, pH 6.0; 1 mM BME; 1 mM EDTA; 50 mM NaCl.
[g] 20 mM HEPES, pH 7.5; 100 mM NaCl; 1 mM BME; 1 mM EDTA.
[h] 20 mM HEPES; 150 mM NaCl; pH 8.0.
References: (1) Ladbury et al. (1995)[e]; (2) Chung et al. (1998)[f]; (3) Bradshaw and Waksman (1999)[g]; (4) Lubman and Waksman (2003)[g]; (5) Lubman, and Waksman (2002)[g]; (6) Bradshaw et al. (1998)[f]; (7) Charifson et al. (1997)[h]; (8) Davidson et al. (2002)[g]; (9) Bradshaw et al. (1999)[g]; (10) Henriques and Ladbury (2001)[f].

detrimental effect (Henriques et al., 2000; Henriques and Ladbury, 2001; Lubman and Waksman, 2003). The third residue C-terminal from the pY in the specific sequence is an isoleucine. This is seen in the crystal structure to bind in a deep, largely hydrophobic pocket, thus completing the "two-pronged plug" mode of binding (Figs. 1, 2A).

The complex between the SH2 domain from PLCγ is somewhat different from that for the Src SH2 domain representing the second group of recognition motifs (Fig. 8.2B; Pascal et al., 1994, 1995). Five residues after the pY make a largely hydrophobic interaction with the SH2 domain. Interestingly, solution structural studies reveal that the residues that line the SH2 domain binding site show a significant dynamic disorder that is increased in the presence of the ligand (Farrow et al., 1994). These unusual data suggest that the binding is accompanied by an entropically favorable side chain mobility that undoubtedly will reduce specific noncovalent bonding interactions and hence reduce any favorable enthalpic contribution to the free energy.

The X-ray crystal structure of the SH2 domain of the protein Grb2 in complex with a tyrosyl-phosphopeptide with the sequence KPFpYVNV (derived from the protein BCR-Abl) reveals how this particular SH2 domain exhibits some specificity in its recognition of its binding partners compared to other SH2 domains (Fig. 8.2C;

Rahuel et al., 1996). This mode of recognition is an example of the third type described above. Grb2 plays an important role in tyrosine kinase–mediated signal transduction. It can bind to a number of ligands. It binds directly to phosphorylated membrane-bound receptors (e.g., epidermal growth factor [EGF] receptor), and connects this early signaling event directly to the entry point to the mitogen-activated protein (MAP) kinase signaling response via interaction with the a guanine nucleotide exchange factor for Ras, *Son of Sevenless* (SOS). The Grb2 SH2 domain shows only quite subtle variation from other SH2 domains. The CD loop is five residues shorter than in, for example, Src, and more closely resembles the SH2 domains from Syp-N, p85, Syk-C, and ZAP70. The pTyr and the three residues C-terminal of this make the only specific interactions of the peptide with the binding site of the domain. In contrast to other SH2–tyrosyl-phosphopeptide interactions, the ligand does not adopt an extended conformation, rather it forms a β-turn. This requires an Asn in the pTyr + 2 position and the β-turn is maintained by a hydrogen bond between to carbonyl oxygen of the pTyr and the main chain nitrogen of the Val +3. A Trp residue from the EF loop of the SH2 domain closes the binding cleft C-terminal to the pTyr and thus forces the peptide into the β-turn conformation. The importance of this Trp residue was demonstrated by changing the recognition of a Src SH2 domain (see above) for extended linear peptides, to that of Grb2 by substituting a Trp into the appropriate position in the former SH2's binding site (Kimber et al., 2000). The modified Src SH2 domain bound peptides that could adopt the β-turn conformation. This requirement for the peptide to be able to adopt a β-turn provides some level of specificity in the recognition. Despite this less common mode of recognition there are still a number of peptide sequences that can adopt the required conformation and bind to the Grb2 SH2 domain. Indeed, most sequences with the pTyr and the Asn in the +2 position will bind (e.g., pYINQ from the EGF receptor).

2.4. Quantification of Interactions of SH2 Domains

Although the structures of various SH2 complexes reveal some differences in recognition, investigation of the affinity of tyrosyl-phosphopeptides shows that SH2 domains do not show a high level of specificity in their interactions. In fact, in most cases they bind in a rather promiscuous manner. A number of techniques have been used to determine the dissociation constants (K_D) for the binding of SH2 domains. Generally these studies have revealed that so-called specific tyrosyl-phosphopeptide binding motifs bind no more than a couple of orders of magnitude more tightly than nonspecific sequences.

Binding data derived using isothermal titration calorimetry (ITC) have provided thermodynamic characterization of the interactions of several SH2 domains with a large number of different peptide-based ligands. These data permit a detailed quantitative insight into the level of specificity in these interactions. Table 8.1 reports the data for the interactions of more than 50 ligands binding to the SH2 domain from Src. Several key points emerge from these studies. First, these data clearly demonstrate that the interaction requires the presence of the pY residue. Absence of this results in no detectable, or very weak binding. The substitution of a phosphoserine for a pY within the "specific" sequence results in a drop in binding affinity of greater than four orders of magnitude (Bradshaw et al., 1999). The intimate interaction of the pY residue is fundamental to ligand binding of to SH2 domains. In the interaction with the case of the Src protein this amino acid is estimated to contribute approximately 60% of the value of the total free energy of the domain–peptide interaction (Bradshaw et al., 1999; Henriques et al., 2000). It should be noted, however, that the presence of this interaction in all SH2 domain ligand recognition means that it does not contribute to the specificity.

A second point to emerge from the ITC studies in Table 8.1 is that changing the residues in C-terminal of the pY does not appear to dramatically affect binding. This dataset reveals the apparent promiscuity of the SH2 domain to ligands. The range of K_D determined is limited to approximately two orders of magnitude. Data have been reported in which all of the residues C-terminal to the pTyr have been individually substituted. In the case of the changing the pTyr + 1 residue very little change is observed when the Glu is substituted for a hydrophobic group or another polar/charged residue. A significant difference in binding of at least an order of magnitude occurs when the negatively charged Glu residue is substituted for a positively charged Arg. In addition, removal of a side chain by substituting with a Gly seems to adversely affect binding. The substitutions in the pTyr + 2 position are essentially insensitive, except when a pTyr is in that position. This residue is able to make a significant number of complementary interactions on the SH2 surface. Substitutions in the pTyr + 3 position are, again, largely insensitive. Going from an aliphatic Ile to a bulky hydrophobic Trp, or a small charged Asp has a net effect on the K_D of less than an order of magnitude. The interactions of the Src SH2 domain with doubly and trebly substituted tyrosyl-phosphopeptides lead to loss in affinity of only a couple of orders of magnitude at worst. Thus, it is anticipated that the residues proximal to the pTyr really do not add a level of specificity that would be required to ensure the integrity of one pathway over another under normal cellular conditions. The affinity of the "specific" sequence is only, at worst, a couple of orders of magnitude weaker than that of an essentially random sequence. Thus, an SH2 that recognizes a completely different sequence from that to pYEEI would only have to be at worst 100-fold more concentrated to compete equally for the Src SH2 binding site. Furthermore, the apparent structural basis for specific recognition is somewhat compromised by these data. Although the structural detail gives us a clear insight into the mode of recognition of selected peptides, it also reveals how the SH2 domain of Src can actually accommodate a range of diverse peptide ligands with only limited perturbation of the affinity.

Table 8.1 also includes some data for peptidomimetic compounds that were precursors for drug compounds. These molecules, as has generally been the case with the multitude of rationally designed inhibitors for SH2 domains, show little improvement in affinity over the specific peptide ligands.

In addition to exploring the issue of specificity by using a range of peptide-based ligands, site-directed mutatgenesis of the SH2 domain itself gives further insight. Substituting individual residues in the binding site appears to have a limited effect. For example, Table 8.2 reveals that mutation of the residue ArgβD$'$1 which makes

Table 8.2. Binding of peptides to mutated forms of the Src SH2 domain[a,b]. Residues in bold type correspond to those interacting with the Src SH2 domain binding site.

Mutant	Peptide	K_D (μM)	ΔH (kJ $\cdot$ mol^{-1})	$T\Delta S$ (kJ $\cdot$ mol^{-1})	ΔG (kJ $\cdot$ mol^{-1})	Reference
Argβ**D**'**1**Ala	PQp**YEEIPI**	0.65	−31.1	4.2	35.3	(4)
Argβ**D**'**1**Ala	PQp**YE**p**YIPI**	0.17	−31.1	7.6	−38.6	(4)
Argβ**D**'**1**Ala	PQp**YEYIPI**	0.80	−30.2	4.2	−34.4	(4)
Argβ**D**'**1**Ala	PQp**YI**p**YVPI**	1.80	−8.4	23.9	−32.3	(4)
Argβ**D**'**1**Phe	PQp**YEEIPI**	0.19	−33.2	4.6	−37.8	(4)
Argβ**D**'**1**Phe	PQp**YE**p**YIPI**	0.07	−28.1	12.6	−40.7	(4)
Argβ**D**'**1**Phe	PQp**YEYIPI**	0.62	−30.24	5.46	−35.7	(4)
Argβ**D**'**1**Phe	PQp**YI**p**YVPI**	0.41	−7.1	29.0	−36.1	(4)
Lysβ**D3**Ala	PQp**YEEIPI**	1.78	−23.1	9.6	−32.7	(5)
Lysβ**D3**Ala	PQp**YAEIPI**	0.91	−28.1	1.26	−29.4	(5)
Lysβ**D3**Ala	PQp**YREIPI**	0.62	−39.1	−3.78	−35.28	(5)
Lysβ**D3**Glu	PQp**YEEIPI**	50	−13.4	10.92	−24.36	(5)
Lysβ**D3**Glu	PQp**YAEIPI**	2.4	−29.0	2.94	−31.92	(5)
Lysβ**D3**Glu	PQp**YREIPI**	3.7	−27.3	3.78	−31.08	(5)
Lysβ**D3**Asp	PQp**YEEIPI**	9.1	−23.5	5.04	−28.56	(5)
Lysβ**D3**Asp	PQp**YAEIPI**	2.3	−32.3	0.84	−33.18	(5)
Lysβ**D3**Asp	PQp**YREIPI**	1.3	−31.9	−1.68	−30.24	(5)
Lysβ**D3**Ala /Asp**CD2**Ala	PQp**YEEIPI**	0.13	−31.9	7.1	−39.0	(5)
Lysβ**D3**Ala /Asp**CD2**Ala	PQp**YAEIPI**	0.50	−30.7	5.9	−36.6	(5)
Lysβ**D3**Ala /Asp**CD2**Ala	PQp**YREIPI**	35.7	−29.4	−4.6	−24.8	(5)
Lysβ**D3**Ala /Aspβ**C8**Ala /Asp**CD2**Ala	PQp**YEEIPI**	0.21	−28.1	9.7	−37.8	(5)
Lysβ**D3**Ala /Aspβ**C8**Ala /Asp**CD2**Ala	PQp**YAEIPI**	0.6	−29.8	5.9	−35.7	(5)
Lysβ**D3**Ala /Aspβ**C8**Ala /Asp**CD2**Ala	PQp**YREIPI**	1.8	−37.8	−4.6	−33.2	(5)
Lysvβ**D3**Ala	PQp**YEAIPI**	8.3	−17.22	11.76	−28.98	(5)
Lysβ**D3**Ala	PQp**YAAIPI**	5.9	−19.74	10.50	−30.24	(5)
Lysβ**D3**Ala	PQp**YKAIPI**	17.2	−19.74	7.14	−26.88	(5)
Lysβ**D3**Ala	PQp**YRAIPI**	4.9	−28.56	2.10	−30.66	(5)
Lysβ**D3**Glu	PQp**YEAIPI**	227	−5.46	15.54	−21.00	(5)
Lysβ**D3**Glu	PQp**YAAIPI**	14.1	−16.38	12.18	−28.56	(5)
Lysβ**D3**Glu	PQp**YKAIPI**	25.0	−16.38	9.24	−25.62	(5)
Lysβ**D3**Glu	PQp**YRAIPI**	25.0	−19.74	6.30	−26.04	(5)
Lysβ**D3**Asp	PQp**YEAIPI**	38.4	−14.28	11.34	−25.62	(5)
Lysβ**D3**Asp	PQp**YAAIPI**	7.7	−21.00	7.98	−28.98	(5)
Lysβ**D3**Asp	PQp**YKAIPI**	10.0	−23.94	5.46	−29.40	(5)
Lysβ**D3**Asp	PQp**YRAIPI**	10.0	−25.20	3.78	−28.98	(5)
Tyrβ**D5**Ile	PQp**YEEIPI**	8.92	−10.1	18.9	−29.0	(11)
Lysβ**D3**Ala	PQp**YEEIPI**	1.77	−23.1	9.7	−32.8	(11)
Argβ**D**'**1**Ala	PQp**YEEIPI**	0.65	−31.1	4.2	−35.3	(11)
Argβ**D**'**1**Phe	PQp**YEEIPI**	0.41	−32.8	3.8	−36.6	(11)

Table 8.2. (continued)

Mutant	Peptide	K_D (μM)	ΔH (kJ · mol^{-1})	$T\Delta S$ (kJ · mol^{-1})	ΔG (kJ · mol^{-1})	Reference
LeuBG4Ala	PQpYEEIPI	0.70	−36.5	−1.3	−35.2	(11)
Ileβ E4Ala	PQpYEEIPI	0.44	−37.4	−0.8	−36.6	(11)
ThrEF1Ala	PQpYEEIPI	0.30	−26.5	10.9	−37.4	(11)
ArgEF3Ala	PQpYEEIPI	0.34	−28.6	8.4	−37.0	(11)
AspBG2Ala	PQpYEEIPI	0.18	−31.9	6.7	−38.6	(11)
Cysβ C3Ala	PQpYEEIPI	38	−31.9	11.3	−43.3	(9)
Cysβ C3Ser	PQpYEEIPI	21	−32.3	9.7	−42.0	(9)
Argβ B5Ala	PQpYEEIPI	0.021	−15.1	10.1	−25.2	(9)
Lysβ D6Ala	PQpYEEIPI	0.56	−26.0	6.7	−32.8	(9)
Argα A2Ala	PQpYEEIPI	0.95	−19.7	14.3	−34.0	(9)
Argα A2Ala	PQpYAEIPI	1.47	−23.5	9.7	−33.2	(9)
Argα A2Ala	PQpYEAIPI	8.93	−13.9	15.1	−29.0	(9)
Argα A2Ala	PQpYEEAPI	9.62	−13.4	15.1	−28.5	(9)
ThrBC2Ala	PQpYEEIPI	1.0	−33.2	1.7	−34.9	(9)
Serβ B7Ala	PQpYEEIPI	1.3	−39.1	−4.2	−34.9	(9)
Serβ B7Ala	PQpYAEIPI	1.78	−37.8	−5.0	−32.8	(9)
Serβ B7Ala	PQpYEAIPI	6.14	−29.4	0.4	−29.8	(9)
Serβ B7Ala	PQpYEEAPI	9.83	−28.6	0.0	−28.6	(9)
GluBC1Ala	PQpYEEIPI	2.7	−29.8	7.1	−37.0	(9)
Hisβ D4Ala	PQpYEEIPI	2.7	−35.3	2.1	−37.4	(9)
Serβ C5Ala	PQpYEEIPI	4.3	−28.1	9.7	−37.8	(9)
ThrBC3Ala	PQpYEEIPI	7.6	−38.2	1.3	−39.5	(9)

[a] All data derived from isothermal titration calorimetry.
[b] Data determined at 25°C.
Buffers as for Tables 1.
References as for Table 1. (11) Bradshaw et al. (2000)[g].

direct contact with the pTyr has a limited effect on binding of ligands with various residues in the PTyr +1 to 3 positions (Lubman and Waksman, 2003). A study that determined the importance of amino acid the SH2 domain in terms of their effects on ligand recognition revealed that only two residues, when mutated to Ala, had significant effects on binding of the pEEI motif. The largest effects seem to derive from inserting a reversed charge mutant into the binding site, which, as might be expected, sets up a repulsive interaction.

Casting attention to the interactions of SH2 domains other than that from the Src protein (Table 8.3) it is clear that the affinities of SH2 domains for their cognate ligands do not get much below the micromolar to 100 nM range. Nonspecific interactions tend not to be more than two orders of magnitude greater than this. One interesting investigation into the specificity of the SH2 domain from Grb2 (McNemar et al., 1997) revealed that, although binding was rather promiscuous when residues in the pTyr + 1, + 3, and + 4 were substituted for Ala, a dramatic change in affinity (three orders of magnitude) resulted on attempting the same mutation at the pTyr +2 position (Table 8.3). These data fit well with the structural data that seem to reveal the importance of

Table 8.3. Binding of peptides and to SH2 domains[a]

SH2	Peptide/ligand	$T(°C)$	$K_D(\mu M)$	ΔH (kJ · mol^{-1})	$T\Delta S$ (kJ · mol^{-1})	ΔG (kJ · mol^{-1})	Reference
Lck	TATEGQpYQPGP	25	4.24	−35.2	−4.4	−30.8	(1)
Lck[c]	TATEGQpYQPGP	25	2.92	−33.7	−1.3	−32.4	(1)
Lck	KTAENPEpYLGL DVPV	10	71.43	−7.9	15.4	−23.3	(12)
Lck	KTAENAEpYLRV APQS	10	4.46	−5.5	24.8	−30.3	(12)
Lck	TATEGQpYQPQP	10	2.96	−22.9	7.5	−30.4	(12)
Lck	EPQpYEEIpIYL	25	1.85	−55.6	−22.9	−32.7	(13)
Lck	EPQpYEEVpIYL	25	1.72	−50.6	−17.7	−32.9	(13)
Lck	EPQpYEEDpIYL	25	5.00	−45.6	−15.4	−30.2	(13)
Grb2	SpYVNVQ	20	0.19	−31.7	−6.1	−25.6	(14)
Grb2	ApYVNVQ	20	0.2	−30.2	−7.4	−22.8	(14)
Grb2	SpYANVQ	20	1.63	−35.9	3.4	−39.3	(14)
Grb2	SpYVAVQ	20	359	−8.2	−11.2	3.0	(14)
Grb2	SpYVNAQ	20	0.77	−25.1	−9.3	−15.8	(14)
Grb2	SpYVNVA	20	0.3	−30.0	−6.6	−23.4	(14)
p85 N-	SVDpYVDMSK	25	0.47	−39.4	3.1	−42.5	(1)
Fyn	EPQpYEEIPIYL	25	0.74	−36.6	−1.61	−35.0	(15)
Fyn	ATEPQpYQPGEN	25	24.39	−18.1	8.16	−26.3	(15)
Fyn	EPQpYEEIPIYL	30	1.43	−42.2	−8.3	−33.9	(15)
Fyn	ATEPQpYQPGEN	30	47.84	−20.2	4.9	−25.1	(15)

[a] All data derived from isothermal titration calorimetry.
[i] 50 mM MOPS, pH 6.8; 100 mM NaCl, 1 mM DTT;[j] 25 mM MOPS, pH 6.8; 100 mM NaCl; 1 mM DTT; [k]50 mM HEPES; 150 mM NaCl, pH 7.5;[l]10 mM KP$_i$, pH 6.0; 30 mM NaCl, 5 mM DTT.
References as in Table 8.1. Lettered footnotes continues from Table 8.1. (12) Lemmon and Ladbury (1994); (13) Renzoni et al. (1997)[j]; (14) McNemar et al. (1997)[k]; (15) Ladbury et al. (1996)[l].

the Asn at this position with respect to maintaining the β-turn binding motif of the peptide (see above).

Although in vitro studies may be considered unrepresentative of the physiological conditions experienced in the cell, the important observation from these data is that the relative affinities of specific versus nonspecific peptide-based interactions are surprisingly close.

2.5. Specificity Investigated In Vivo

In vivo studies confirm the absolute requirement for the pTyr residue, but reveal a distinct lack of the high levels of specificity expected to maintain the expected integrity in signal transduction in the residues proximal to this. For example, in experiments in mice substitution of the two closely spaced pTyr sites in the carboxy-tail of the mammalian Met receptor tyrosine kinase for Phe resulted in embryonic lethality (Maina et al., 1996). However, in substituting only the Asn in the +2 position on the receptor, which was deemed crucial for the binding of the protein Grb2, a significantly milder phenotype was observed that led to defective muscle development.

This suggests that whereas in the absence of a pTyr residue no signal is transduced, in the substitution of a residue C-terminal of the pTyr is still capable of signaling. This implies a level of "leakiness" in the recognition. This lack of specificity is exemplified further in experiments designed to alter the specificity of SH2 domains.

Normal vulval development in *C. elegans* involves the interaction the Grb2 homologue, Sem-5. A single amino acid substitution in the SH2 domain of this protein produces a vulvaless phenotype. Microinjection of a chimeric form of this Sem-5 with its SH2 domain replaced by that from the protein Src resulted in the rescue of approximately 30% of the vulval development observed for the wild-type Sem-5 (Merengere et al., 1994). This experiment demonstrated that different SH2 domains could function adequately in vivo, albeit with somewhat reduced effectiveness. This would require a significant level of degeneracy in the ligand recognition process to occur.

Although for the most part SH2 domains have high levels of primary, secondary, and tertiary structural homology, the N-terminal domain of the protein Cbl binds to tyrosyl-phosphopeptides. The pTyr binding site exhibits similarity to the canonical SH2 domain (Meng et al., 1999); however, there are some structural differences. For example, this domain lacks one of the characteristic β-strands. The discovery of this domain perhaps hints that there may be other variations on the SH2 domain theme that also recognize the pTyr residues in tyrosine kinase–mediated signaling, adding a further complication to the issue of specificity (Schlessinger and Lemmon, 2003).

In considering the above data on interactions of SH2 domains the overriding conclusion appears to be that they are not sufficient to ensure mutually exclusive pathways in respect of linear processing of tyrosine kinase–mediated signaling (Ladbury and Arold, 2000). Thus, it appears that alternative modes of ensuring that signal transduction does not result in "crossed wires" have to be inferred.

3. PHOSPHOTYROSINE BINDING (PTB) DOMAINS

3.1. Introduction

Coincident with the high activity in the investigation of the structure and binding characteristics of SH2 domains was the discovery of another domain capable of interaction with pY with similar affinity (Blaikie et al., 1994; Kavanaugh and Williams, 1994). In screening studies against the phosphorylated EGF receptor the protein Shc was pulled down even in the absence of its SH2 domain. The domain responsible for this pY interaction was identified and was denoted as the phosphotyrosine binding (PTB) domain. This domain contains approximately 200 amino acids. A series of studies revealed that the sequence dependence for specific recognition was quite different from that of SH2 domains. Investigation of binding of the Shc PTB domain to the polyoma middle tumor antigen (Campbell et al., 1994; Dilworth et al., 1994) as well as to Trk (Obermeier et al., 1994) showed that residues N-terminal, rather than the C-terminal (as in the case of SH2 domains) of the pY were important. The Shc

PTB domain recognizes the sequence NPXpY (where X corresponds to any amino acid). PTB domains were subsequently discovered in a number of other proteins (currently there are believed to be as many as 30 domains of similar structure to the PTB domain from Shc in humans; Schlessinger and Lemmon, 2003).

Around the same time the insulin receptor substrates IRS-1 and IRS-2 were shown to bind to the NPXpY motif via a domain that had some sequence homology to the Shc PTB domain (Gustafson et al., 1995; Sun et al., 1995). Despite the limited sequence homology, the structures of the PTB domains from Shc and IRS-1 show clear similarity (Zhou et al., 1995; Dhalluin et al., 2000).

3.2. Implications of Structure in PTB Domain Interactions

The structures of PTB domains from both the Shc-like family and the IRS-like family show a three- and a four- stranded sheet forming a nearly orthogonal β-sandwich structure (similar to that seen in pleckstrin homology domains) as shown in Figure 8.3. An α-helix is positioned at one end of this sandwich completing the ligand binding site (Zhou et al., 1995; Eck et al., 1996; Zwahlen et al., 2000; Stolt et al., 2003). Structural studies have revealed that the recognition sequence adopts a

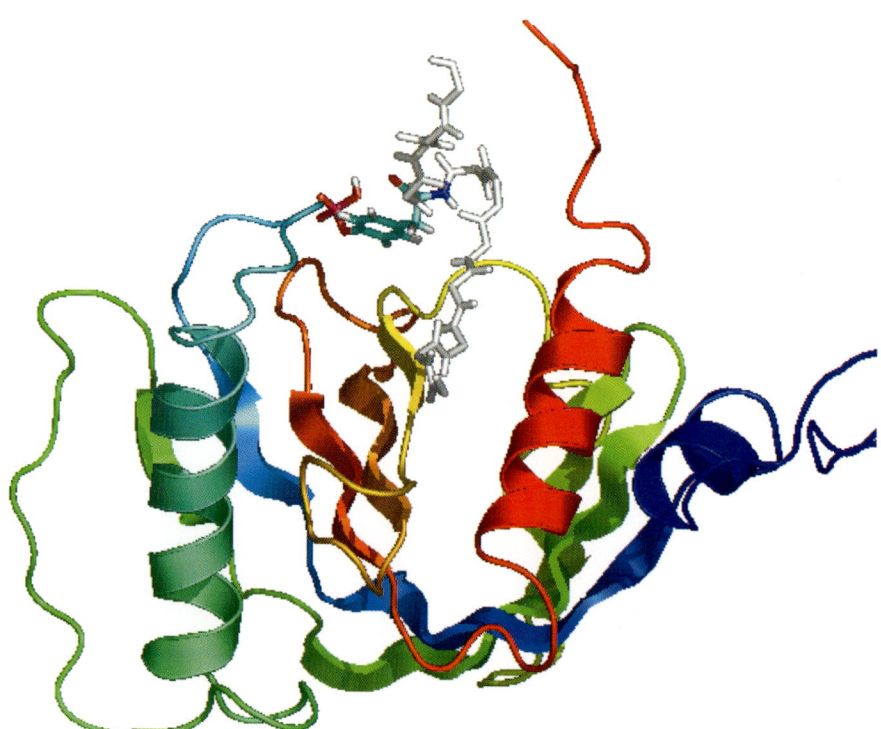

Figure 8.3. The structure of the PTB domain from Shc bound to a cognate ligand.

β-turn conformation. The residues N-terminal of the proline residue form an extended, hydrogen bonded conformation that binds as an additional strand of the domain itself. Interestingly, and in contrast with the situation in SH2 domains, the residues involved in pTyr binding are not conserved between PTB domains. Indeed, the pTyr binding appears to play a minor role in the overall recognition of ligands by PTB domains.

3.3. Specificity in PTB Domain Binding

Although it was established that Shc can bind to phosphoproteins in the absence of its SH2 domain, and thus it could be assumed that PTB domains play a primary role in recruiting these proteins to their targets, more recent data have demonstrated that the presence of a pTyr is not required for protein recognition (Kelley et al., 2002). Structural and biochemical studies from several proteins have demonstrated high-affinity binding of nonphosphorylated target sequences (Zhang et al., 1997; Dhalluin et al., 2000; Zwahlen et al., 2000). The presence of the phosphate moiety has no significant effect on the binding of a peptide from the amyloid precursor protein, βAPP bearing the sequence NPTY to the PTB domain from the protein X11 (Zhang et al., 1997). The PTB domain from *Drosophila* Numb protein has been reported to bind selectively ($K_D = 5.8 \mu M$) to proteins containing the YIGPYΨ (where Ψ is a hydrophobic residue; Li et al., 1997). However, only a slightly tighter interaction could be invoked on phosphorylation of the second tyrosine in the sequence ($K_D = 1.4 \mu M$)

The lack of specificity is highlighted in several observations of binding of PTB domains to distinctly different ligands, both phosphorylated and nonphosphorylated. For example, the PTB domain from fibroblast receptor substrate, FRS2, can bind with similar affinity to the NPXpY sequence (also recognized by Shc) as well as a sequence, AVHKLAKSIPLRRQVTVS, derived from the juxtamembrane region of the fibroblast growth factor receptor (Xu et al., 1998; Ong et al., 2000).

Whereas the binding of SH2 domains to phosphopeptides is dictated primarily by the presence of the pY moiety, the binding of PTB domains seems to rest largely in the propensity of the surrounding amino acid sequence to form a β-strand antiparallel to the $\beta5$ and the C-terminal α-helix of the PTB core β-sandwich. The pY moiety is thus only of secondary importance in PTB domain interactions.

3.4. Significance of the Specificity SH2 Domain and PTB Domain Interactions in Cellular Signaling

The description of the structures and interactions of both SH2 domains and PTB domains above reveals that these domains show levels of promiscuity in binding that would likely compromise the mutual exclusivity of a signal in vivo. The data for SH2 domains (in the accompanying Tables 8.1, 8.2, and 8.3) show only approximately two orders of magnitude difference in affinity between, so called "specific" interactions, and those with apparently random sequences. This rather limited specificity is further exemplified on looking at the effects of single mutations to the ligand (or the SH2 domain). In these data even dramatic changes in the SH2 domain–peptide ligand interface usually results in small differences in the overall affinity. As a result the whole issue of linear processing of tyrosine kinase–mediated signaling has to be questioned. In other words, the idea that a signal can be transduced by the sequential interactions of single proteins in a relay cannot be correct. As a result, other mechanisms for this type of signaling have to be invoked.

3.5. The Effect of Multiple Signaling Domains

It is quite apparent that many signaling proteins are comprised of more than one domain. For example, Src family proteins have one SH2 domain, one SH3 domain (which can bind to the same ligand, e.g., focal adhesion kinase [in the case of the Src family protein Fyn]), and a kinase domain. On the other hand Grb2 has one SH2 domain and two SH3 domains (which are known to both bind to the protein SOS). This common feature among the majority of these proteins suggests that the combined effects of multiple domain interactions could add an additional level to the specificity either through enhanced affinity and/or improved steric selectivity. Thus, an interaction that involves two domains recognizing two distinct sites on a cognate ligand could result in an additive effect toward the total free energy of the interaction, that is, increase the affinity. Futhermore, the juxtaposition of the two domains would have to be correct for docking onto the respective sites on the cognate ligand; otherwise the double binding could not occur. This steric effect would improve the selectivity since other competing proteins with similar domains may not be able to conformationally adjust the domains to the appropriate orientation. Investigation of this phenomenom has suggested that in the cases in which two domains are required for binding there is no additive effect of the free energy of binding (O'Brien et al., 2000; Arold et al., 2001). Indeed, in some cases the interaction of the two domains has an anticooperative effect. For example, the binding of the tandem SH2 domains of the p85 subunit of PI3-kinase to a peptide with the two appropriately positioned pY motifs (derived from a cognate protein ligand) binds with a change in free energy significantly below what would be expected from the addition of the interactions of the individual domains (O'Brien et al., 2000). It was hypothesized that this is a result of incurring an entropic penalty from the conformational change associated with having to orientate the domains appropriately for docking onto the ligand.

3.6. Different Modes of Signaling Required for Specificity

Assuming that the linear processing of tyrosine kinase–mediated signal transduction cannot explain the required level of specificity required to effect a distinct downstream response, other mechanisms have to be considered. There are several aspects of signaling that have largely been ignored in the general focus on linear pathway recognition. For example, there is evidence that the integrity of signaling could be based on more than one pathway. For example, the stimulation of a given receptor or series of receptors could result in more than one linear response. This "parallel" processing of a signal could ensure specificity in that the downstream response would occur only in the presence of the multiple pathways arriving at a given juncture in the cell, or producing a combined effect (from secondary messenger molecules) that could be sensed prior to gene transcription (Fambrough et al., 1999). This parallel processing model could even have evolved into a more complex network system in which there are a series of "junction proteins" forming checkpoints for the various pathways that gate downstream response (Dueber et al., 2003). In addition, the aspect of temporal control of the signaling process appears to be important. For example, the longevity of activation of a signal could control the ultimate downstream response (O'Rourke and Ladbury, 2003).

The understanding of tyrosine kinase–mediated signal transduction is still rather naïve; however, it is clear that advances in this area will help to unveil the intricate nuances of this fundamental biology and provide a roadmap for possible pharmaceutical interventions when these pathways lead to disease.

Acknowledgments

J.E.L. is a Wellcome Trust Senior Fellow. The help of Jonathan Taylor in preparation of the figures for this chapter is acknowledged.

REFERENCES

Arold, S.T., Ulmer, T.S., Mulhern, T.D., Werner, J.M., Ladbury, J.E., Campell, I.D., and Noble, M.E.M. (2001). The role of the Src homology 3–Src homology 2 interface in the regulation of the Src kinases. *J. Biol. Chem.* 276:17199–17205.

Blaikie, P., Immanuel, D., Wu, J., li, N., Yajnik, V., and Margolis, B. (1994) A region in Shc distinct from the SH2 domain can bind tyrosine-phosphorylated growth factor receptors. *J. Biol. Chem.* 269:32031–32034.

Bradshaw, J.M., and Waksman, G. (1999). Calorimetric examination of high-affinity Src SH2 domain-tyrosyl phosphopeptide binding: dissection of the phosphopeptide sequence specificity and coupling energetics. *Biochemistry* 38:5147–5154.

Bradshaw, J.M., Grucza, R.A., Ladbury, J.E., and Waksman, G. (1998). Probing the "two-pronged plug two-holed socket" model for the mechanism of binding of the Src SH2 domain to phosphotyrosyl peptides: a thermodynamic study. *Biochemistry* 37:9083–9090.

Bradshaw, J.M., Mitaxov, V., and Waksman, G. (1999). Investigation of phosphotyrosine recognition by the SH2 domain of the Src kinase. *J. Mol. Biol.* 293:971–985.

Bradshaw, J.M., Mitaxov, V., and Waksman, G. (2000). Mutational investigation of the specificity determining region of the Src SH2 domain. *J. Mol. Biol.* 299:521–535.

Campbell, K.S., Ogris, E., Burke, B., Su, W., Auger, K.R., Druker, B.J., Schaffhausen, B.S., Roberts, T.M., and Pallas, D.C. (1994) Polyoma middle tumor antigen interacts with SHC protein via NPTY (Asn-Pro-Thr-Tyr) motif in middle tumor antigen. *Proc. Natl. Acad. Sci. USA* 91:6344–6348.

Charifson, P.S., Shewchuk, L.M., Rocque, W., Hummel, C.W., Jordan, S.R., Mohr, C., Pacofsky, G.J., Peel, M.R., Rodriguez, M., Sternbach, D.D., and Consler, T.G. (1997). Peptide ligands of pp60(c-src) SH2 domains: A thermodynamic and structural study. *Biochemistry* 36:6283–6293.

Chung, E., Henriques, D., Renzoni, D., Zvelebil, M., Bradshaw, J.M., Waksman, G., Robinson, C.V., and Ladbury, J.E. (1998). Mass spectrometric and thermodynamic studies reveal the role of water molecules in complexes formed between SH2 domains and tyrosyl phosphopeptides. *Struct. Fold. Des.* 6:1141–1151.

Davidson, J.P., Lubman, O., Rose, T., Waksman, G., and Martin, S.F. (2002). Calorimetric and structural studies of 1,2,3-trisubstituted cyclopropanes as conformationally constrained peptide inhibitors of Src SH2 domain binding. *J. Am. Chem. Soc.* 124:205–215.

Dhalluin, C., Yan, K., Plotnikova, O., Lee, K.W., Zeng, L., Kuti, M., Mujtaba, S., Goldfarb, M.P., and Zhou, M.M. (2000). Structural basis of SNT PTB domain interactions with distinct neurotrophic receptors. *Mol. Cell.* 6:921–929.

Dilworth, S.M., Brewster, C.E., Jones, M.D., Lanfrancome, L., Pelicci, G., and Pelicci, P.G. (1994). Transformation by polyoma virus middle T-antigen involves the binding and tyrosine phosphorylation of Shc. *Nature* 367:87–90.

Dueber, J.E., Yeh, B.J., Chak, K., and Lim, W.A. (2003). Reprogramming control of an allosteric switch through modular recombination. *Science* 301:1904–1908.

Eck, M.J., Dhe-Paganon, S., Trüb, T., Nolte, R., and Shoelson, S.E. (1996). Structure of the IRS-1 PTB domain bound to the juxtamembrane region of the insulin receptor. *Cell* 85:695–705.

Fambrough, D., McClure, K., Kazlauskas, A., and Lander, E.S. (1999) Diverse signaling pathways activated by growth factor receptors indue broadly overlapping, rather than independent, sets of genes. *Cell* 97:727–741.

Farrow, N.A., Muhandiram, R., Singer, A.U., Pascal, S.M., Kay, C.M., Gish, G., Shoelson, S.E., pawson, T., Forman-Kay, J.D., and Kay, L.E. (1994). Backbone dynamics of a free and a phosphopeptide-complexed Src homology 2 domain studied by ^{15}N NMR relaxation. *Biochemistry* 33:5984–6003.

Gustafson, T.A., He, W., Craparo, A., Schaub, C.D., and O'Neill, T.J. (1995). Phosphotyrosine-dependent interaction of SHC and insulin receptor substrate 1 with the NPEY motif of the insulin receptor via a novel non-SH2 domain. *Mol. Cell Biol.* 15:2500–2508.

Henriques, D.A., and Ladbury, J.E. (2001). Inhibitors to the Src SH2 domain: a lesson in thermodynamic structural correlation in drug design. *Arch. Biochem. Biophys.* 390:158–168.

Henriques, D.A., Ladbury, J.E., and Jackson, R.M. (2000). Comparison of binding energies of SrcSH2-phosphotyrosyl peptides with structure-based prediction using surface area based empirical parameterization. Prot. Sci. 9:1975–1985.

Hunter, T. (2000). Signalling—2000 and Beyond. *Cell* 100:113–127.

Kavanaugh, W.M., and Williams, L.T. (1994). An alternative to SH2 domains for binding tyrosine-phosphorylated proteins. *Science* 266:1862–1865.

Kimber, M.S., Nachman, J., Cunningham, A.M., Gish, G.D., Pawson, T., and Pai, E.F. (2000). Structural basis for specificity switching of the Src SH2 domain. *Mol. Cell* 5:1043–1049.

Kuriyan, J., and Cowburn, D. (1993). Structures of SH2 and SH3 domains. *Curr. Opin. Struct. Biol.* 3:828–837.

Kuriyan, J., and Cowburn, D. (1997). Modular peptide recognition domains in eukaryotic signalling. *Annu. Rev. Biophys. Biomol. Struct.* 26:259–288.

Ladbury, J.E., and Arold, S. (2000). Searching for specificity in SH domains. *Chem. and Biol.* 7:R3–R8.

Ladbury, J.E., Wright, J.G., Sturtevant, J.M., and Sigler, P.B. (1994). A thermodynamic study of the *trp* repressor-operator interaction. *J. Mol. Biol.* 238:669–681.

Ladbury, J.E., Lemmon, M.A., Zhou, M., Green, J., Botfield, M.C., and Schlessinger, J. (1995). Measurement of the binding of tyrosyl phosphopeptides to SH2 domains—a reappraisal. *Proc. Natl. Acad. Sci. USA.* 92:3199–3203.

Ladbury, J.E., Hensmann, M., Panayotou, G., and Campbell, I.D. (1996). Alternative modes of tyrosyl phosphopeptide binding to a Src family SH2 domain: implications for regulation of tyrosine kinase activity. *Biochemistry* 35:11062–11069.

Lemmon, M.A., and Ladbury, J.E. (1994). Thermodynamic studies of tyrosyl-phosphopeptide binding to the Sh2 domain of P56(Lck). *Biochemistry* 33:5070–5076.

Li., C., Songyang, Z., Vincent, S.J., Zwahlen, C., Wiley, S., Cantley, L. Kay, L.E., Forman-Kay, J., and Pawson, T. (1997). High-affinity binding of the Drosophila Numb phosphotyrosine-binding domain to peptides containing a Gly-Pro-(p)Tyr motif. *Proc. Natl. Acad. Sci. USA* 94:7204–7209.

Lubman, O.Y., and Waksman, G. (2002). Dissection of the energetic coupling across the Src SH2 domain-tyrosyl phosphopeptide interface. *J. Mol. Biol.* 316:291–304.

Lubman, O.Y., and Waksman, G. (2003). Structural and thermodynamic basis for the interaction of the Src SH2 domain with the activated form of the PDGF beta- receptor. *J. Mol. Biol.* 328:655–668.

Maina, F., Casagranda, F., Andero, E., Simeone, A., Comoglio, P.M., Klein, and Ponzetto, G., (1996). Uncoupling of Grb2 from the Met receptor in vivo reveals complex roles in muscle development. *Cell* 87:531–542.

McNemar, C., Snow, M.E., Windsor, W.T., Prongay, A., Mui, P., Zhang, R.M., Durkin, J., Le, H.V., and Weber, P.C. (1997). Thermodynamic and structural analysis of phosphotyrosine polypeptide binding to Grb2-SH2. *Biochemistry* 36:10006–10014.

Meng, W., Sawasdikosol, S., Burakoff, S.J., and Eck (1999). Structure of the amino-terminal domain of Cbl complexed to its binding site on ZAP-70 kinase. *Nature* 398:84–90.

Merengere, L.E.M., Songyang, Z., Gish, G.D., Schaller, M.D., Parsons, J.T., Stern, M.J., Cantley, L.C., and Pawson, T. (1994). SH2 domain specificity and activity modified by a single residue. *Nature* 369:502–505.

Obermeier, A., Bradshaw, R.A., Seedorf, K., Choidas, A., Schlessinger, J., and Ullrich, A. (1994). Neuronal differentiation signals are controlled by nerve growth factor receptor/Trk binding sites for SHC and PLCγ. *EMBO J.* 13:1585–1590.

O'Brien, R., Rugman, P., Renzoni, D., Layton, M., Handa, R., Hilyard, K., Waterfield, M.D., Driscoll, P.C., and Ladbury, J.E. (2000). Alternative modes of binding of proteins with tandem SH2 domains. *Protein Sci.* 9:570–579.

Ong, S.H., Guy, G.R., Hadari, Y.R., Laks, S., Gotoh, N., Schlessinger, J. and Lax, I. (2000). FRS2 proteins recruit intercellular signalling pathways by binding to diverse targets on fibroblast growth factor and nerve cell growth factor receptors. *Mol. Cell. Biol.* 20:979–989.

O'Rourke, L., and Ladbury, J.E. (2003). Specificity iscomplex and time consuming: mutual exclusivity in tyrosine kinase-mediated signalling. *Acc. Chem. Res.* 36:410–416.

Pascal, S.M., Singer, A.U., Gish G., Yamazaki, T., Shoelson, S.E., Pawson, T., Kay, L.E., and Farman-Kay, J.D. (1994). Nuclear magnetic resonance structure of an SH2 domain of phospholipase C-gamma-1 complexed with a high affinity binding peptide. *Cell* 77:461–472.

Pascal S.M., Yamazaki, T., Singer, A.U., Kay, L.E., and Forman-Kay, J.D. (1995). Structural and dynamic characterization of the binding region of a Src homology domain-phosphopeptide complex by NMR relaxation, proton exchange, and chemical shift approaches. *Biochemistry* 34:11353–11362.

Pawson T., and Nash, P. (2000) Protein–protein interactions define specificity in signal transduction. *Genes and Dev.* 14:1027–1047.

Pawson T., Gish, G.D., and Nash, P. (2001). SH2 domains, interaction modules and cellular wiring. *Trends Cell Biol.* 11:504–511.

Rahuel J., Gay, B., Erdmann, D., Strauss, A., Garcia-Echeverria, C., Furet, P., Caravatti, G., Fretz, H., Schoepfer, J., and Grtter, M.G. (1996). Structural basis for specificity of Grb2-SH2 revealed by a novel ligand binding mode. *Nat. Struct. Biol.* 3:586–589.

Renzoni, D.A., Zvelebil, M.J., Lundback, T., and Ladbury, J.E. (1997). Exploring uncharted waters: water molecules in drug design strategies. In *Structure-based drug design: thermodynamics, modelling and strategy* (Ladbury, J.E. & Connelly, P.R., eds.), Vol. Chapter 6.

Schlessinger, J. (2000). Cell signalling by receptor tyrosine kinases. *Cell* 103:211–225.

Schlessinger, J., and Lemmon, M.A. (2003). SH2 and PTB domains in tyrosine kinase signaling. *Science STKE* 191:1–9.

Songyang, Z., Shoelson, S.E., Chaudhuri, M., Gish, G., Pawson, T., Haser, W.G., King, F., Roberts, T., Ratnofsky, S., Lechleider, R.J., Neel, B.G., Birge, R.B., Fajardo, J.E., Chou, M.M., Hanafusa, H., Schaffhausen, B., and Cantley, L.C. (1993). SH2 domains recognize specific phosphopeptide sequences. *Cell* 72:767–778.

Stolt, P.C., Jeon, H., Song, H.K., Herz, J., Eck, M.J., and Blacklow, S.C. (2003) Origins of peptide selectivity and phosphoinositide binding revealed by structures of disabled-1 PTB domain complexes. *Structure* 11:569–579.

Sun, X.J., Wang, L.M., Zhang, Y., Yenush, L., Myers M.G., Jr., Glasheen, E., Lane, W.S., Pierce, J.H., and White, M.F. (1995). Role of IRS-2in insulin and cytokine signalling. *Nature* 377:173–177.

Tessari, M., Gentile, L.N., Taylor, S.J., Shalloway, D.I., Nicholson, L.K., and Vuister, G.W. (1997). Heteronuclear NMR studies of the combined Src homology domains 2 and 3 of pp60-c-src: effects of phosphopeptide binding. *Biochemistry* 36:14561–14571.

Waksman, G., Kominos, D., Robertson, S.C., Pant, N., Baltimore, D., Birge, R.B., Cowburn, D., Hanafusa, H., Mayer, B.J., Overduin, M., Resh, M.D., Rios, C.B., Silverman, L., and Kuriyan, J. (1992). Crystal structure of the phosphotyrosine recognition domain SH2 of *v-src* complexed with tyrosine-phosphorylated peptides. *Nature* 358:646–653.

Waksman, G., Shoelson, S.E., Pant, N., Cowburn, D., and Kuriyan, J. (1993). Binding of a high affinity phosphotyrosyl peptide to the Src SH2 domain: Crystal structures of the complexed and peptide-free forms. *Cell* 72:779–790.

Williams, J.C., Weijland, A., Gonfloni, S., Thompson, A., Courtneidge, S.A., Superti-Furga, G., and Wierenga, R.K. (1997). The 2.35 Å crystal structure of the inactivated form of chicken Src: a dynamic molecule with multiple regulatory interactions. *J. Mol. Biol.* 274:757–775.

Xu, H., Lee, K.W. and Goldfarb, M. (1998). Novel recognition motif on fibroblast growth factor receptor mediates direct association and activation of SNT adaptor proteins. *J. Biol. Chem.* 273:17987–17990.

Xu, R.X., Word, J.M., Davis, D.G., Rink, M.J., Willard, D.H., Jr., and Gampe, R.T. (1995). Solution structure of the human pp60^{c-src} SH2 domain complexed with a phosphorylated tyrosine pentapeptide. *Biochemistry* 34:2107–2121.

Yan, K.S., Kuti, M., and Zhou, M.M. (2002). PTB or not PTB—that is the question. *FEBS Lett.* 513:67–70.

Zhang, Z., Lee, C.-H., Mandiyan, V., Borg, J.-P., Margolis, B., Schlessinger, J., and Kuriyan, J. (1997). Sequence-specific recognition of the internalization motif of the Alzheimer's amyloid precursor protein by the X11 PTB domain. *EMBO J* 16:6141–6150.

Zhou, M.M., Ravichandran, K.S., Olejniczak, E.F., Petros, A.M., Meadows, R.P., Sattler, M., Harlan, J.E. ,Wade, W.S., Burakoff, S.J., and Fesik, S.W. (1995). Structure and ligand recognition of the phosphotyrosine binding domain of Shc. *Nature* 378:584–592.

Zwahlen, C., Li, S.C., Kay, L.E., Pawson, T., and Forman-Kay, J.D. (2000). Multiple modes of peptide recognition by the PTB domain of the cell fate determinant Numb. *EMBO J.* 19:1505–1515.

9

Competitive Binding of Proline-Rich Sequences by SH3, WW, and Other Functionally Related Protein Domains

Marius Sudol and Mark T. Bedford

ABSTRACT

Protein domains or modules are families of small (35 to 100 amino acids), conserved globular folds that bind DNA, RNA, phosphoinositides, and protein motifs (≈ 10 amino acids). A subset of at least five different domain types possesses the ability to bind proline-rich sequences; these include SH3, WW, EVH1, GYF, and UEV domains. Some of these domain types recognize the same or overlapping proline-rich motifs, thus generating competitive pressure for motif binding. In addition, the phosphorylation and methylation of residues within a proline-rich motif can regulate domain binding. Here, we highlight which domains likely compete for the same ligands, and how some of these interactions are regulated by posttranslational modifications.

MARIUS SUDOL • Department of Medicine, Mount Sinai School of Medicine, One Gustave Levy Place, New York, NY 10029, USA. Current Affiliation • Weis Center for Research, Geisinger Clinic, Danville, PA 17822, USA
MARK T. BEDFORD • Department of Carcinogenesis, M.D. University of Texas Anderson Cancer Center, Smithville, TX, 78957, USA.

Proteomics and Protein–Protein Interactions: Biology, Chemistry, Bioinformatics, and Drug Design, edited by Waksman. Springer, New York, 2005.

1. INTRODUCTION

It has been more than a decade since the discovery that families of protein do-mains bind short, recognizable motifs within their ligands (Anderson et al., 1990). Since that time it has become a scientific mind-set that protein domains play key roles in all aspects of cell function, including signal transduction, the structural integrity of the cell, vesicular trafficking, and chromatin remodeling. After the Src homology 2 (SH2) domain (Sadowski et al., 1986), the SH3 domain was the second modular protein–protein interaction unit to be recognized (Lehto et al., 1988; Mayer et al., 1988; Stahl et al., 1988), and the first modular domain identified to display affinity to-ward proline-rich motifs (Ren et al., 1993). The WW domain (a domain harboring two conserved tryptophan residues) was the second structurally distinct domain identified that can bind proline-rich ligands (Bork and Sudol, 1994; Chen and Sudol, 1995). Both these early studies used recombinant SH3 and WW domains to screen expression libraries for interacting partners, which turned out to be proline rich. When reciprocal experiments were performed, using a proline-rich sequence as probes, it became clear that subsets of SH3 and WW domains could bind the same short stretches of amino acids (Chan et al., 1996). In this chapter we focus on the specificity of binding of WW and SH3 domains, the ability of these structurally distinct domains (and others) to recognize common proline-rich motifs, and the ways in which these interactions are regulated.

2. PROLINE-RICH SEQUENCES ACT AS RECOGNITION MOTIFS FOR PROTEIN–PROTEIN INTERACTIONS

The importance of proline-rich regions is clearly demonstrated with the recent availability of an abundance of proteomic information, derived from various genome-sequencing projects—proline-rich regions are among the most common motifs identi-fied (Zarrinpar et al., 2003). Indeed, in the *Drosophila* proteome proline-rich regions represent the most common sequence motifs (Rubin and Lewis, 2000) and in the predicted proteomes of *Caenorhabditis elegans* and human genomes, proline-rich regions are also widely distributed (Hu et al., 2003). The properties of this distinctive residue have been harnessed by evolution, spawning multiple proline-based recogni-tion domains.

2.1. Biophysical Properties of Proline-Rich Sequences

Proline differs from the other amino acids in that it has an aliphatic side chain that is bound to both the nitrogen and α-carbon atoms. This cyclic side chain imposes conformational constraints that markedly influence protein structure. Proline residues often generate a kink in a protein fold. As a consequence of these unique properties of proline, a nonrepetitive proline-rich regions adopt a type II polyproline (PPII) helix. A PPII helix is a left-handed helix composed of three residues per turn. This structure resembles a triangular prism with the proline residues on the same face and the side chains and backbone carbonyls are often displayed in pseudo-symmetry, which permits this structure to be bound in two possible orientations. The carbonyl and amide groups within a PPII helix are available for hydrogen bonding and the prolines form a ridged hydrophobic patch. An analysis of a group of more than 250 nonhomologous protein structures reveals that more than half contain one or more PPII helix (Stapley and Creamer, 1999). Thus, PPII helices are common structural motifs that allow sequence-specific recognition by protein interaction domains.

2.2. Domains that Bind Proline-Rich Motifs

2.2.1. SH3 Domains

The first domain type to be identified among the now known superfamily of proline binding modules was the SH3 domain. The domain was noted as a region of similarity between Src family tyrosine kinases and other signaling proteins, hence its name SH3—Src Homology 3 region (Mayer, 2001). The SH3 domains are approximately 60 residues long and form a well-defined structure composed of β-strands (Macias et al., 2002). The domain was shown to bind proline-rich ligands with a general core: PxxP, where x denoted any amino acid (Ren et al., 1993). Subsequently, the basic core was refined further into PpΨP consensus, where p meant most frequently proline and Ψ denoted aliphatic amino acid (Kay et al., 2000; Aasland et al., 2002). Soon after, the flanking sequences of the basic core were also defined and the role of positively charged amino acids such as arginine (R) and lysine (K) became evident. The Rs and Ks found in the flanks dictated specificity and the orientation of the ligand with respect to the binding pocket of the domain (reviewed in Cesareni et al., 2002). The position of the R/K residue(s) relative to the proline-rich tract of SH3 domain ligands is used to classify these domains into two larger classes. Class I SH3 domains bind ligands harboring RxxPxxP motifs and Class II SH3 domains proteins with PxxPxR motifs. The Abl SH3 domain does not require the presence of a charged residue (R or K) from its ligands; instead it often harbors a leucine residue within or close to the PxxP motif (Musacchio, 2002). The SH3 domain, through its presence in several major adaptor proteins, participates in many important signaling routes (e.g., role of Grb2 in receptor signaling) and its regulatory role in maintaining autoinhibitory conformation of Src and Abl kinases is a convincing testimony to the functional plasticity of the module (Zarrinpar et al., 2003).

2.2.2. WW Domains

WW domains are small modular units (35 to 40 amino acids) that bind a variety of proline-rich sequences (Bork and Sudol, 1994; Chen and Sudol, 1995; Chan et al., 1996). This domain type is characterized by the presence of two highly conserved tryptophan residues that are spaced 20 to 22 amino acids apart. A broad spectrum of proteins harbor WW domains, including splicing factors, transcription factors, E3 ubiquitin ligases, dystrophin, and prolyl isomerases. A demonstration of the specificities of proline-rich peptide recognition can be visualized in Figure 9.1. Here different proline-rich peptides bind different groups of WW domains with very little

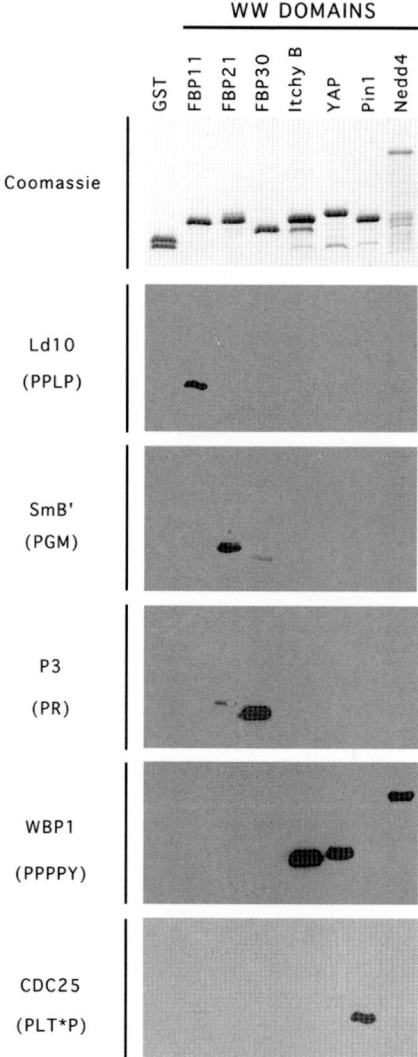

Figure 9.1. WW domain binding specificity. Blot overlay analysis of a set of different WW domains. GST alone or GST fusion proteins containing the WW domains of FBP11, FBP21, FBP30, YAP, Pin1, Nedd4, and Itch 3 and 4 were separated by SDS-PAGE. Six identical gels were run. One gel was Coomassie-stained as a loading control and the other five gels were transfered onto PVDF membranes. These filters were probed with the indicated biotinylated peptides. The presence of the bound peptides was visualized using streptavidin-horseradish peroxidase, followed by chemiluminescence. The peptide sequences are: Ld10—biotin-SGSGAPPTPPPLPP; SmB′—biotin-PPGMRPPPPGMRRGPPPPGMRPPRP; P3—biotin-GVSVRGRGAAPPPPP-VPRGRGVGP; WBP1—biotin-SGSGGTPPPPYTVG; CDC25—biotin-SGSGEQPLT*PVTDL (T* = phosphothreonine).

cross-recognition. The nature of the proline-rich ligands that bind specifically to subsets of WW domains is used to classify these subsets (Espanel and Sudol, 1999; Bedford et al., 2000b). Following this cataloging format, group I WW domains recognize a PPxY binding motif, group II WW domains bind PPLP motifs, group III WW domains ligand that harbor a PPR motif, and group IV WW domains bind a motif that contains a proline residue preceded by a phosphoserine or phosphothreonine residue (S*/T*-P).

Recently a WW domain interaction map was generated for the human proteome (Hu et al., 2003). Fifty-seven human WW domains expressed as (glutathione S-transferase) GST fusions were probed with almost 2000 proline-rich peptides containing cores recognized by three major classes of WW domains (Sudol and Hunter, 2000) and each peptide corresponding to the known protein or open reading frame (ORF) in the human proteome. A network of more that 69,000 interactions was deciphered and it now serves as a blue print for identification of new signaling pathways that utilize WW domains.

2.2.3. EVH1 Domains

EVH1 domains (Ena/VASP homology 1) are present in two profilin binding proteins, VASP and Mena, and play a role in actin cytoskeleton remodeling (Niebuhr et al., 1997). EVH1 domains are also found in the Wiskott–Aldrich syndrome (WASP) family of proteins that regulate actin polymerization (Volkman et al., 2002). The motif recognized by the EVH1 domain of Mena is DFPPPPT (Niebuhr et al., 1997) and the EVH1 domain from WASP recognizes a similar motif (DLPPPEP) (Volkman et al., 2002). As with profilin, SH3 and WW domains, EVH1 domains bind their ligands in both N- and C-terminal orientations and targets are classified as either class I (FPxxP) or class II (PPxxF) ligands accordingly (Ball et al., 2002).

2.2.4. GYF Domains

The GYF domain was originally found in the T-cell signaling protein, CD2BP2. These domains are named after a highly conserved signature Gly-Tyr-Phe motif within the domain core (Nishizawa et al., 1998). This domain is a unique fold that binds a PPPPGHR motif in the cytoplasmic tail of CD2 cell surface receptor (Freund et al., 1999). Twenty-six GYF domain-containing proteins have been identified in eukaryotic proteins and it is likely that CD2BP2 harbors a subclass of GYF domain (Freund et al., 2002). Further studies with the other classes of GYF domains will likely identify ligands with variant proline-rich motifs.

2.2.5. UEV Domains

The REPTAPP motif (the late region) of the HIV Gag protein adopts a PPII helical conformation (Pornillos et al., 2002a) and interacts with the N-terminal ubiquitin E2 variant (UEV) domain of Tsg101 (Garrus et al., 2001). The Tsg101 protein is involved in vesicular sorting and is localized to the endosomal membrane through an interaction with a PSAP motif within the cellular endosomal protein, Hrs (Pornillos et al., 2003). The UEV domain of Tsg101 also binds ubiquitin, although this interaction does not prevent binding to its proline-rich ligands (Pornillos et al., 2002b). Although the UVE domain is not found in many proteins it does fold into a novel structure that resembles the binding pockets of SH3 and WW domains.

2.2.6. Profilin

Profilin is an actin binding protein that regulates actin polymerization. In addition to binding actin, profilin also interacts with proline-rich regions in its ligands through a patch of aromatic amino acids on its surface (Mahoney et al., 1997). The proline-rich binding region of profilin is not a classic protein interacting domain, in that it is not a modular unit found in many proteins, it is, however, a prominent binder of proline-rich proteins. Profilin, like WW, SH3, and EVH1 domains, is able to bind proline-rich ligands in both amide backbone orientations (Mahoney et al., 1999). The recognition motif for profilin is xPPPP, where x is often a leucine residue but can also be isoleucine, serine, or glycine (Holt and Koffer, 2001). This recognition motif is similar to that bound by subsets of WW and SH3 domains (Bedford et al., 1997).

SH3 and WW domains have emerged as the primary two families of proline-rich binding domains. In humans there are upwards of 300 SH3 domains and 100 WW domains. These estimates reflect the current number of the SH3 and WW domains found in public databases supplemented by the numbers resulting from mining of the human proteome with more advanced but also in some cases relaxed algorithms.

3. STRUCTURAL COMPARISON OF PROLINE-RICH BINDING DOMAINS

High-resolution crystal structures are available for representative members of all the above-mentioned proline recognition domains. Moreover, several comprehensive reviews have been published recently discussing the structures of the domain–ligand complexes, identifying common structural denominators and molecular mechanisms of the complex formation (Kay et al., 2000; Macias et al., 2002; Zarrinpar et al., 2003). As with other modular domains, the proline binding domains have their N- and C-termini located in close proximity, allowing the domain to be "inserted" into polypeptide chain of the host protein while being exposed and available for interaction with target proteins. The most prominent feature of all proline binding domains is the presence of conserved and exposed clusters of aromatic residues. Actually, in two instances these conserved aromatics even prompted the coining of the domain's name: WW, GYF (Bork and Sudol, 1994; Freund et al., 1999). One or more of the conserved aromatic side chains of the domain directly participate in molecular interactions with prolines of cognate ligands (Macias et al., 2002; Zarrinpar et al., 2003). Examining the interface of the domain–ligand complexes, we observe a landscape that is generally hydrophobic with sandwiches between proline rings of the ligands and aromtics rings of the domains. In an apparent contrast, yet in concert within this hydrophobic "milieu" there is at least one dominant hydrogen bond formed by the carbonyl group of the proline and a proton-donating aromatic ring—frequently from the conserved tryptophan as in complexes of SH3 or WW domains with their ligands (Huang et al., 2000). The dominance of the "proline-aromatic" hydrogen bond stems also from one of the unique features of proline having a carbonyl group that is the most electron rich among all amino acids and therefore having strong propensity for hydrogen bond formation (Hagerman and Butler, 1981). The concerted arrangement of the very hydrogen bond between proline and aromatic residue and the hydophobic stacks of proline and aromatic rings were noticed as common structural arrangements in SH3, WW, EVH1 domains, and profilin protein and named an "aromatic cradle"(Huang et al., 2000).

More synthetic view of proline binding domains is known as "xP" hypothesis of Lim and colleagues (Zarrinpar et al., 2003) proposing that aromatics on the binding surface of domains form a groove (or grooves) into which "xP" dipeptide of the ligand fits snuggly. The "xP" dipeptide (where x is often a hydrophobic amino acid and P is proline) forms a continuous ridge built from the C-substituted residue followed by the N-substituted proline. With the exception of EVH1 domain where the "xP" groove is not clearly visible, other proline binding domains contain distinct "xP" grooves as major sites engaged in ligand binding.

As always, Nature is complex and a small number of proline binding modules evolved to bind noncanonical motifs (Mongiovi et al., 1999) or even peptide motifs without prolines (Berry et al., 2002). However, these are exceptions to general rules.

In sum, numerous structural studies increased our understanding of molecular aspects of proline binding domains. This knowledge is being harnessed to develop small nonpeptide inhibitors (Oneyama et al., 2002, 2003) aimed at modulation of the complexes in experimental settings that frequently represent cell culture and animal models of human diseases.

4. THE SAME PROLINE-RICH MOTIFS RECOGNIZED BY DIFFERENT DOMAINS

Nonproline residues within proline-rich motifs dictate the specificity and orientation of domain binding. It is often the case that proline-rich sequences can harbor superimposed domain binding motifs. This phenomenon was first observed with a 10-amino-acid proline-rich stretch, termed Ld10, from the protein formin (Chan et al., 1996). The Ld10 peptide can bind both the Abl SH3 domain and the FBP11 WW domains. Alanine scanning experiments demonstrated that the key residues within this 10-amino-acid ligand were not the same for the SH3 and the WW domain. Instead, two overlapping motifs within this short proline-rich ligand were exposed. Further expression library screens with the WW domains of FBP11 identified multiple protein ligands that are also bound by the Abl SH3 domain, and the relative equilibrium binding constants of the two domains for the same ligands are similar (Bedford et al., 1997). Thus the Abl SH3 domain and group II WW domains both bind PPLP motifs.

Group III WW domains bind proline-rich ligands that are peppered with arginine residues (Bedford et al., 1998, 2000b). Arginine (and in some instances lysine) residues on the edge of a proline-rich sequence are also critical for Src-like SH3 domain binding to their ligands (Musacchio, 2002). Proline-rich ligands use the flanking arginine/lysine residue to orientate themselves in WW and SH3 domain binding grooves. The requirement for arginine is different for the two domain types: WW domains use the charged nature of the arginine residue for binding (Bedford et al., 2000b), and SH3 domains require the arginine residue to generate a key salt bridge with a highly conserved aspartic acid or glutamic acid residue within the domain (Weng et al., 1995; Arold et al., 1997). Expression library screening revealed that a subset of proline-rich proteins arc bound by both group III WW domains and Src-like SH3 domain (Bedford et al., 2000b). One such common ligand is the c-Src kinase substrates, Sam68 (Fumagalli et al., 1994; Richard et al., 1995). Isolated proline-rich sequences form Sam68 are bound by both SH3 (Src-like) and WW domains (Group III) (Bedford et al., 2000a). By probing arrayed protein domains with proline-rich peptides it is apparent that SH3 and WW domains share ligands (Fig. 9.2).

The actin binding protein profilin interacts with the proline-rich region of the Ena/VASP family of proteins (Gertler et al., 1996). SH3 and WW domains (Gertler et al., 1995; Ermekova et al., 1997) also bind the same proline-rich region. This raises the possibility of competitive binding, and indeed this is the case. Increasing concentrations of profilin can compete off n-Src SH3 domain binding to this region in vitro (Lambrechts et al., 2000). These competitive interactions are further regulated by protein kinase A (PKA) phosphorylation of EVL, which results in the loss of Abl and nSrc SH3 domain binding but has no effect on profilin binding (Lambrechts et al., 2000).

Yet another example of different domains binding the same proline-rich stretch are the GYF domain of CD2BP2 and the SH3 domain of Fyn, which can both bind a motif in the cytoplasmic tail of the T-cell adhesion molecule, CD2 (Freund et al., 2002). Although these two distinct domains compete for the same CD2 binding site in vitro, it seem unlikely to happen in vivo. This is because CD2BP2 and Fyn are localized in different subcellular compartments. Fyn probably both bind CD2 in lipid rafts and CD2BP2 is the cytosolic binding partner for CD2 (Freund et al., 2002).

Finally, the tyrosine residue that is part of the group I WW domain recognition motif can be phosphorylated within certain contexts. Such a phosphorylation event can simultaneously prevent WW domain binding and facilitate SH2 domain binding. An example of this type of regulation is seen with the tyrosine phosphorylation of a PPxY motif in the C-terminal tail of β-dystroglycan (Sotgia et al., 2001), which prevents interactions with the WW domain of dystrophin and utrophin (Ilsley et al., 2001) and stimulates binding with a number of SH2 domain-containing proteins, including Src, Fyn, Csk, Nck, and Shc (Sotgia et al., 2001).

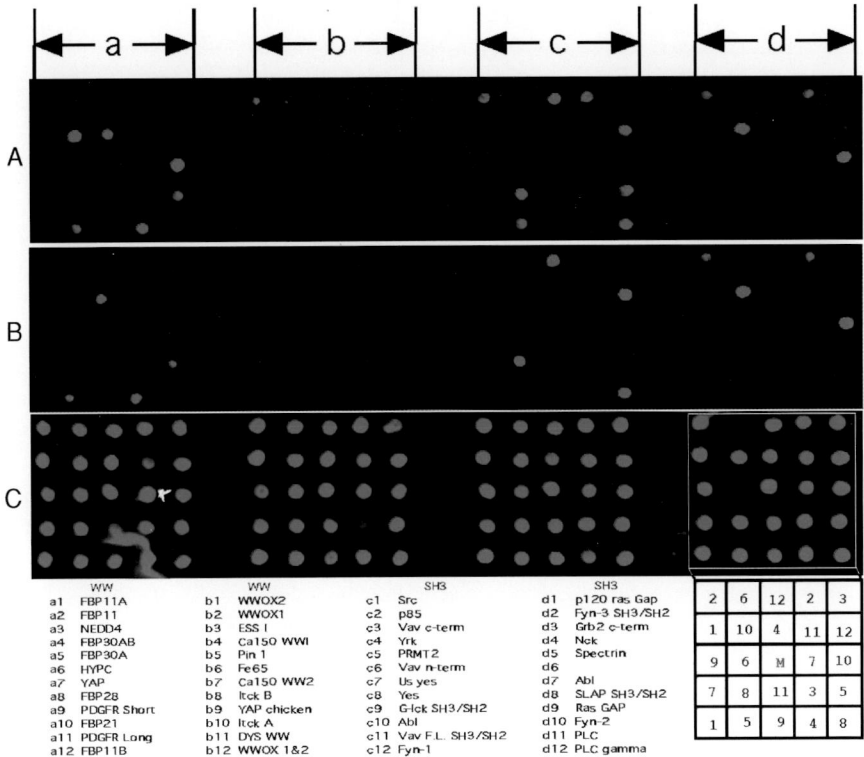

WW		WW		SH3		SH3						
a1	FBP11A	b1	WWOX2	c1	Src	d1	p120 ras Gap	2	6	12	2	3
a2	FBP11	b2	WWOX1	c2	p85	d2	Fyn-3 SH3/SH2	1	10	4	11	12
a3	NEDD4	b3	ESS I	c3	Vav c-term	d3	Grb2 c-term					
a4	FBP30AB	b4	Ca150 WWI	c4	Yrk	d4	Nck	9	6	M	7	10
a5	FBP30A	b5	Pin 1	c5	PRMT2	d5	Spectrin					
a6	HYPC	b6	Fe65	c6	Vav n-term	d6		7	8	11	3	5
a7	YAP	b7	Ca150 WW2	c7	Us yes	d7	Abl					
a8	FBP28	b8	Itck B	c8	Yes	d8	SLAP SH3/SH2	1	5	9	4	8
a9	PDGFR Short	b9	YAP chicken	c9	G-Ick SH3/SH2	d9	Ras GAP					
a10	FBP21	b10	Itck A	c10	Abl	d10	Fyn-2					
a11	PDGFR Long	b11	DYS WW	c11	Vav F.L. SH3/SH2	d11	PLC					
a12	FBP11B	b12	WWOX 1&2	c12	Fyn-1	d12	PLC gamma					

Figure 9.2. Proline and arginine-rich peptide sequences bind both WW and SH3 domains. GST fusion proteins were arrayed in duplicate on nitrocellulose slides. Protein domain microarrays were probed with two different proline and arginine-rich peptides. (**A**) The array was probed with a Cy3-labeled peptide from the splicing factor, SmB′ (biotin-PPGMRPPPPGMRRGPPPPGMRPPRP). (**B**) The array was probed with a Cy3-labeled peptide from the signaling molecule, Sam68 (biotin-GVSVRGRGAAPPPPPVPRGRGVGP). (**C**) The array was probed with an anti-GST primary antibody and detected with a FITC-conjugated secondary antibody. A key to the arrayed domains is displayed below. 24 WW domains (*a, b*) and 23 SH3 domains (*c, d*) are arrayed. Both peptides (**A, B**) bind both domain types.

The competitive nature of many of these interactions is depicted in a proline-rich peptide binding wheel (Fig. 9.3).

5. REGULATION OF WW AND SH3 BINDING TO SUBSTRATES

The posttranslational modifications within SH3 and WW domains, as well as the modification within the vicinity of domain binding motifs, has proved to be a way of regulating interactions with proline-rich regions. Unlike SH2 domains that demonstrate phosphotyrosine-dependent ligand binding, modifications of SH3 domains, WW domains, and their targets generally interfere with protein–protein interactions—group IV WW domains are the one exception to this rule.

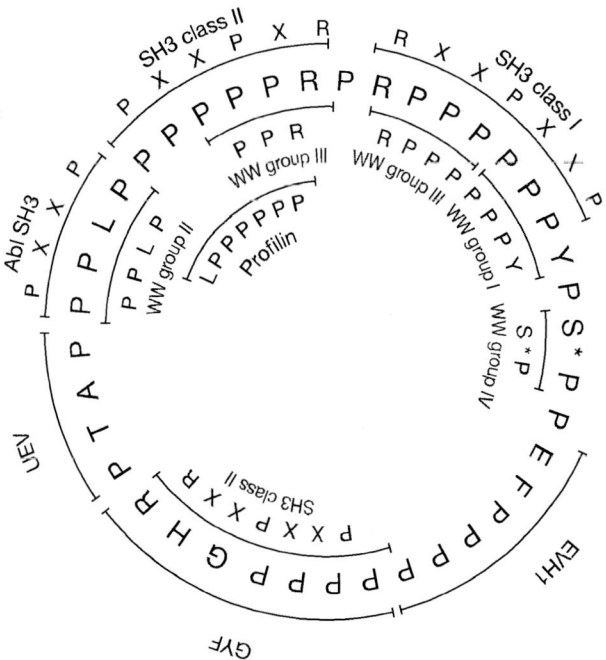

Figure 9.3. Proline-rich peptide binding wheel. The ability of different domains to bind similar or over-lapping proline-rich motifs is depicted in a circular design.

5.1. Serine/Threonine Phosphorylation of Group IV WW Domain Ligands

A small number of protein–protein interaction modules bind protein motifs in serine/threonine phosphorylation-dependent fashion. They include Forkhead-associated (FHA) domain, polo-box domains, and group IV WW domains (14-3-3s proteins also fall into this category even though they are not domains as such). The prolyl isomerase Pin1 harbors the best-characterized phospho-dependent motif binding WW domain (Lu et al., 1996). Proline-directed protein kinases phosphorylate serine or threonine residues that precede proline (S^*/T^*–P). These kinases play an important role in cell cycle progression. Pin1 is a peptidyl-prolyl *cis/trans* isomerase that binds phosphorylated ligands through its WW domain and then isomerizes the same S^*/T^*–P bond. This results in a structural and concomitant functional change of the ligand. Protein ligands that are bound and regulated by Pin1 include the mitotic phosphatases Cdc25 (Zhou et al., 2000), the transcription factor nuclear factor of activated T cells (NFAT) (Liu et al., 2001), the protein kinases never in Mitosis A (NIMA) and CK2 (Lu et al., 1996; Messenger et al., 2002), and p53 (Zacchi et al., 2002; Zheng et al., 2002). Pin1 also binds the phosphorylated C-terminal domain (CTD) of RNA polymerase II and the structural basis of this interaction has been established (Verdecia et al., 2000). The Pin1 WW domain has evolved a rather unique ability, among WW domains, to bind in a phosphospecific fashion.

5.2. Tyrosine Phosphorylation of Group I WW Domain Ligands

Group I WW domain ligands harbor an obligatory tyrosine residue (PPxY), which raises the possibility that tyrosine phosphorylation could play a role in the regulation of group I WW domain–mediated ligand binding. This has indeed been found to be a method of regulation in the case of β-dystroglycan–dystrophin WW domain interactions. The C-terminal tail of β-dystroglycan harbors a PPxY motif that binds dystrophin predominantly through its WW domain (Rentschler et al., 1999). β-Dystroglycan is a substrate for the c-Src tyrosine kinase and the site of phosphorylation is the tyrosine residue within the C-terminal PPxY motif (Sotgia et al., 2001). Most importantly, tyrosine phosphorylation of β-dystroglycan prevents its interaction with the WW domain of dystrophin (Ilsley et al., 2001). Phosphospecific antibodies have been generated to the C-terminal tail of β-dystroglycan to demonstrate the dependence on c-Src and the in vivo relevance of this modification (Sotgia et al., 2003). Not only does tyrosine phosphorylation of the C-terminal tail of β-dystroglycan inhibit dystrophin WW domain binding, it also generates binding sites for SH2 domain binding (Sotgia et al., 2001). So, in this case, either a WW domain or a SH2 domain, depending on its phosphorylation status of the PPxY sequence, can bind the same motif.

5.3. Arginine Methylation of Group III WW Domain Ligands

Arginine residues form key components of both SH3 and WW domain binding motifs. In the case of Src-like SH3 domains, structural analysis with complexed ligands has revealed the formation of a salt bridge between a conserved acidic residue within the SH3 domain and an arginine residue either N-terminal (class I ligand) or C-terminal (class II ligand) to the PxxP motif of the ligand (Weng et al., 1995; Arold et al., 1998). Oriented peptide libraries and expression cloning have demonstrated that group III WW domains also bind proline-rich motifs flanked with arginine residues (Bedford et al., 1998, 2000b; Komuro et al., 1999). Structural predictions suggested that the positively charged arginine residues dock with negatively charged patches on the WW domain (Bedford et al., 2000b).

Arginine residues can be mono- or dimethylated within an arginine and glycine rich motif (GAR motif) (Gary and Clarke, 1998). Arginine methyltransferases modify a large number of proteins that are involved in RNA processing, signal transduction, and the regulation of transcription (McBride and Silver, 2001). This modification does not change the charge of the arginine residue, but the addition of one or two methyl groups does add steric bulk to a ligand. On occasion, a GAR domain overlaps with a proline-rich motif and key arginine residues (for SH3/WW binding) can be posttranslationally modified. The signaling molecule, Sam68, is one such protein that is bound by both SH3 and WW domains, and is also arginine methylated (Bedford et al., 2000a). SH3 and WW domain binding activity as well as arginine methylation is restricted to a few proline/arginine/glycine-rich regions of fewer than 20 amino acids in size. This raised the possibility that arginine methylation within these regions could regulate either SH3 or WW domain binding, or both. Indeed, it was shown that arginine methylation selectively modulates the in vitro binding of SH3 domains but not WW domains (Bedford et al., 2000a). Thus, competitive binding of the same proline-rich region by SH3 and WW domains can be driven toward WW domain binding as a result of arginine methylation. This phenomenon of selective inhibition of SH3 domain binding by arginine methylation is clearly demonstrated in Figure 9.4.

5.4. Posttranslational Modifications Within Domains Themselves

Phosphorylation events within the SH3 domain itself have been shown to play a role in the regulation of domain binding. The cytoskeleton-associated protein, PSTPIP (Proline, Serine, Threonine phosphatase interacting protein) harbors an SH3 domain that binds the proline-rich motif of the Wiskott–Aldrich syndrome protein (WASP). A tyrosine residue within the PSTPIP SH3 domain is a target for the non–membrane-bound tyrosine kinase, c-Src. This modification negatively regulates the binding between PSTPIP and WASP (Wu et al., 1998).

Bruton's tyrosine kinase (Btk), which is a member of the Tec family of protein tyrosine kinases, is also phosphorylated at a specific residue in the SH3 domain (Morrogh et al., 1999). This is an autophosphorylation event. The Btk SH3 domain has been shown to bind WASP and c-Cbl. On autophosphorylation the Btk SH3 domain is inhibited from WASP binding, but still binds c-Cbl. Thus, phosphorylation of a SH3 domain in this case selectively regulates ligand binding. This study has recently been expanded to demonstrate that additional Tec family members (Itk, Tec, and Bmx) are autophosphorylated within their SH3 domains (Nore et al., 2003). In addition, transphosphorylation of one family member by another is also possible. Thus, this method of regulating SH3 domain mediated protein–protein interactions has been described for a number of different proteins and may prove to be fairly common.

Not only SH3 domain, but also WW domain binding is regulated by domain phosphorylation. The WW domain of Pin1 is phosphorylated by PKA (Lu et al., 2002). The phosphorylation site is a serine residue positioned five amino acids after the first highly conserved tryptophan of the Pin1 WW domain. Phosphorylation of the Pin1 WW domain prevents it from interacting with substrates in vivo.

In this chapter we discussed how phosphorylation and methylation can regulate SH3–WW domain interactions with proline-rich sequences. In addition, proline itself can be modified in two rather unique ways: by prolyl hydroxylases and proline isomerases. Proline is the only amino acid that exists in both a *cis* and a *trans* peptide bond configuration under physiological conditions within a cell (Grathwohl and Wuthrich, 1976). The *cis-trans* conversion is catalyzed by a family of enzymes—the proline isomerases. With the aid of proline isomerases a protein conformation can be altered to regulate its activity or perhaps even its propensity for protein–protein interactions (Andreotti, 2003). In addition, proline residues within a LxxLAP motif can be posttranslationally modified by prolyl hydroxylases (Huang et al., 2002). It is conceivable that prolyl hydroxylation could regulate SH3–WW domain binding affinities for modified proline-rich sequences, although this has yet to be experimentally established.

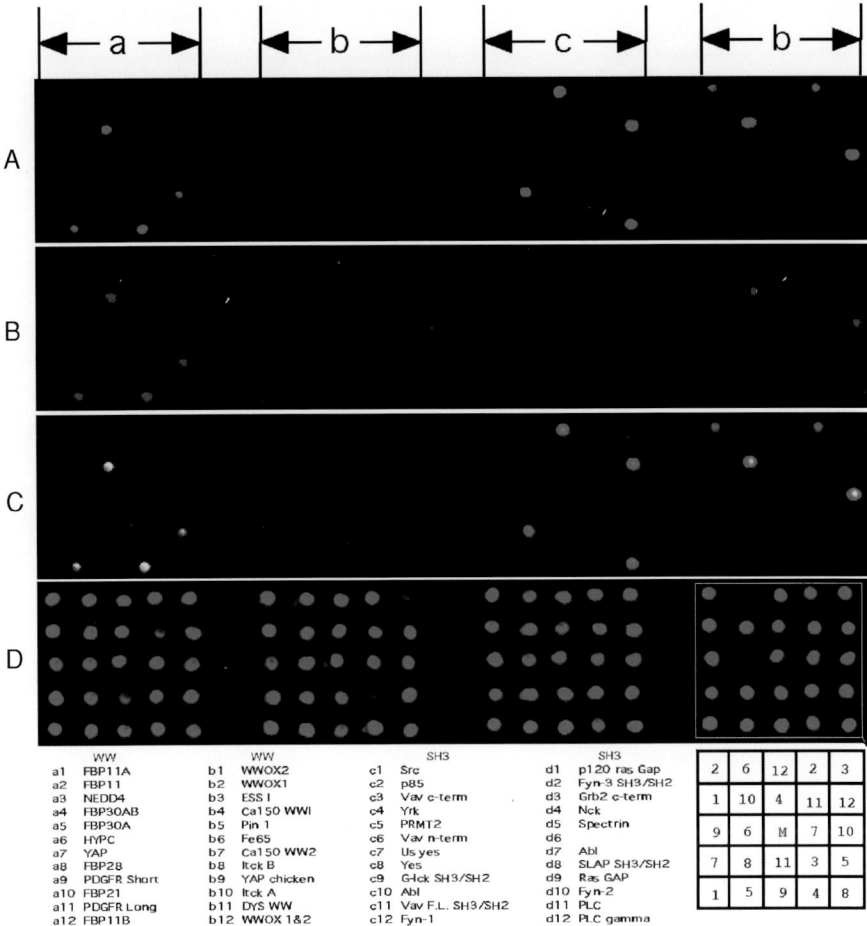

		WW				SH3				SH3						
a1	FBP11A	b1	WWOX2	c1	Src	d1	p120 ras Gap	2	6	12	2	3
a2	FBP11	b2	WWOX1	c2	p85	d2	Fyn-3 SH3/SH2	1	10	4	11	12
a3	NEDD4	b3	ESS I	c3	Vav c-term	d3	Grb2 c-term					
a4	FBP30AB	b4	Ca150 WWI	c4	Yrk	d4	Nck	9	6	M	7	10
a5	FBP30A	b5	Pin 1	c5	PRMT2	d5	Spectrin					
a6	HYPC	b6	Fe65	c6	Vav n-term	d6		7	8	11	3	5
a7	YAP	b7	Ca150 WW2	c7	Us yes	d7	Abl					
a8	FBP28	b8	Itck B	c8	Yes	d8	SLAP SH3/SH2	1	5	9	4	8
a9	PDGFR Short	b9	YAP chicken	c9	G-ick SH3/SH2	d9	Ras GAP					
a10	FBP21	b10	Itck A	c10	Abl	d10	Fyn-2					
a11	PDGFR Long	b11	DyS WW	c11	Vav F.L. SH3/SH2	d11	PLC					
a12	FBP11B	b12	WWOX 1&2	c12	Fyn-1	d12	PLC gamma					

Figure 9.4. Posttranslational modifications regulate binding of proline-rich peptides to their associated domains. GST fusion proteins were arrayed in duplicate on nitrocellulose slides. Protein domain microarrays were probed with a proline-rich peptide that was either unmodified or harbored methylated arginine residues. (**A**) The array was probed with a Cy3-labeled peptide from the signaling molecule, Sam68 (biotin-GVSVRGRGAAPPPPPVPRGRGVGP). (**B**) The array was probed with a Cy5-labeled Sam68 peptide that is arginine methylated (biotin-GVSVR*GR*GAAPPPPPVPR*GR*GVGP; *asterisks* denote asymmetrically dimethylated arginine residues). (**C**) The Cy3 and Cy5 signals were superimposed. The *yellow* signal indicates protein interactions that are insensitive to arginine methylation, and the *green* signals mark protein interactions that are inhibited by arginine methylation. (**D**) The array was probed with an anti-GST primary antibody and detected with a fluorescein isothiocyanate (FITC)-conjugated secondary antibody. A key to the arrayed domains is displayed below. 24 WW domains (*a, b*) and 23 SH3 domains (*c, d*) are arrayed. Both peptides bind WW domains but only unmethylated peptides bind SH3 domains.

Acknowledgments

We thank Alexsandra Espejo for generating Figures 9.2 and 9.4. Mark T. Bedford is supported by NIH Grant DK62248-01, the Welch Foundation (G-1495), and in part by ES07784 and ES01104. Marius Sudol is supported by grants from NIH (AR45626) and from Human Frontier Science Program Organization (RG0234).

REFERENCES

Aasland, R., Abrams, C., Ampe, C., Ball, L.J., Bedford, M.T., Cesareni, G., Gimona, M., Hurley, J.H., Jarchau, T., Lehto, V.P., Lemmon, M.A., Linding, R., Mayer, B.J., Nagai, M., Sudol, M., Walter, U., and Winder, S.J. (2002). Normalization of nomenclature for peptide motifs as ligands of modular protein domains. *FEBS Lett.* 513:141–144.

Anderson, D., Koch, C.A., Grey, L., Ellis, C., Moran, M.F., and Pawson, T. (1990). Binding of SH2 domains of phospholipase C gamma 1, GAP, and Src to activated growth factor receptors. *Science* 250:979–982.

Andreotti, A.H. (2003). Native state proline isomerization: an intrinsic molecular switch. *Biochemistry* 42:9515–9524.

Arold, S., Franken, P., Strub, M.P., Hoh, F., Benichou, S., Benarous, R., and Dumas, C. (1997). The crystal structure of HIV-1 Nef protein bound to the Fyn kinase SH3 domain suggests a role for this complex in altered T cell receptor signaling. *Structure* 5:1361–1372.

Arold, S., O'Brien, R., Franken, P., Strub, M.P., Hoh, F., Dumas, C., and Ladbury, J.E. (1998). RT loop flexibility enhances the specificity of Src family SH3 domains for HIV-1 Nef. *Biochemistry* 37:14683–14691.

Ball, L.J., Jarchau, T., Oschkinat, H., and Walter, U. (2002). EVH1 domains: structure, function and interactions. *FEBS Lett.* 513:45–52.

Bedford, M.T., Chan, D.C., and Leder, P. (1997). FBP WW domains and the Abl SH3 domain bind to a specific class of proline-rich ligands. *EMBO J.* 16:2376–2383.

Bedford, M.T., Reed, R., and Leder, P. (1998). WW domain-mediated interactions reveal a spliceosome-associated protein that binds a third class of proline-rich motif: the proline glycine and methionine-rich motif. *Proc. Natl. Acad. Sci. USA* 95:10602–10607.

Bedford, M.T., Frankel, A., Yaffe, M.B., Clarke, S., Leder, P., and Richard, S. (2000a). Arginine methylation inhibits the binding of proline-rich ligands to Src homology 3, but not WW, domains. *J. Biol. Chem.* 275:16030–16036.

Bedford, M.T., Sarbassova, D., Xu, J., Leder, P., and Yaffe, M.B. (2000b). A novel pro-Arg motif recognized by WW domains. *J. Biol. Chem.* 275:10359–10369.

Berry, D.M., Nash, P., Liu, S.K., Pawson, T., and McGlade, C.J. (2002). A high-affinity Arg-X-X-Lys SH3 binding motif confers specificity for the interaction between Gads and SLP-76 in T cell signaling. *Curr. Biol.* 12:1336–1341.

Bork, P., and Sudol, M. (1994). The WW domain: a signalling site in dystrophin? *Trends Biochem. Sci.* 19:531–533.

Cesareni, G., Panni, S., Nardelli, G., and Castagnoli, L. (2002). Can we infer peptide recognition specificity mediated by SH3 domains? *FEBS Lett.* 513:38–44.

Chan, D.C., Bedford, M.T., and Leder, P. (1996). Formin binding proteins bear WWP/WW domains that bind proline-rich peptides and functionally resemble SH3 domains. *EMBO J.* 15:1045–1054.

Chen, H.I., and Sudol, M. (1995). The WW domain of Yes-associated protein binds a proline-rich ligand that differs from the consensus established for Src homology 3-binding modules. *Proc. Natl. Acad. Sci. USA* 92:7819–7823.

Ermekova, K.S., Zambrano, N., Linn, H., Minopoli, G., Gertler, F., Russo, T., and Sudol, M. (1997). The WW domain of neural protein FE65 interacts with proline-rich motifs in Mena, the mammalian homolog of *Drosophila* enabled. *J. Biol. Chem.* 272:32869–32877.

Espanel, X., and Sudol, M. (1999). A single point mutation in a group I WW domain shifts its specificity to that of group II WW domains. *J. Biol. Chem.* 274:17284–17289.

Freund, C., Dotsch, V., Nishizawa, K., Reinherz, E.L., and Wagner, G. (1999). The GYF domain is a novel structural fold that is involved in lymphoid signaling through proline-rich sequences. *Nat. Struct. Biol.* 6:656–660.

Freund, C., Kuhne, R., Yang, H., Park, S., Reinherz, E.L., and Wagner, G. (2002). Dynamic interaction of CD2 with the GYF and the SH3 domain of compartmentalized effector molecules. *EMBO J.* 21:5985–5995.

Fumagalli, S., Totty, N.F., Hsuan, J.J., and Courtneidge, S.A. (1994). A target for Src in mitosis. *Nature* 368:871–874.

Garrus, J.E., von Schwedler, U.K., Pornillos, O.W., Morham, S.G., Zavitz, K.H., Wang, H.E., Wettstein, D.A., Stray, K.M., Cote, M., Rich, R.L., Myszka, D.G., and Sundquist, W.I. (2001). Tsg101 and the vacuolar protein sorting pathway are essential for HIV-1 budding. *Cell* 107:55–65.

Gary, J.D., and Clarke, S. (1998). RNA and protein interactions modulated by protein arginine methylation. *Prog. Nucleic Acid Res. Mol. Biol.* 61:65–131.

Gertler, F.B., Comer, A.R., Juang, J.L., Ahern, S.M., Clark, M.J., Liebl, E.C., and Hoffmann, F.M. (1995). Enabled, a dosage-sensitive suppressor of mutations in the *Drosophila* Abl tyrosine kinase, encodes an Abl substrate with SH3 domain-binding properties. *Genes Dev.* 9:521–533.

Gertler, F.B., Niebuhr, K., Reinhard, M., Wehland, J., and Soriano, P. (1996). Mena, a relative of VASP and Drosophila Enabled, is implicated in the control of microfilament dynamics. *Cell* 87:227–239.

Grathwohl, C., and Wuthrich, K. (1976). Nmr studies of the molecular conformations in the linear oligopeptides H-(L-Ala)n-L-Pro-OH. *Biopolymers* 15:2043–2057.

Hagerman, A.E., and Butler, L.G. (1981). The specificity of proanthocyanidin-protein interactions. *J. Biol. Chem.* 256:4494–4497.

Holt, M.R., and Koffer, A. (2001). Cell motility: proline-rich proteins promote protrusions. *Trends Cell Biol.* 11:38–46.

Hu, H., Columbus, J., Zhang, Y., Wu, D., Lian, L., Yang, S., Goodwin, J., Luczak, C., Carter, M., Chen, L., James, M., Davis, R., Sudol, M., Rodwell, J., and Herrero, H. (2003). A map of WW domain family interactions. *Proteomics* (in press).

Huang, J., Zhao, Q., Mooney, S.M., and Lee, F.S. (2002). Sequence determinants in hypoxia-inducible factor-1alpha for hydroxylation by the prolyl hydroxylases PHD1, PHD2, and PHD3. *J. Biol. Chem.* 277:39792–39800.

Huang, X., Poy, F., Zhang, R., Joachimiak, A., Sudol, M., and Eck, M.J. (2000). Structure of a WW domain containing fragment of dystrophin in complex with beta-dystroglycan. *Nat. Struct. Biol.* 7:634–638.

Ilsley, J.L., Sudol, M., and Winder, S.J. (2001). The interaction of dystrophin with beta-dystroglycan is regulated by tyrosine phosphorylation. *Cell Signal.* 13:625–632.

Kay, B.K., Williamson, M.P., and Sudol, M. (2000). The importance of being proline: the interaction of proline-rich motifs in signaling proteins with their cognate domains. *FASEB J.* 14:231–241.

Komuro, A., Saeki, M., and Kato, S. (1999). Association of two nuclear proteins, Npw38 and NpwBP, via the interaction between the WW domain and a novel proline-rich motif containing glycine and arginine. *J. Biol. Chem.* 274:36513–36519.

Lambrechts, A., Kwiatkowski, A.V., Lanier, L.M., Bear, J.E., Vandekerckhove, J., Ampe, C., and Gertler, F.B. (2000). cAMP-dependent protein kinase phosphorylation of EVL, a Mena/VASP relative, regulates its interaction with actin and SH3 domains. *J. Biol. Chem.* 275:36143–36151.

Lehto, V.P., Wasenius, V.M., Salven, P., and Saraste, M. (1988). Transforming and membrane proteins. *Nature* 334:388.

Liu, W., Youn, H.D., Zhou, X.Z., Lu, K.P., and Liu, J.O. (2001). Binding and regulation of the transcription factor NFAT by the peptidyl prolyl cis-trans isomerase Pin1. *FEBS Lett.* 496:105–108.

Lu, K.P., Hanes, S.D., and Hunter, T. (1996). A human peptidyl-prolyl isomerase essential for regulation of mitosis. *Nature* 380:544–547.

Lu, P.J., Zhou, X.Z., Liou, Y.C., Noel, J.P., and Lu, K.P. (2002). Critical role of WW domain phosphorylation in regulating phosphoserine binding activity and Pin1 function. *J. Biol. Chem.* 277:2381–2384.

Macias, M.J., Wiesner, S., and Sudol, M. (2002). WW and SH3 domains, two different scaffolds to recognize proline-rich ligands. *FEBS Lett.* 513:30–37.

Mahoney, N.M., Janmey, P.A., and Almo, S.C. (1997). Structure of the profilin-poly-L-proline complex involved in morphogenesis and cytoskeletal regulation. *Nat. Struct. Biol.* 4:953–960.

Mahoney, N.M., Rozwarski, D.A., Fedorov, E., Fedorov, A.A., and Almo, S.C. (1999). Profilin binds proline-rich ligands in two distinct amide backbone orientations. *Nat. Struct. Biol.* 6:666–671.

Mayer, B.J. (2001). SH3 domains: complexity in moderation. *J. Cell. Sci.* 114:1253–1263.

Mayer, B.J., Hamaguchi, M., and Hanafusa, H. (1988). A novel viral oncogene with structural similarity to phospholipase C. *Nature* 332:272–275.

McBride, A.E., and Silver, P.A. (2001). State of the arg: protein methylation at arginine comes of age. *Cell* 106:5–8.

Messenger, M.M., Saulnier, R.B., Gilchrist, A.D., Diamond, P., Gorbsky, G.J., and Litchfield, D.W. (2002). Interactions between protein kinase CK2 and Pin1. Evidence for phosphorylation-dependent interactions. *J. Biol. Chem.* 277:23054–23064.

Mongiovi, A.M., Romano, P.R., Panni, S., Mendoza, M., Wong, W.T., Musacchio, A., Cesareni, G., and Di Fiore, P.P. (1999). A novel peptide-SH3 interaction. *EMBO J.* 18:5300–5309.

Morrogh, L.M., Hinshelwood, S., Costello, P., Cory, G.O., and Kinnon, C. (1999). The SH3 domain of Bruton's tyrosine kinase displays altered ligand binding properties when auto-phosphorylated in vitro. *Eur. J. Immunol.* 29:2269–2279.

Musacchio, A. (2002). How SH3 domains recognize proline. *Adv. Protein Chem.* 61:211–268.

Niebuhr, K., Ebel, F., Frank, R., Reinhard, M., Domann, E., Carl, U.D., Walter, U., Gertler, F.B., Wehland, J., and Chakraborty, T. (1997). A novel proline-rich motif present in ActA of Listeria monocytogenes and cytoskeletal proteins is the ligand for the EVH1 domain, a protein module present in the Ena/VASP family. *EMBO J.* 16:5433–5444.

Nishizawa, K., Freund, C., Li, J., Wagner, G., and Reinherz, E.L. (1998). Identification of a proline-binding motif regulating CD2-triggered T lymphocyte activation. *Proc. Natl. Acad. Sci. USA* 95:14897–14902.

Nore, B.F., Mattsson, P.T., Antonsson, P., Backesjo, C.M., Westlund, A., Lennartsson, J., Hansson, H., Low, P., Ronnstrand, L., and Smith, C.I. (2003). Identification of phosphorylation sites within the SH3 domains of Tec family tyrosine kinases. *Biochim. Biophys. Acta* 1645:123–132.

Oneyama, C., Nakano, H., and Sharma, S.V. (2002). UCS15A, a novel small molecule, SH3 domain-mediated protein-protein interaction blocking drug. *Oncogene* 21:2037–2050.

Oneyama, C., Agatsuma, T., Kanda, Y., Nakano, H., Sharma, S.V., Nakano, S., Narazaki, F., and Tatsuta, K. (2003). Synthetic inhibitors of proline-rich ligand-mediated protein–protein interaction. Potent analogs of UCS15A. *Chem. Biol.* 10:443–451.

Pornillos, O., Alam, S.L., Davis, D.R., and Sundquist, W.I. (2002a). Structure of the Tsg101 UEV domain in complex with the PTAP motif of the HIV-1 p6 protein. *Nat. Struct. Biol.* 9:812–817.

Pornillos, O., Alam, S.L., Rich, R.L., Myszka, D.G., Davis, D.R., and Sundquist, W.I. (2002b). Structure and functional interactions of the Tsg101 UEV domain. *EMBO J.* 21:2397–2406.

Pornillos, O., Higginson, D.S., Stray, K.M., Fisher, R.D., Garrus, J.E., Payne, M., He, G.P., Wang, H.E., Morham, S.G., and Sundquist, W.I. (2003). HIV Gag mimics the Tsg101-recruiting activity of the human Hrs protein. *J. Cell. Biol.* 162:425–434.

Ren, R., Mayer, B.J., Cicchetti, P., and Baltimore, D. (1993). Identification of a ten-amino acid proline-rich SH3 binding site. *Science* 259:1157–1161.

Rentschler, S., Linn, H., Deininger, K., Bedford, M.T., Espanel, X., and Sudol, M. (1999). The WW domain of dystrophin requires EF-hands region to interact with beta-dystroglycan. *Biol. Chem.* 380:431–442.

Richard, S., Yu, D., Blumer, K.J., Hausladen, D., Olszowy, M.W., Connelly, P.A., and Shaw, A.S. (1995). Association of p62, a multifunctional SH2- and SH3-domain-binding protein, with src family tyrosine kinases, Grb2, and phospholipase C gamma-1. *Mol. Cell. Biol.* 15:186–197.

Rubin, G.M., and Lewis, E.B. (2000). A brief history of *Drosophila's* contributions to genome research. *Science* 287:2216–2218.

Sadowski, I., Stone, J.C., and Pawson, T. (1986). A noncatalytic domain conserved among cytoplasmic protein-tyrosine kinases modifies the kinase function and transforming activity of Fujinami sarcoma virus P130gag-fps. *Mol. Cell. Biol.* 6:4396–4408.

Sotgia, F., Lee, H., Bedford, M.T., Petrucci, T., Sudol, M., and Lisanti, M.P. (2001). Tyrosine phosphorylation of beta-dystroglycan at its WW domain binding motif, PPxY, recruits SH2 domain containing proteins. *Biochemistry* 40:14585–14592.

Sotgia, F., Bonuccelli, G., Bedford, M., Brancaccio, A., Mayer, U., Wilson, M.T., Campos-Gonzalez, R., Brooks, J.W., Sudol, M., and Lisanti, M.P. (2003). Localization of phospho-beta-dystroglycan (pY892) to an intracellular vesicular compartment in cultured cells and skeletal muscle fibers in vivo. *Biochemistry* 42:7110–7123.

Stahl, M.L., Ferenz, C.R., Kelleher, K.L., Kriz, R.W., and Knopf, J.L. (1988). Sequence similarity of phospholipase C with the non-catalytic region of src. *Nature* 332:269–272.

Stapley, B.J., and Creamer, T.P. (1999). A survey of left-handed polyproline II helices. *Protein Sci.* 8:587–595.

Sudol, M., and Hunter, T. (2000). New Wrinkles for an old domain. *Cell* 103:1001–1004.

Verdecia, M.A., Bowman, M.E., Lu, K.P., Hunter, T., and Noel, J.P. (2000). Structural basis for phosphoserine-proline recognition by group IV WW domains. *Nat. Struct. Biol.* 7:639–643.

Volkman, B.F., Prehoda, K.E., Scott, J.A., Peterson, F.C., and Lim, W.A. (2002). Structure of the N-WASP EVH1 domain-WIP complex: insight into the molecular basis of Wiskott-Aldrich syndrome. *Cell* 111:565–576.

Weng, Z., Rickles, R.J., Feng, S., Richard, S., Shaw, A.S., Schreiber, S.L., and Brugge, J.S. (1995). Structure-function analysis of SH3 domains: SH3 binding specificity altered by single amino acid substitutions. *Mol. Cell. Biol.* 15:5627–5634.

Wu, Y., Spencer, S.D., and Lasky, L.A. (1998). Tyrosine phosphorylation regulates the SH3-mediated binding of the Wiskott-Aldrich syndrome protein to PSTPIP, a cytoskeletal-associated protein. *J. Biol. Chem.* 273:5765–5770.

Zacchi, P., Gostissa, M., Uchida, T., Salvagno, C., Avolio, F., Volinia, S., Ronai, Z., Blandino, G., Schneider, C., and Del Sal, G. (2002). The prolyl isomerase Pin1 reveals a mechanism to control p53 functions after genotoxic insults. *Nature* 419:853–857.

Zarrinpar, A., Bhattacharyya, R.P., and Lim, W.A. (2003). The structure and function of proline recognition domains. *Sci STKE* 2003:RE8.

Zheng, H., You, H., Zhou, X.Z., Murray, S.A., Uchida, T., Wulf, G., Gu, L., Tang, X., Lu, K.P., and Xiao, Z.X. (2002). The prolyl isomerase Pin1 is a regulator of p53 in genotoxic response. *Nature* 419:849–853.

Zhou, X.Z., Kops, O., Werner, A., Lu, P.J., Shen, M., Stoller, G., Kullertz, G., Stark, M., Fischer, G., and Lu, K.P. (2000). Pin1-dependent prolyl isomerization regulates dephosphorylation of Cdc25C and tau proteins. *Mol. Cell* 6:873–883.

10

The Structure and Molecular Interactions of the Bromodomain

Kelley S. Yan and Ming-Ming Zhou

ABSTRACT

The bromodomain is a structurally conserved protein module that is present in a large number of chromatin-associated proteins and in many nuclear histone acetyltransferases. The bromodomain functions as an acetyl-lysine binding domain and has recently been shown to play an important role in regulating protein–protein interactions in chromatin-mediated cellular gene transcription as well as in viral transcriptional activation. Recent structural analyses of bromodomains in complex with acetyl-lysine–containing biological ligands provide insights into the molecular basis of differences in ligand selectivity of the bromodomain family, and reinforce the concept that functional diversity of a conserved protein structure is achieved by evolutionary changes of amino acid sequences in the ligand binding site.

1. INTRODUCTION

Given a limited number of genes within its genome, the functional complexity of an organism requires more than just a sum of the products directly encoded by its genes. This implies that mechanisms exist to diversify the functions of gene products.

KELLEY S. YAN AND MING-MING ZHOU • Structural Biology Program, Department of Physiology and Biophysics, Mount Sinai School of Medicine, New York University, New York, NY 10029-6574, USA.

Proteomics and Protein–Protein Interactions: Biology, Chemistry, Bioinformatics, and Drug Design, edited by Waksman. Springer, New York, 2005.

This notion is supported by the vertebrate immune system, in which massive numbers of different antibodies with distinct specificities are generated by somatic mutation or recombination of a relatively small number of genes (Baltimore, 1981). Likewise, functional diversification of gene products may be achieved by combinatorial shuffling of a limited number of modules to produce distinct proteins as well as by modifications of these building blocks through various mechanisms. In this manner, a finite number of protein building blocks can be amplified to produce an almost infinite array of different biological outcomes. One illustration of this general principle is the posttranslational modification of individual amino acids in proteins, a common regulatory mechanism that expands the functional diversity of these proteins. Such modifications of a protein can result in switching of its molecular function or creation of docking sites for molecular interactions in myriad processes ranging from signaling at cell surface receptors (Keyse, 2000; Chang and Karin, 2001; Johnson and Lapadat, 2002; Pawson and Nash, 2003) to remodeling of chromatin structure in the nucleus (Strahl and Allis, 2000; Jenuwein and Allis, 2001; Turner, 2002; Fischle et al., 2003). Thus, the combinatorial arrangement of modular domains within proteins and the posttranslational modifications of those proteins represent two central cellular mechanisms that amplify the functional repertoire of the genome through the generation of large networks of molecular interactions that regulate cellular processes.

Covalent modifications of histones play a pivotal role in the control of chromatin structure, which in turn regulates a wide array of DNA-templated processes including transcription, replication, recombination and segregation (John and Workman, 1998; Wolffe and Hayes, 1999; Strahl and Allis, 2000; Jenuwein and Allis, 2001; Turner, 2002; Fischle et al., 2003). Eukaryotic DNA is packaged in the form of chromatin, which is composed of a repeating nucleoprotein unit called the nucleosome. Within each nucleosome, chromosomal DNA of approximately 146 base pairs is wrapped around a histone octamer comprised of two copies of each histone protein H2A, H2B, H3, and H4 (Wolffe and Hayes, 1999). Nucleosome cores are connected by short stretches of DNA bound to the linker histones H1 and H5 to form a nucleosomal filament, which is further folded into the higher-order chromatin fiber structure. Such dense packing and precise organization of DNA within chromatin structure are necessary for compaction of the genome into the nucleus of a eukaryotic cell. However, the question of how transcription or replication machineries gain access to the chromosomal DNA has been a subject of intense investigation. A rapidly growing body of knowledge has provided direct mechanistic links between the modification-induced dynamic modulation of chromatin architecture and the regulation of gene transcription (Mizzen et al., 1998; Struhl, 1998; Jenuwein and Allis, 2001; Fischle et al., 2003). These modifications, including acetylation, methylation, phosphorylation, ubiquitination, ribosylation, and glycosylation, can occur on the conserved amino acid residues in the flexible N- and C-terminal sequences of histones, and are directly linked to gene transcriptional activation and repression (Grunstein, 1997; Luger et al., 1997; Mizzen et al., 1998). Such posttranslational modifications, alone or in combination, of nucleosomal histones have been shown to be associated with a broad spectrum of distinct biological outcomes, a phenomenon that has been referred to as the "histone code hypothesis" (Strahl and Allis, 2000; Jenuwein and Allis, 2001; Turner, 2002; Fischle et al., 2003).

2. THE BROMODOMAIN FAMILY

Although it was known that histones could be acetylated on specific lysine residues (Marcus et al., 1994; Brownell and Allis, 1996; Brownell et al., 1996; Filetici et al., 1998), the consequences of these modifications were not understood at a molecular level until more recently. Site-specific lysine acetylation of histones could serve as docking sites for the recruitment of chromatin remodeling complexes or simply alter electrostatic interactions between histones and DNA. The discovery by Dhalluin et al. (Dhalluin et al., 1999) that the bromodomain may function as an acetyl-lysine binding domain by interacting with lysine-acetylated peptides derived from histones provided the first supporting evidence for the former possibility. The bromodomain is an evolutionarily conserved region of approximately 110 amino acids, which was first identified in the *Drosophila* protein brahma (Haynes et al., 1992; Tamkun et al., 1992) and named by analogy to the chromo domain, a chromatin-associated protein module (Jeanmougin et al., 1997). The extensive bromodomain family contains members from more than 500 eukaryotic chromatin-associated proteins and nuclear histone acetyltransferases (HATs), including approximately 128 human proteins (Schultz et al., 1998; Letunic et al., 2002). The suggested biological function of bromodomain binding to lysine-acetylated histones (Dhalluin et al., 1999; Winston and Allis, 1999) is analogous to Src homology 2 (SH2) (Schlessinger and Lemmon, 2003) and phosphotyrosine binding (PTB) (Yan et al., 2002a) domains of adaptor proteins binding to tyrosine-phosphorylated receptor tyrosine kinases in signal transduction (Pawson and Nash, 2003). Thus, histone lysine acetylation may serve as a regulatory modification to promote acetylation-dependent recruitment of proteins for chromatin remodeling or gene transcription. This mechanism agrees well with the histone code hypothesis, which postulates that distinct patterns of posttranslational histone modifications function as a recognition code for the recruitment of different chromatin remodeling complexes (Strahl and Allis, 2000; Jenuwein and Allis, 2001; Turner, 2002; Fischle et al., 2003).

The discovery of the bromodomain as an acetyl-lysine binding domain hinted at a mechanism for regulating protein–protein interactions via lysine acetylation. Such a mechanism has broad implications for the molecular events underlying a wide variety of cellular processes, including chromatin remodeling and transcriptional activation (Winston and Allis, 1999; Strahl and Allis, 2000; Zeng and Zhou, 2001). This mechanism also suggests that bromodomains may contribute to the observed hyperacetylated state of histones during transcription by tethering enzymatic activity of HATs to target chromosomal sites (Brownell and Allis, 1996; Travers, 1999; Manning et al., 2001) in order to propagate the acetylation to neighboring nucleosomes. Moreover, the bromodomain may also assist in the directed assembly and activity of multiprotein chromatin remodeling complexes such as Spt-Ada-Gcn5 acetyltransferase (SAGA),

Remodeling the Structure of Chromatin (RSC), SWI/SNF, and NuA4 (Sterner et al., 1999; Brown et al., 2001). Bromodomain disruption or deletion in various proteins across organisms results in pleiotrophic effects and may provide insights into the function of this module in vivo. For example, it has been shown that this module is indispensable for the function of GCN5p in the catalytic activity of the SAGA complex in *Saccharomyces cerevisiae* (Georgakopoulos et al., 1995; Syntichaki et al., 2000). Deletion of a bromodomain in HBRM, a protein component of the human SWI/SNF remodeling complex, causes both decreased stability and loss of nuclear localization (Muchardt et al., 1998; Muchardt and Yaniv, 1999). Bromodomains of Bdf1p, a *S. cerevisiae* protein, are required for sporulation and normal mitotic growth (Chua and Roeder, 1995). Finally, bromodomain deletion in Sth1, Rsc1, and Rsc2, three members of the nucleosome remodeling complex RSC, can cause a conditional lethal phenotype (in Sth1) (Du et al., 1998) or a strong phenotypic inhibition on cell growth (in Rsc1 and Rsc2) (Cairns et al., 1999). Notably, the phenotypic effect observed in Rsc1 and Rsc2 results from deletion of only the second but not the first bromodomain, suggesting that these two bromodomains serve distinct functions through interactions with different biological ligands (Cairns et al., 1999).

3. THE BROMODOMAIN STRUCTURE

The first three-dimensional structure of the large bromodomain family was solved using the bromodomain from the transcriptional coactivator p300/CBP-associated factor (PCAF), determined by nuclear magnetic resonance (NMR) spectroscopy (Dhalluin et al., 1999). The PCAF bromodomain adopts a left-handed four-helix bundle comprised of amphipathic helices α_Z, α_A, α_B, and α_C (Fig. 10.1A) (Dhalluin et al., 1999). At one end of the helical bundle, the N- and C-termini come together, emphasizing the modular architecture of this domain and underscoring the idea that the bromodomain acts as an independent functional unit for protein interactions (Jeanmougin et al., 1997; Dhalluin et al., 1999; Zeng and Zhou, 2001). At the opposite end of the bundle, a long intervening segment connecting helices α_Z and α_A (the "ZA loop") packs against the relatively short segment connecting helices α_B and α_C (the "BC loop") to form a surface-accessible hydrophobic pocket. Site-directed mutational analysis demonstrates that hydrophobic and aromatic tertiary contacts between the ZA and BC interhelical loops are important for stabilizing the three-dimensional (3D) structure of this protein (Dhalluin et al., 1999).

This left-handed four-helix bundle structural fold is highly conserved within the bromodomain family, as confirmed by other more recently determined structures of bromodomains from human GCN5 (Hudson et al., 2000), *S. cerevisiae* GCN5p (Owen et al., 2000), the double bromodomain module of human TAF$_{II}$ 250 (Jacobson et al., 2000), and the human transcriptional coactivator CREB-binding protein (CBP) (Mujtaba et al., 2004). Although the structural similarity shared by these bromodomains is very high for the four helices at the backbone level, structural differences do exist and are localized to the loop regions, particularly in the ZA and BC loops, which

A

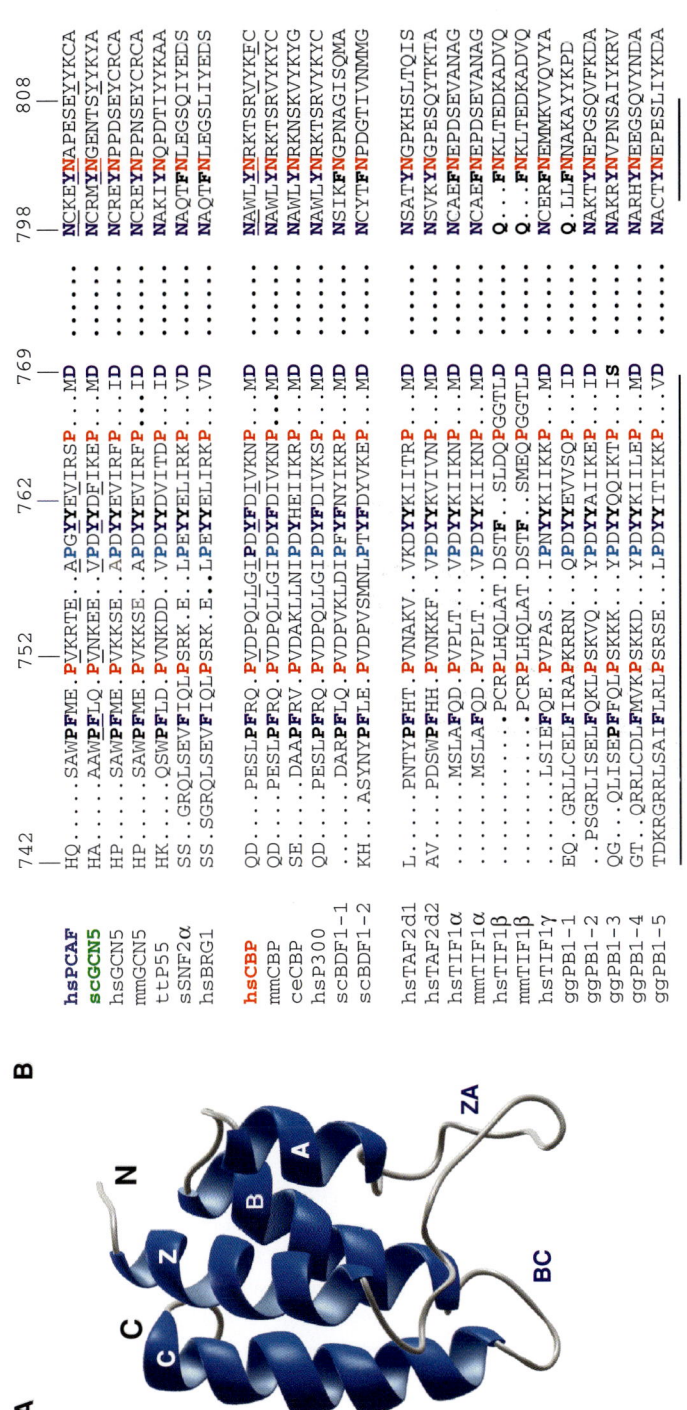

N B A Z C C

ZA   BC

B

```
              742        752          762              769        798    808
hsPCAF    HQ....SAWPFME.PVKRTE. APGYYEVIRSP....MD  ......  NCKEYNAPESEYYKCA
scGCN5    HA....AAWPFLQ.PVNKEE. VPDYYDFIKEP....MD  ......  NCRMYNGENTSYYKYA
hsGCN5    HP....SAWPFME.PVKKSE. APDYYEVIRFP....ID  ......  NCREYNPPDSEYCRCA
mmGCN5    HP....SAWPFME.PVKKSE. APDYYEVIRFP....ID  ......  NCREYNPPNSEYCRCA
ttP55     HK....QSWPFLD.PVNKDD. VPDYYDVITDP....ID  ......  NAKIYNQPDTIYYKAA
sSNF2α    SS..GRQLSEVFIQLPSRK.E. LPEYYELIRKP....VD  ......  NAQTFNLEGSQIYEDS
hsBRG1    SS.SGRQLSEVFIQLPSRK.E. LPEYYELIRKP....VD  ......  NAQTFNLEGSLIYEDS

hsCBP     QD....PESLPFRQ.PVDPQLLGIPDYFDIVKNP....MD  ......  NAWLYNRKTSRVYKFC
mmCBP     QD....PESLPFRQ.PVDPQLLGIPDYFDIVKNP....MD  ......  NAWLYNRKTSRVYKYC
ceCBP     SE....DAAPFRV.PVDAKLLNIPDYHEIIKRP....MD  ......  NAWLYNRKNSKVYKYG
hsP300    QD....PESLPFRQ.PVDPQLLGIPDYFDIVKSP....MD  ......  NAWLYNRKTSRVYKYC
scBDF1-1  .....DARPFLQ.PVDPVKLDIPFYFNYIKRP....MD  ......  NSIKFNGPNAGISQMA
scBDF1-2  KH..ASYNYPFLE.PVDPVSMNLPTYFDYVKEP....MD  ......  NCYTFNPDGTIVNMMG

hsTAF2d1  L....PNTYPFHT.PVNAKV..VKDYYKIITRP....MD  ......  NSATYNGPKHSLTQIS
hsTAF2d2  AV....PDSWPFHH.PVNKKF..VPDYYKVIVNP....MD  ......  NSVKYNGPESQYTKTA
hsTIF1α   .....MSLAFQD.PVPLT...VPDYYKIIKNP....MD  ......  NCAEFNEPDSEVANAG
mmTIF1α   .....MSLAFQD.PVPLT...VPDYYKIIKNP....MD  ......  NCAEFNEPDSEVANAG
hsTIF1β   .....PCRPLHQLAT.DSTF..SLDQPGGTLD       ......  Q...FNKLTEDKADVQ
mmTIF1β   .....PCRPLHQLAT.DSTF..SMEQPGGTLD       ......  Q...FNKLTEDKADVQ
hsTIF1γ   .....LSIEFQE.PVPAS..IPNYYKIIKKP....MD  ......  NCERFNEMMKVVQVYA
ggPB1-1   EQ.GRLLCELFIRAPKRRN..QPDYYEVVSQP....ID  ......  Q.LLFNNAKAYYKPD
ggPB1-2   .PSGRLISELFQKLPSKVQ..YPDYYQQIKTP....ID  ......  NAKTYNEPGSQVFKDA
ggPB1-3   QG..QLISEPFFQLPSKK...YPDYYQQIKTP....IS  ......  NAKRYNVPNSAIYKRV
ggPB1-4   GT..QRRLCDLFMVKPSKKD..YPDYYKIILEP....MD  ......  NARHYNEEGSQVYNDA
ggPB1-5   TDKRGRRLSAIFLRLPSRSE..LPDYYITIKKP....VD  ......  NACTYNEPESLIYKDA
```

ZA loop   BC loop

Figure 10.1. The bromodomain as an acetyl-lysine binding domain. (**A**) The three-dimensional structure of the PCAF bromodomain in the free form, as determined by NMR spectroscopy (Dhalluin et al., 1999). (**B**) Sequence alignment of bromodomains highlighting amino acid variations in the ZA and BC loops. Bromodomains are grouped according to sequence similarities. Sequence numbers of PCAF are shown above the sequence. Absolutely and highly conserved residues in bromodomains are colored in *red* and *blue*, respectively. Residues in the PCAF or CBP bromodomain that interact with p53 or HIV-1 Tat peptide, as shown by intermolecular NOEs, are *underlined*. Similarly, the residues of the yeast GCN5 bromodomain that directly contact the histone H4 peptide, as defined in the crystal structure, are also *underlined*.

correspond to the segments of high amino acid sequence divergence (Jeanmougin et al., 1997) (Fig. 10.1B). The conservation of 3D structure, as seen for the bromodomain fold, is another illustration of nature's general principle of exploiting a simple 3D scaffold to generate biological diversity. Although there may be only a limited number of evolutionarily conserved 3D structural folds, the functional use of these scaffolds can be amplified through amino acid sequence changes at their ligand binding sites to recognize a multitude of different biological targets.

4. THE BROMODOMAIN AS AN ACETYL-LYSINE BINDING DOMAIN

The discovery of acetyl-lysine recognition by bromodomains is attributed to the unique ability of NMR spectroscopy to measure changes in local chemical environment and/or conformation of a protein induced on binding to a ligand. Weak but highly specific interactions between a protein and a ligand (with the dissociation constant K_D in the micromolar to millimolar range) can be reliably detected with NMR whereas most other techniques are limited to higher affinity binding (Hajduk et al., 1999). Furthermore, NMR spectroscopy provides insights into the location of the ligand-binding site within the protein through chemical shift mapping techniques, which were used to study the bromodomain from PCAF (Dhalluin et al., 1999). Ligand concentration-dependent NMR titrations of the PCAF bromodomain revealed that the protein can bind in a highly specific manner to lysine-acetylated peptides derived from major known acetylation sites on histones H3 or H4 (Dhalluin et al., 1999). The PCAF bromodomain failed to bind with either ligand in the absence of acetylation, demonstrating that the interaction is indeed dependent on lysine acetylation. Chemical shift mapping using the titration data and NMR structural analysis of the PCAF bromodomain in complex with acetyl-histamine, a chemical analog of acetyl-lysine, showed that the acetyl-lysine binding site is localized to the hydrophobic cavity between the ZA and BC loops (Dhalluin et al., 1999). The methyl and methylene groups of acetyl-histamine make extensive contacts with the side chains of Val 752, Ala 757, Tyr 760, Tyr 802, Asn 803, and Tyr 809, which are highly conserved among the large bromodomain family (Jeanmougin et al., 1997; Dhalluin et al., 1999). The observed acetyl-lysine dependence of the interactions and the location of the ligand-binding site were supported by another NMR study of the human GCN5 bromodomain binding to lysine-acetylated histone H4 peptides (Hudson et al., 2000).

A crystal structure of *S. cerevisiae* GCN5p bromodomain solved in complex with an acetylated peptide derived from histone H4 at Lys 16 (A-AcK-RHRKILRNSIQGI, where AcK represent N^ε—acetyl lysine) reveals the molecular details of its acetyl-lysine recognition (Fig. 10.2A) (Owen et al., 2000). In addition to binding to the conserved hydrophobic and aromatic residues seen in the PCAF bromodomain, the acetyl-lysine forms a specific hydrogen bond between the oxygen of the acetyl carbonyl group and the side-chain amide nitrogen of an invariant asparagine residue in the bromodomain, Asn 407 (corresponding to Asn 803 in PCAF). A network of

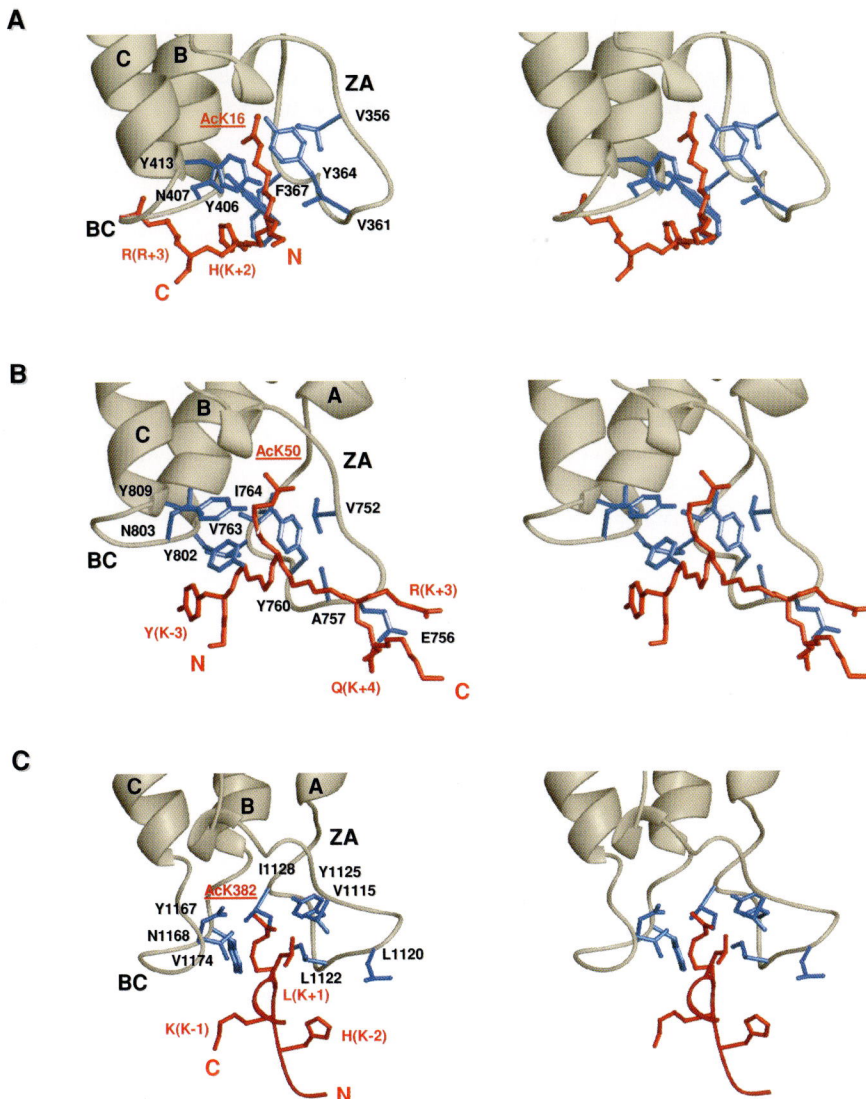

Figure 10.2. Differences in ligand selectivity of bromodomains. Stereoviews of the three-dimensional structures of (**A**) scGCN5, (**B**) PCAF, or (**C**) CBP bromodomain in complex with the acetyl-lysine–containing peptide derived from histone H4 at Lys 16 (A-AcK-RHRKILRNSIQGI), HIV-1 Tat at Lys 50 (SYGR-AcK-KRRQR), or p53 at Lys 382 (SHLKSKKGQSTSRHK-AcK-LMFK), respectively, showing interactions of the protein residues (*blue*) and the lysine-acetylated petpide residues (*red*) in the ligand binding sites. In all three bromodomain–ligand complex structures, the protein residues (*blue*) are numbered according to the sequences, and the peptide residues (*red*) are annotated according to their position with respect to the acetyl-lysine.

water-mediated hydrogen bonds with protein backbone carbonyl groups at the base of the cleft also contributes to acetyl-lysine binding. Site-directed mutagenesis confirmed the critical role of these amino acid residues in binding to acetyl-lysine, suggesting that acetyl-lysine recognition is a general feature of bromodomains (Dhalluin et al., 1999).

5. MOLECULAR DETERMINANTS OF LIGAND SPECIFICITY

5.1. The GCN5p Bromodomain–Histone H4 Complex

While the major binding determinant in the GCN5p bromodomain-H4 peptide complex is the acetylated lysine itself, which sits in a deep hydrophobic pocket, the protein also has a limited number of contacts with residues C-terminal to the AcK at (AcK+2) and (AcK+3) in the H4 peptide that act as prongs plugged in two separate, shallower pockets (Owen et al., 2000) (Fig. 10.2A). Specifically, the aromatic ring of a histidine at (AcK+2) interacts directly with aromatic side chains of Tyr 406 and Phe 367, which are conserved in the bromodomain family. In addition to the GCN5p bromodomain–H4-AcK16 complex structure, the understanding of ligand specificity of bromodomains is further enhanced by the recent structural studies of two bromodomains in complex with biologically relevant, nonhistone protein ligands. The first one is the highly selective association between the PCAF bromodomain and a lysine-acetylated *trans*-activator protein Tat of human immunodeficiency virus type 1 (HIV-1) (Mujtaba et al., 2002). The second is the interaction between the bromodomain of the coactivator CBP and a lysine-acetylated region in the C-terminal segment of the tumor suppressor protein p53 (Mujtaba et al., 2004). These structures also provide the first glimpses into structural features of bromodomain interactions with nonhistone proteins.

5.2. The PCAF Bromodomain–HIV-1 Tat Complex

The viral Tat protein stimulates transcriptional activation of the integrated HIV-1 genome and promotes viral replication in infected host cells (Adams et al., 1994; Cullen, 1998; Garber and Jones, 1999; Jeang et al., 1999; Karn, 1999). Tat transactivation activity is dependent on acetylation at Lys 50 by the HAT activity of p300/CBP and on its subsequent association with PCAF through a bromodomain-mediated interaction (Hottiger and Nabel, 1998; Kiernan et al., 1999; Ott et al., 1999). This bromodomain interaction results in the release of lysine-acetylated Tat from its association with the viral TAR RNA, leading to activation of HIV-1 transcription (Benkirane et al., 1998; Wei et al., 1998; Deng et al., 2000). Deletion of the PCAF C-terminal region comprising the bromodomain potently abrogated Tat transactivation of integrated, but not unintegrated HIV-1 provirus (Benkirane et al., 1998).

The NMR structure of the PCAF bromodomain in complex with an acety-lated Tat Lys 50 peptide (SYGR-AcK-KRRQR) showed that in addition to the acetyl-lysine, flanking residues both N- and C-terminal to the acetyl-lysine are important for this bromodomain interaction (Fig. 10.2B). The Tat peptide adopts an extended conformation in the complex, in which its N-terminal Y(AcK-3) residue contacts Tyr 802 and Val 763, and its C-terminal R(AcK+3) and Q(AcK+4) residues interact with E756. These specific interactions involving the acetyl-lysine moiety and its flanking residues, confirmed by site-directed mutagenesis, confer a highly selective association between the PCAF bromodomain and Tat (Mujtaba et al., 2002). The extensive number of contact points involved in ligand interactions may explain why the PCAF bromodomain binds to the Tat AcK50 peptide with a binding affinity ($K_D \sim 10 \ \mu M$) about 30-fold higher than that for a histone H4 AcK16 peptide ($K_D \sim 300 \ \mu M$) (Dhalluin et al., 1999; Mujtaba et al., 2002).

The PCAF and GCN5p bromodomains share a high degree of sequence identity ($\sim 40\%$), yet the structures of these modules in their complexes with different ligands suggest that they possess different binding specificities. The differences in ligand selectivity are striking in both the location and orientation of the bound peptides in the PCAF and GCN5p bromodomains. The backbones of the Tat and H4 peptides both adopt an extended conformation, but are antiparallel in the two corresponding structures with their N- and C-termini oriented nearly opposite to each other. Despite these differences, it is interesting to note that GCN5p binding of H4 H(AcK+2) residue is reminiscent of PCAF bromodomain recognition of Tat Y(AcK-3) residue via residues Tyr 802 and Val 763, which are equivalent to residues Tyr 406 and Phe 367 in GCN5p. Because of this similar mode of molecular interaction, the two aromatic residues, which are located in very different positions in the Tat and H4 peptides with respect to the acetyl-lysine, are found surprisingly to be in a nearly identical position in the corresponding bromodomain complex structures. The high level of conservation of these ligand recognition residues in bromodomains suggests that selection of an aromatic or hydrophobic residue neighboring the acetyl-lysine is possibly a common mechanism used by this subgroup of the bromodomain family, and that the ligand may be maneuvered into an orientation to accommodate this selection.

5.3. The CBP Bromodomain–p53 Complex

The human tumor suppressor p53 is another nonhistone protein that requires acetylation of its C-terminal lysine residues Lys 320, Lys 373, Lys 382 and to a lesser extent Lys 372 and Lys 381 for its activity (Barlev et al., 2001; Ito et al., 2001, 2002; Li et al., 2002). Recent in vivo studies show that acetylation-induced p53 activation as a transcription factor in response to DNA damage does not result from an increase of its DNA binding activity as hypothesized previously (Gu and Roeder, 1997; Sakaguchi et al., 1998; Liu et al., 1999), but rather from its recruitment of coactivators and subsequent histone acetylation (Barlev et al., 2001). Despite the identification of these multiple acetylation sites, specific effects of single or combined acetylation of these lysine residues on p53 activity remain elusive. A recent study using structure-based functional analysis demonstrates that the bromodomain of CBP binds selectively to p53 at the acetylated Lys 382 (Mujtaba et al., 2004). This molecular interaction is responsible for p53 acetylation-dependent coactivator recruitment after DNA damage, which is an essential step for p53-induced transcriptional activation of the cyclin-dependent kinase inhibitor p21 in G_1 cell cycle arrest (Mujtaba et al., 2004).

The structure of the CBP bromodomain–p53 AcK382 peptide complex extends our knowledge on ligand selectivity of the bromodomain family (Fig. 10.2C). Structural comparison of the CBP bromodomain–p53 AcK382 peptide (SHLKSKKGQSTSRHK-AcK-LMFK), the PCAF bromodomain/Tat AcK50 peptide (Mujtaba et al., 2002) and the GCN5p bromodomain–H4 AcK16 peptide (Owen et al., 2000) complexes further confirms that the mechanism of acetyl-lysine recognition is conserved. AcK recognition involves a nearly identical set of conserved residues in these different bromodomains, corresponding to Val 1115, Tyr 1167, Asn 1168, and Val 1174 in CBP. However, a different set of residues are used in the CBP bromodomain to recognize different amino acids flanking the AcK, including L(AcK+1), K(AcK-1), and H(AcK-2), to achieve its specificity. Notably, Val 763 in PCAF interacts with Y(AcK-3) in HIV-1 Tat (Mujtaba et al., 2002), whereas mutation of the corresponding Ile 1128 to alanine in CBP has only a partial reduction in p53 peptide binding. Moreover, Glu 756 in PCAF, which is important for interactions with R(AcK+3) and Q(AcK+4) at the AcK50 in Tat, is changed to Leu 1119 followed by the unique two-amino-acid insertion in CBP. The hydrophobic residues near this insertion are involved in CBP bromodomain binding to the L(AcK+1) and H(AcK-2) at the p53 AcK382 site. These distinct intermolecular interactions confer the binding preference of the CBP bromodomain for the AcK382 over AcK373 or AcK320 site in p53. Finally, the conformation of the bound peptides in CBP and PCAF bromodomains is also different due to the differences in their modes of ligand interactions. The p53 peptide forms a β-turn-like conformation (Mujtaba et al., 2004), whereas the HIV-1 Tat peptide adopts an extended conformation (Mujtaba et al., 2002). Taken together, these structural features of bromodomain–ligand complexes reinforce the notion that differences in ligand selectivity are attributed to a few but important differences in bromodomain sequences, mostly in the ZA loop.

6. EMERGING DEVELOPMENTS

The structural and biochemical understanding of bromodomain–acetyl-lysine binding also facilitates the recent investigations of bromodomain functions in vivo. For instance, during transcription, p300 has been shown to bind directly to histones, preferentially to histone H3 (Manning et al., 2001; An et al., 2002). p300–CBP has been reported to interact through its bromodomain with lysine-acetylated myogenic factor MyoD (Polesskaya et al., 2001). Bromodomains of the catalytic subunits of (Spt-Ada-Gens) SAGA and SWI/SNF may anchor these chromatin remodeling complexes to lysine-acetylated promoter nucleosomes (Hassan et al., 2002). Moreover, the fundamental importance of the bromodomain is further highlighted in a systematic study that demonstrates its functional role in the recruitment of transcription complexes to lysine-acetylated histones (Agalioti et al., 2002). The bromodomain modules from different proteins, or even from within the same proteins that contain two bromodomains, as exemplified by Bdf1p, are frequently found not to be biologically equivalent (Matangkasombut et al., 2000; Ladurner et al., 2003; Matangkasombut and Buratowski, 2003). Using fluorescence resonance energy transfer, Kanno and coworkers have recently demonstrated that bromodomain-containing proteins recognize different patterns of acetylated histones in the intact nuclei of living cells (Kanno et al., 2004). Specifically, they show that the bromodomain protein Brd2 selectively interacts with acetylated K12 on histone H4, whereas TAF_{II} 250 and PCAF recognize H3 and other acetylated histones, indicating a high degree of specificity toward histone recognition exhibited by different bromodomains.

However, in other cases, these double bromodomains found within the same protein may operate together to form a functional unit to act as a supermodule. Brd4, one such protein containing tandem bromodomains, requires both bromodomains to interact with acetylated chromatin during mitosis and in transmitting transcriptional memory to daughter cells (Dey et al., 2003). In the case of human TAF_{II} 250, it has been proposed that the tandem bromodomains operate together in the cooperative binding of two neighboring acetyl-lysine sites in a histone protein that are separated by a distance of 25 Å, as suggested by the crystal structure of the bromodomains (Jacobson et al., 2000). In addition, some other bromodomains may function in combination with other modules to form heteromeric supermodules, such as the TIF1β bromodomain that is implicated to form a functional unit with its adjacent PHD finger (Aasland et al., 1995), in transcriptional repression (Schultz et al., 2001). A number of other modular domains including the BAH (bromo-adjacent homology) domain (Callebaut et al., 1999) are also frequently found to be adjacent to the bromodomain, and the juxtaposition of these two domains suggests that they could also operate as supermodules. Taken together, these findings suggest that while bromodomains share a common basic biochemical function in acetyl-lysine binding that has been conserved throughout evolution, in vivo biological functions of individual bromodomains may be further modulated by the biological contexts in which they are found.

7. CONCLUDING REMARKS

Like the SH2 domain (Schlessinger and Lemmon, 2003) and PTB domain (Yan et al., 2002b) recognition of tyrosine-phosphorylated proteins in signal transduction (Pawson and Nash, 2003), the bromodomain also binds with high selectivity to lysine-acetylated proteins through interactions with amino acid residues flanking the acetyl-lysine (Zeng and Zhou, 2001). Since the residues important for acetyl-lysine recognition are largely conserved in bromodomains, binding of lysine-acetylated proteins is likely a general biochemical function for this family. The three-dimensional, left-handed four-helix bundle architecture provides a molecular framework for acetyl-lysine interaction using a hydrophobic cleft formed by the ZA and BC loops at one end of the bundle. Structural variations in the binding site are encoded by amino acid sequence variations in these loop regions to allow for discrimination of different interaction targets. Differences in ligand selectivity may be attributed to a few but important differences in bromodomain sequences. These include variations in the ZA and BC loops, which contain relatively low sequence conservation and amino acid deletion or insertion in different bromodomains. These sequence variations enable individual bromodomains to use distinct sets of amino acids to interact with residues flanking the acetyl-lysine in a target protein. Because of the limited number of biologically relevant bromodomain–ligand complex structures currently available, a consensus understanding of ligand binding specificity of different bromodomains is still lacking. Such new insights into ligand specificity will undoubtedly require additional structural analysis, which will help understand how functional diversity of this conserved structural fold is achieved through evolutionary modification of amino acid sequences that comprise the ligand-binding site. The emerging knowledge of the structure and function relationships of the bromodomain will enhance our mechanistic understanding of specific biological functions of bromodomain-containing proteins that have been implicated in human diseases including Williams syndrome (Lu et al., 1998; Bochar et al., 2000), lymphoma, and leukemia (Greenwald et al., 2004), as well as in control of a large network of molecular interactions that regulate chromatin remodeling and gene transcription.

Acknowledgments

We thank all past and present members of the Zhou laboratory who have contributed to the studies discussed in this chapter. This work is supported by grants from the National Institutes of Health.

REFERENCES

Aasland, R., Gibson, T.J., and Stewart, A.F. (1995). The PHD finger: implications for chromatin-mediated transcriptional regulation. *Trends Biochem. Sci.* 20:56–59.

Adams, M., Sharmeen, L., Kimpton, J., Romeo, J.M., Garcia, J.V., Peterlin, B.M., Groudine, M., and Emerman, M. (1994). Cellular latency in human immunodeficiency virus-infected individuals with high CD4 levels can be detected by the presence of promoter-proximal transcripts. *Proc. Natl. Acad. Sci. USA* 91:3862–3866.

Agalioti, T., Chen, G., and Thanos, D. (2002). Deciphering the transcriptional histone acetylation code for a human gene. *Cell* 111:381–392.

An, W., Palhan, V.B., Karymov, M.A., Leuba, S.H., and Roeder, R.G. (2002). Selective requirements for histone H3 and H4 N termini in p300-dependent transcriptional activation from chromatin. *Mol. Cell* 9:811–821.

Baltimore, D. (1981). Gene conversion: some implications for immunoglobulin genes. *Cell* 24:592–594.

Barlev, N.A., Liu, L., Chehab, N.H., Mansfield, K., Harris, K.G., Halazonetis, T.D., and Berger, S.L. (2001). Acetylation of p53 activates transcription through recruitment of coactivators/histone acetyltransferases. *Mol. Cell* 8:1243–1254.

Benkirane, M., Chun, R.F., Xiao, H., Ogryzko, V.V., Howard, B.H., Nakatani, Y., and Jeang, K.-T. (1998). Activation of integrated provirus requires histone acetyltransferase: p300 and P/CAF are co-activators for HIV-1 Tat. *J. Biol. Chem.* 273:24898–24905.

Bochar, D.A., Savard, J., Wang, W., Lafleur, D.W., Moore, P., Cote, J., and Shiekhattar, R. (2000). A family of chromatin remodeling factors related to Williams syndrome transcription factor. *Proc. Natl. Acad. Sci. USA* 97:1038–1043.

Brown, C.E., Howe, L., Sousa, K., Alley, S.C., Carozza, M.J., Tan, S., and Workman, J.L. (2001). Recruitment of HAT complexes by direct activator interactions with the ATM-related Tra1 subunit. *Science* 292:2333–2337.

Brownell, J.E., and Allis, C.D. (1996). Special HATs for special occasions: linking histone acetylation to chromatin assembly and gene activation. *Curr. Opin. Genet. Dev.* 6:176–184.

Brownell, J.E., Zhou, J., Ranalli, T., Kobayashi, R., Edmondson, D.G., Roth, S.Y., and Allis, C.D. (1996). Tetrahymena histone acetyltransferase A: a homolog to yeast Gcn5p linking histone acetylation to gene activation. *Cell* 84:843–851.

Cairns, B.R., Schlichter, A., Erdjument-Bromage, H., Tempst, P., Kornberg, R.D., and Winston, F. (1999). Two functionally distinct forms of the RSC nucleosome-remodeling complex, containing essential AT hook, BAH, and bromodomains. *Mol. Cell* 4:715–723.

Callebaut, I., Courvalin, J.C., and Mornon, J.P. (1999). The BAH (bromo-adjacent homology) domain: a link between DNA methylation, replication and transcriptional regulation. *FEBS Lett.* 446:189–193.

Chang, L., and Karin, M. (2001). Mammalian MAP kinase signaling cascades. *Nature* 410:37–40.

Chua, P., and Roeder, G.S. (1995). Bdf1, a yeast chromosomal protein required for sporulation. *Mol. Cell Biol.* 15:3685–3696.

Cullen, B.R. (1998). HIV-1 auxiliary proteins: making connections in a dying cell. *Cell* 93:685–692.

Deng, L., Fuente, C.d.l., Fu, P., Wang, L., Donnelly, R., Wade, J.D., Lambert, P., Li, H., Lee, C.-G., and Kashanchi, F. (2000). Acetylation of HIV-1 Tat by CBP/P300 increases transcription of integrated HIV-1 genome and enhances binding to core histones. *Virology* 277:278–295.

Dey, A., Chitsaz, F., Abbasi, A., Misteli, T., and Ozato, K. (2003). The double bromodomain protein Brd4 binds to acetylated chromatin during interphase and mitosis. *Proc. Natl. Acad. Sci. USA* 100:8758–8763.

Dhalluin, C., Carlson, J.E., Zeng, L., He, C., Aggarwal, A.K., and Zhou, M.-M. (1999). Structure and ligand of a histone acetyltransferase bromodomain. *Nature* 399:491–496.

Du, J., Nasir, I., Benton, B.K., Kladde, M.P., and Laurent, B.C. (1998). Sth1p, a Saccharomyces cerevisiae Snf2p/Swi2p homolog, is an essential ATPase in RSC and differs from Snf/Swi in its interactions with histones and chromatin-associated proteins. *Genetics* 150:987–1005.

Filetici, P., Aranda, C., Gonzalez, A., and Ballario, P. (1998). GCN5, a yeast transcriptional co-activator, induces chromatin reconfiguration of HIS3 promotor *in vivo. Biochem. Biophys. Res. Commun.* 242:84–87.

Fischle, W., Wang, Y., and Allis, C.D. (2003). Binary switches and modification cassettes in histone biology and beyond. *Nature* 425:475–479.

Garber, M.E., and Jones, K.A. (1999). HIV-1 Tat: coping with negative elongation factors. *Curr. Opin. Immunol.* 11:460–465.

Georgakopoulos, T., Gounalaki, N., and Thireos, G. (1995). Gentic evidence for the interaction of the yeast transcriptional co-activator proteins GCN5 and ADA2. *Mol. Gen. Genet.* 246:723–728.

Greenwald, R.J., Tumang, J.R., Sinha, A., Currier, N., Cardiff, R.D., Rothstein, T.L., Faller, D.V., and Denis, G.V. (2004). E mu-RD2 transgenic mice develop B-cell lymphoma and leukemia. *Blood* 103:1475–1484.

Grunstein, M. (1997). Histone acetylation in chromatin structure and transcription. *Nature* 389:349–352.

Gu, W., and Roeder, R.G. (1997). Activation of p53 sequence-specific DNA binding by acetylation of the p53 C-terminal domain. *Cell* 90:595–606.

Hajduk, P.J., Measdows, R.P., and Fesik, S.W. (1999). NMR-based screening in drug discovery. *Q. Rev. Biophys.* 32:211–240.

Hassan, A.H., Prochasson, P., Neely, K.E., Galasinski, S.C., Chandy, M., Carrozza, M.J., and Workman, J.L. (2002). Function and selectivity of bromodomains in anchoring chromatin-modifying complexes to promoter nucleosomes. *Cell* 111:369–379.

Haynes, S.R., Dollard, C., Winston, F., Beck, S., Trowsdale, J., and Dawid, I.B. (1992). The bromodomain: a conserved sequence found in human, *Drosophia* and yeast proteins. *Nucleic Acids Res.* 20:2603–2603.

Hottiger, M.O., and Nabel, G.J. (1998). Interaction of human immunodeficiency virus type 1 Tat with the transcriptional coactivators p300 and CREB binding protein. *J. Virol.* 72:8252–8256.

Hudson, B.P., Martinez-Yamout, M.A., Dyson, H.J., and Wright, P.E. (2000). Solution structure and acetyl-lysine binding activity of the GCN5 bromodomain. *J. Mol. Biol.* 304:355–370.

Ito, A., Lai, C.H., Zhao, X., Saito, S., Hamilton, M.H., Appella, E., and Yao, T.P. (2001). p300/CBP-mediated p53 acetylation is commonly induced by p53-activating agents and inhibited by MDM2. *EMBO J.* 20:1331–1340.

Ito, A., Kawaguchi, Y., Lai, C.H., Kovacs, J.J., Higashimoto, Y., Appella, E., and Yao, T.P. (2002). MDM2-HDAC1-mediated deacetylation of p53 is required for its degradation. *EMBO J.* 21:6236–6245.

Jacobson, R.H., Ladurner, A.G., King, D.S., and Tjian, R. (2000). Structure and function of a human TAFII250 double bromodomain module. *Science* 288:1422–1425.

Jeang, K.-T., Xiao, H., and Rich, E.A. (1999). Multifaceted activities of the HIV-1 transactivator of transcription, Tat. *J. Biol. Chem.* 274:28837–28840.

Jeanmougin, F., Wurtz, J.M., Le Douarin, B., Chambon, P., and Losson, R. (1997). The bromodomain revisited. *Trends Biochem. Sci.* 22:151–153.

Jenuwein, T., and Allis, C.D. (2001). Translating the histone code. *Science* 293:1074–1080.

John, S., and Workman, J.L. (1998). Just the facts of chromatin transcription. *Science* 282:1836–1837.

Johnson, G.L., and Lapadat, R. (2002). Mitogen-activated protein kinase pathways mediated by ERK, JNK, and p38 protein kinases. *Science* 298:1911–1912.

Kanno, T., Kanno, Y., Siegel, R.M., Jang, M.K., Lenardo, M.J., and Ozato, K. (2004). Selective recognition of acetylated histones by bromodomain proteins visualized in living cells. *Mol. Cell* 13:33–43.

Karn, J. (1999). Tackling Tat. *J. Mol. Biol.* 293:235–254.

Keyse, S.M. (2000). Protein phosphatases and the regulation of mitogen-activated protein kinase signalling. *Curr. Opin. Cell. Biol.* 12:186–192.

Kiernan, R.E., Vanhulle, C., Schiltz, L., Adam, E., Xiao, H., Maudoux, F., Calomme, C., Burny, A., Nakatani, Y., Jeang, K.-T., and Van Lint. C. (1999). HIV-1 Tat transcriptional activity is regulated by acetylation. *EMBO J.* 18:6106–6118.

Ladurner, A.G., Inouye, C., Jain, R., and Tjian, R. (2003). Bromodomains mediate an acetyl-histone encoded antisilencing function at heterochromatin boundaries. *Mol. Cell* 11:365–376.

Letunic, I., Goodstadt, L., Dickens, N.J., Doerks, T., Schultz, J., Mott, R., Ciccarelli, F., Copley, R.R., Ponting, C.P., and Bork, P. (2002). Recent improvements to the SMART domain-based sequence annotation resource. *Nucleic Acids Res* 30:242–244.

Li, M., Luo, J., Brooks, C.L., and Gu, W. (2002). Acetylation of p53 inhibits its ubiquitination by Mdm2. *J. Biol. Chem.* 277:50607–50611.

Liu, L., Scolnick, D.M., Trievel, R.C., Zhang, H.B., Marmorstein, R., Halazonetis, T.D., and Berger, S.L. (1999). p53 sites acetylated in vitro by P/CAF and p300 are acetylated in vivo in response to DNA damage. *Mol. Cell Biol.* 19:1202–1209.

Lu, X., Meng, X., Morris, C.A., and Keating, M.T. (1998). A novel human gene, WSTF, is deleted in Williams syndrome. *Genomics* 54:241–249.

Luger, K., Mäder, A.W., Richmond, R.K., Sargent, D.F., and Richmond, T.J. (1997). Crystal structure of the nucleosome core particle at 2.8 Å resolution. *Nature* 389:251–260.

Manning, E.T., Ikehara, T., Ito, T., Kadonaga, J.T., and Kraus, W.L. (2001). p300 forms a stable, template-committed complex with chromatin: role for the bromodomain. *Mol. Cell Biol.* 21:3876–3887.

Marcus, G.A., Silverman, N., Berger, S.L., Horiuchi, J., and Guarente, L. (1994). Functional similarity and physical association between GCN5 and ADA2: putative transcriptional adaptors. *EMBO J.* 13:4807–4815.

Matangkasombut, O., and Buratowski, S. (2003). Different sensitivities of bromodomain factors 1 and 2 to histone H4 acetylation. *Mol. Cell* 11:353–363.

Matangkasombut, O., Buratowski, R.M., Swilling, N.W., and Buratowski, S. (2000). Bromodomain factor 1 corresponds to a missing piece of yeast TFIID. *Genes Dev.* 14:951–962.

Mizzen, C., Kuo, M.-H., Smith, E., Brownell, J., Zhou, J., Ohba, R., Wei, Y., Monaco, L., Sassone-Corsi, P., and Allis, C.D. (1998). Signaling to chromatin through histone modifications: How clear is the signal? *Cold Spring Harbor Symp. Quant. Biol.* LXIII:469–481.

Muchardt, C., and Yaniv, M. (1999). The mammalian SWI/SNF complex and the control of cell growth. *Semin. Cell Dev. Biol.* 10:189–195.

Muchardt, C., Bourachot, B., Reyes, J.C., and Yaniv, M. (1998). ras transformation is associated with decreased expression of the brm/SNF2alpha ATPase from the mammalian SWI-SNF complex. *EMBO J.* 17:223–231.

Mujtaba, S., He, Y., Zeng, L., Farooq, A., Carlson, J.E., M.Ott, Verdin, E., and Zhou, M.-M. (2002). Structural basis of lysine-acetylated HIV-1 Tat recognition by P/CAF bromodomain. *Mol. Cell* 9:575–586.

Mujtaba, S., He, Y., Zeng, L., Yan, S., Plotnikova, O., Sanchez, R., Zeleznik-Le, N., Ronai, Z., and Zhou, M.-M. (2004). Structural mechanism of the bromodomain of the coactivator CBP in p53 transcriptional activation. *Mol. Cell* 13:251–263.

Ott, M., Schnolzer, M., Garnica, J., Fischle, W., Emiliani, S., Rackwitz, H.-R., and Verdin, E. (1999). Acetylation of the HIV-1 Tat protein by p300 is important for its transcriptional activity. *Curr. Biol.* 9:1489–1492.

Owen, D.J., Ornaghi, P., Yang, J.C., Lowe, N., Evans, P.R., Ballario, P., Neuhaus, D., Eiletici, P., and Travers, A.A. (2000). The structural basis for the recognition of acetylated histone H4 by the bromodomain of histone acetyltransferase gcn5p. *EMBO J.* 19:6141–6149.

Pawson, T., and Nash, P. (2003). Assembly of cell regulatory systems through protein interaction domains. *Science* 300:445–452.

Polesskaya, A., Naguibneva, I., Duquet, A., Bengal, E., Robin, P., and Harel-Bellan, A. (2001). Interaction between acetylated MyoD and the bromodomain of CBP and/or p300. *Mol. Cell Biol.* 21:5312–5320.

Sakaguchi, K., Herrera, J.E., Saito, S., Miki, T., Bustin, M., Vassilev, A., Anderson, C.W., and Appella, E. (1998). DNA damage activates p53 through a phosphorylation-acetylation cascade. *Genes Dev.* 12:2831–2841.

Schlessinger, J., and Lemmon, M.A. (2003). SH2 and PTB domains in tyrosine kinase signaling. *Sci. STKE* 2003:RE12.

Schultz, D.C., Friedman, J.R., and Rauscher, F.J., 3rd (2001). Targeting histone deacetylase complexes via KRAB-zinc finger proteins: the PHD and bromodomains of KAP-1 form a cooperative unit that recruits a novel isoform of the Mi-2alpha subunit of NuRD. *Genes Dev.* 15:428–443.

Schultz, J., Milpetz, F., Bork, P., and Ponting, C.P. (1998). SMART, a simple modular architecture research tool: identification of signaling domains. *Proc. Natl. Acad. Sci. USA* 95:5857–5864.

Sterner, D.E., Grant, P.A., Roberts, S.M., Duggan, L.J., Belotserkovskaya, R., Pacella, L.A., Winston, F., Workman, J.L., and Berger, S.L. (1999). Functional organization of the yeast SAGA complex: distinct components involved in structural integrity, nucleosome acetylation, and TATA-binding protein interaction. *Mol. Cell Biol.* 19:86–98.

Strahl, B.D., and Allis, C.D. (2000). The language of covalent histone modifications. *Nature* 403:41–45.

Struhl, K. (1998). Histone acetylation and transcriptional regulatory mechanisms. *Genes Dev.* 12:599–606.

Syntichaki, P., Topalidou, I., and Thireos, G. (2000). The Gcn5 bromodomain coordinates nucleosome remodelling. *Nature* 404:414–417.

Tamkun, J.W., Deuring, R., Scott, M.P., Kissinger, M., Pattatucci, A.M., Kaufman, T.C., and Kennison, J.A. (1992). brahma: a regulator of *Drosophila* homeotic genes structurally related to the yeast transcriptional activator SNF2/SWI2. *Cell* 68:561–572.

Travers, A. (1999). Chromatin modification: how to put a HAT on the histones. *Curr. Biol.* 9:23–25.

Turner, B.M. (2002). Cellular memory and the histone code. *Cell* 111:285–291.

Wei, P., Garber, M.E., Fang, S.M., Fischer, W.H., and Jones, K.A. (1998). A novel CDK9-associated C-type cyclin interacts with HIV-1 Tat and mediates its high-affinity, loop-specific binding to TAR RNA. *Cell* 92:451–462.

Winston, F., and Allis, C.D. (1999). The bromodomain: a chromatin-targeting module? *Nat. Struct Biol.* 6:601–604.

Wolffe, A.P., and Hayes, J.J. (1999). Chromatin disruption and modification. *Nucleic Acids Res.* 27:711–720.

Yan, K.S., Kuti, M., Mujtaba, S., Farooq, A., Goldfarb, M.P., and Zhou, M.-M. (2002a). SNT PTB domain conformation regulates interactions with divergent neurotrophic receptors. *J. Biol. Chem.* 277:17088–17094.

Yan, K.S., Kuti, M., and Zhou, M.-M. (2002b). PTB or not PTB—that is the question. *FEBS Lett.* 513:67–70.

Zeng, L., and Zhou, M.-M. (2001). Bromodomain: an acetyl-lysine binding domain. *FEBS Lett.* 513:124–128.

11

SMART Drug Design: Novel Phosphopeptide and ATP Mimetic-Based Small Molecule Inhibitors of the Oncogenic Protein Kinase pp60^{src} (Src)

T.K. Sawyer, R.S. Bohacek, W.C. Shakespeare, C.A. Metcalf, III, Y. Wang, R. Sundaramoorthi, T. Keenan, S. Narula, and D.C. Dalgarno

ABSTRACT

Over the past two decades, the oncogenic protein kinase pp60^{src} (Src) has been the focus of tremendous biological investigations that have identified it to be a promising therapeutic target for both cancer and bone disease drug discovery. The molecular, cellular and in vivo functional properties of Src provide a detailed framework for strategies to advance small molecule inhibitors relative to both its noncatalytic (e.g., SH2) and catalytic (i.e., kinase) domains. This chapter illustrates phoshopeptide mimetic-based small molecule Src SH2 inhibitors and ATP

T.K. SAWYER, R.S. BOHACEK, W.C. SHAKESPEARE, C.A. METCALF III, Y. WANG, R. SUNDARAMOORTHI, T. KEENAN, S. NARULA, AND D.C. DALGARNO • ARIAD Pharmaceuticals, Inc., 26 Landsdowne Street, Cambridge, MA 02139-4234, USA.

This chapter is dedicated to Manfred Weigele, our friend, colleague and outstanding chemist who inspires us by his passion for structure-based drug design, synthetic chemistry and the interdisciplinary development of novel, small-molecule therapeutics at ARIAD Pharmaceuticals.

Proteomics and Protein–Protein Interactions: Biology, Chemistry, Bioinformatics, and Drug Design, edited by Waksman. Springer, New York, 2005.

mimetic-based, small molecule Src kinase inhibitors. Key lead compounds exemplifying Src SH2 and Src kinase inhibitors are described with respect to structural biology, drug design and biological activity (in vitro and in vivo). The term SMART refers to small molecule ARIAD therapeutics that has been particularly focused on generating and optimizing novel lead compounds such as AP22408 and AP23464. AP22408 is a prototype bone-targeted Src SH2 inhibitor that blocks binding to phosphorylated ligands and was first to achieve in vivo proof-of-concept in a bone disease model. AP23451 is a second-generation, bone-targeted Src inhibitor and determined to be effective in both osteolytic bone metastasis and osteoporosis in vivo models. AP23464 is a prototype Src kinase inhibitor that is competitive to ATP and is extraordinarily potent in vitro and provides proof-of-concept in Src-dependent, cell assays representing both bone degrading osteoclasts and cancer cells. X-ray crystallographic structures of the aforementioned Src SH2 and Src kinase inhibitors provide insight to SMART drug design strategies. Second-generation Src kinase inhibitors are amidst preclinical and clinical drug development, and such small molecules illustrate varying template classes.

1. SRC AND NOVEL SRC INHIBITOR DRUG DISCOVERY

1.1. Src as a Therapeutic Target for Cancer and Bone Diseases

The pp60^{c-Src} (Src) tyrosine kinase is a nonreceptor tyrosine kinase that, by virtue of a plethora of molecular and cellular investigations to understand its complex signal transduction roles, has emerged as a promising therapeutic target for drug discovery for cancer and bone diseases (Brugge and Erikson, 1977; Collet and Erikson, 1978; Levinson et al., 1978; Hunter and Sefton, 1980; Soriano et al., 1991; Boyce et al., 1992; Talamonti et al., 1993; Maa et al., 1995; Lowell and Soriano, 1996; Tanaka et al., 1996; Verbeek et al., 1996; Abu-Amer et al., 1997; Mao et al., 1997; Staley et al., 1997; Thomas and Brugge; 1997; Duong et al., 1998; Ellis et al., 1998; Jeschke et al., 1998; Lutz et al., 1998; Turkson et al., 1998; Van Oijen et al., 1998; Biscardi et al., 1999; Catlett-Falcone et al., 1999; Egan et al., 1999; Irby et al., 1999; Karni et al., 1999; Wong et al., 1999; Marzia et al., 2000; Susa et al., 2000; Tsai et al., 2000; Martin, 2001; Avizienyte et al., 2002; Frame, 2002; Metcalf et al., 2002; Kauffman et al., 2003; Russello and Shore, 2003; Sawyer et al., 2003; Shakespeare et al., 2003; Summy and Gallick, 2003; Warmuth et al., 2003a). In retrospect, milestone studies of Rous sarcoma virus led to the identification of the first oncogene, v-*src*, which then provided impetus to research linking the constitutively activated tyrosine kinase gene product of v-*src* to its enhanced cell signaling and oncogenic transforming properties (Brugge and Erikson, 1977; Collet and Erikson, 1978; Levinson et al., 1978; Hunter and Sefton, 1980). Collectively, the above pioneering discoveries on v-Src and its cellular homolog, c-Src, have unveiled the critical mechanistic roles of this oncogenic protein kinase in a number of cellular processes, including proliferation, cell/cell adhesion, cell/matrix adhesion, cell migration, cell survival, vascular permeability and bone remodeling. Thus, Src stands among the first protein kinases to be comprehensively characterized by functional genomics, structural biology, cellular biology, and biochemical studies to understand its role in signal transduction pathways as well as its role in disease processes.

Elevated Src expression and/or activity has been correlated with tumor growth in specific cancers having ErbB2 or c-Met receptors (Maa et al., 1995; Mao et al., 1997; Biscardi et al., 1999), by studies using Src-specific antisense DNA (Staley et al., 1997; Ellis et al., 1998), and the recent identification of activating Src mutations in advanced human colon cancer (Irby et al., 1999). Specifically, elevated Src expression and/or activity has been found in breast cancer cell lines and malignant breast tumors (Verbeek et al., 1996; Egan et al., 1999). Src has been implicated in metastatic colon cancer (Talamonti et al., 1993), head and neck cancers (Van Oijen et al., 1998), and pancreatic cancer (Lutz et al., 1998). Recently, Src has been implicated in malignant transformations for certain cancers, such as breast cancer and multiple myeloma, via epidermal growth factor receptor (EGF-R) or interleukin-6 receptor (IL6-R) signaling pathways, respectively, that commonly activate the transcription factor known as signal transducer and activator of transcription-3 (STAT3) (Turkson et al., 1998; Karni et al., 1999; Tsai et al., 2000). The STAT3 pathway is different from the previously described integrin and receptor tyrosine kinase signal transduction pathways, and Src inhibitors have been shown to reverse the transformed cell phenotypes in specific examples of breast cancer and multiple myeloma (Turkson et al., 1998; Karni et al., 1999; Tsai et al., 2000). Aberrant activation of STAT signaling pathways have been linked to oncogenesis with respect to the prevention of apoptosis (Catlett-Falcone et al., 1999).

Studies involving genetically engineered *src* (–/–) knockout mice have provided compelling evidence for critical role of Src in bone remodeling (Lowell and Soriano, 1996), hence implicating the therapeutic opportunity of Src inhibitors for osteoporosis, Paget's disease, osteolytic bone metastasis and hypercalcemia associated with malignancy. In osteoclasts (bone-resorbing cells), Src is important for functional activity (ruffled border formation), survival, motility, and adhesion through various signal transduction pathways (Tanaka et al., 1996; Abu-Amer et al., 1997; Duong et al., 1998; Jeschke et al., 1998; Wong et al., 1999). Other key proteins, including the adapter protein Cbl and Pyk2, have been implicated as Src substrates and in Src-dependent signal transduction pathways in osteoclasts. In osteoblasts (bone-forming cells), Src recently has been implicated as a negative regulator of osteoblast functional activity (Marzia et al., 2000). However, a detailed understanding of Src signaling pathways in osteoblasts has not yet been elucidated.

1.2. Src Family Kinases (SFKs) Molecular and Cellular Biology

Src has been determined to be the prototype member of a group of structurally homologous proteins (Src, Fyn, Yes, Yrk, Lyn, Hck, Fgr, Blk, Lck, and the Frk subfamily Frk/Rak and Iyk/Bsk) that are referred to as Src family kinases (SFKs). Such SFKs are expressed in many different cell types. The highest levels of Src have been found primarily in platelets, neurons, and osteoclasts (Thomas and Brugge, 1997). Structurally, Src and related SFKs are defined by a sequence of common modular domains (Fig.11.1) that include an N-terminal myristoylated "unique" region (~70 amino acids), noncatalytic Src homology 2 (SH2) (~100 amino acids) and SH3

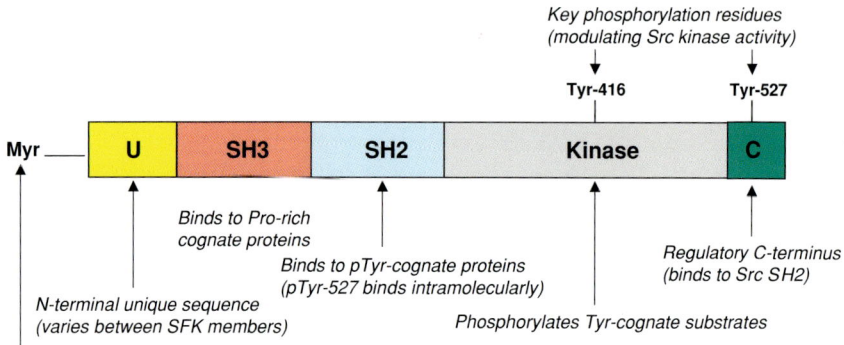

Figure 11.1. Organization of Src family kinases (SFKs) into regulatory domains (SH3 and SH2) and catalytic domain (tyrosine kinase), including additional key regulatory sequences and amino acid residues. See text for detailed discussion and references.

($\sim$60 amino acids) domains, a bilobal catalytic kinase domain ($\sim$300 amino acids), and a tyrosine-containing C-terminal regulatory tail (bound to the SH2 domain on tyrosine phosphorylation). The noncatalytic SH2 and SH3 domains mediate protein–protein interactions in cells through the recognition of sequence-specific p-Tyr containing peptides (Songyang et al., 1993; Sawyer et al., 2001, Vetter and Zhang, 2002; Vidal et al., 2001; Bradshaw and Waksman, 2002; Pawson, 2004), and proline-rich peptide sequences (Dalgarno et al., 1997; Gmeiner and Horita, 2001; Sawyer et al., 2001; Vidal et al., 2001; Musacchio, 2002), respectively. The tyrosine kinase catalytic domain contains N-terminal and C-terminal domains with an ATP and peptide substrate-binding site, and is partially regulated by the conformational status of its activation loop containing an autophosphorylation tyrosine residue (see below).

1.3. Src Regulatory Mechanisms (Intramolecular and Intermolecular)

Mechanistic insight into the regulatory conformational states of Src has been previously obtained through the collective X-ray structures of inactive Src and Hck kinases (SH3-SH2-catalytic domain) (Sicheri et al., 1997; Xu et al., 1997, 1999 Schindler et al., 1999) and active Lck kinase (isolated catalytic domain) (Yamaguchi and Hendrickson, 1996). As observed in the Src and Hck inactive structures, the protein adopts a compact conformation that is stabilized through various low-affinity intramolecular interactions (Fig. 11.2). Such interactions include both the phosphorylated C-terminal tail (Tyr-527 residue) which is bound to the SH2 domain, and the SH3 domain, which forms multiple interactions with the SH2-kinase linker (PPII helical conformation) and the N-terminal catalytic domain. The activation loop in the inactive Src structure adopts a conformation that contributes to the disruption of the active catalytic site (i.e., displacement of C-helix). This inhibitory conformation (via A-loop helix) also blocks access to the peptide substrate-binding site and prevents

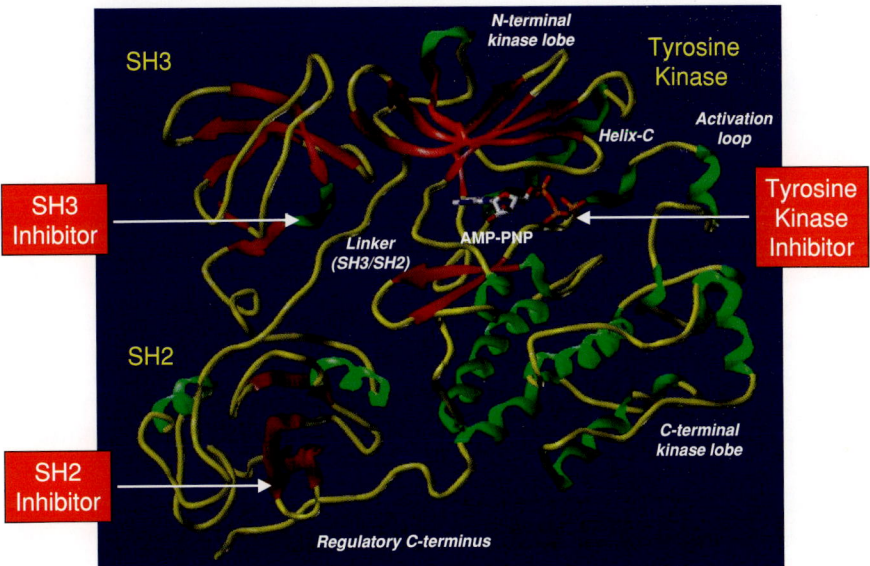

Figure 11.2. Ribbon representation of an X-ray structure of the Src kinase (SH3-SH2-catalytic construct) in the downregulated or inactive conformation. This Src X-ray structure contains a nonhydrolyzable ATP analog (AMP-PNP). Src inhibitors targeting the SH3, SH2, and kinase domain are highlighted by *arrows*. See text for detailed discussion and references.

autophosphorylation of Tyr-416. Indeed, it is proposed that the phosphorylation status of Tyr-416 in the activation loop plays a key role in regulating SFK active and inactive states. This is supported by the active Lck structure, in which the highly reorganized conformation of the activation loop, containing a phosphorylated Tyr-416 residue, is positioned to allow protein substrate access to a fully assembled catalytic site.

A proposed mechanism for activation of Src kinase and SFK proteins involves release of the SH3 and SH2 intramolecular interactions (either through competitive displacement by protein-binding partners or by dephosphorylation of Tyr-527 in the case of the SH2 domain) along with phosphorylation of Tyr-416 and a reorganization of the inhibitory conformation of the activation loop to provide an accessible active catalytic site. Examples of Src activation by interaction of its SH3 and/or SH2 domains with cognate proteins include: autophosphorylated platelet-derived growth factor receptor (PDGF-R) binding via Src SH2 (Erpel and Courtneidge, 1995), focal adhesion kinase (FAK) binding via both Src SH3 and SH2 (Thomas et al., 1998), and Sin binding via both Src SH3 and SH2 (Alexandropoulos and Baltimore, 1996). Such high-affinity, cognate protein ligands for the Src SH2 and/or SH3 domains are capable of out-competing the low-affinity, intramolecular interactions that individually exist within the assembled, inactive conformation of Src. Furthermore, intermolecular engagement of the SH2 and/or SH3 domains with such cognate proteins then likely destabilizes the inhibitory conformations within the catalytic domain to free Tyr-416 for phosphorylation as well as establishing binding sites for substrate and fixing inter-molecular binding interactions with complexed ATP relative to its triphosphate ester functionality to set the stage for Src-catalyzed substrate phosphorylation.

1.4. Strategies for Small-molecule Src Inhibitor Drug Discovery

Both structure-based design and screening-related lead identification approaches have significantly impacted the discovery of Src inhibitors. The scope of known Src inhibitors includes phosphopeptide mimetic-based peptidomimetics and de novo de-signed nonpeptides, ATP mimetic-based analogs, peptide substrate-based analogs, and other small-molecule inhibitors that have been derived from screening of nat-ural products and/or combinatorial libraries (Bridges, 1995, 2001a,b; Levitzki and Gazit, 1995; Dalgarno et al., 1997, 2000; Stankovic et al., 1997; Traxler, 1997, 1998; Lawrence and Niu, 1998; McMahon et al., 1998; Strawn and Shawver, 1998; Stover et al., 1999; Toledo et al., 1999; Cody et al., 2000; Muller, 2000; Sedlacek, 2000; Susa et al., 2000; Susa and Teti, 2000; Tsatsanis and Spandidos, 2000; Vu, 2000; Burke et al., 2001; Dumas, 2001; Sawyer et al., 2001, 2002; Shakespeare, 2001; Metcalf et al. 2002; Burke and Lee, 2003; Shakespeare et al., 2003a; Metcalf and Sawyer, 2004). Such Src inhibitors may be categorized into three major classes: (1) SH3 inhibitors; (2) SH2 inhibitors; and (3) tyrosine kinase inhibitors. In the case of tyrosine kinase inhibitors, both ATP- and/or substrate-binding site inhibitors are in-cluded. SH3 and/or SH2 inhibitors are expected to block intermolecular interactions between Src and its cognate proteins to abrogate SH3- and/or SH2-dependent protein complexes critical in signal transduction pathways. In contrast, tyrosine kinase in-hibitors (i.e., targeting the ATP- and/or peptide substrate-binding sites) are expected to block Src-dependent phosphorylation of cognate substrate proteins critical in signal transduction pathways. Noteworthy achievements to advance novel phosphopeptide mimetic-based SH2 inhibitors and ATP mimetic-based, Src kinase inhibitors illus-trate small molecule drug discovery that has ultimately led to clinical candidates (see below).

2. PHOSPHOPEPTIDE MIMETIC-BASED, SMALL MOLECULE SRC SH2 INHIBITORS

2.1. Phosphopeptide Binding to Src SH2 and Drug Design

Numerous X-ray and/or nuclear magnetic resonance (NMR) structures have been determined for Src SH2 and other SH2 domains (e.g., Lck, Abl, Grb2, Zap) and complexes thereof with phosphopeptide, peptidomimetic, or nonpeptide inhibitors (Waksman et al., 1992, 1993, 2004; Gilmer et al., 1994; Hatada et al., 1995; Plummer et al., 1996, 1997a,b; Rahuel et al., 1996; Lunney et al, 1997; Stankovic et al., 1997; Alligood et al., 1998; Rickles et al., 1998; Buchanan et al., 1999; Para et al., 1999; Shakespeare et al., 2000a,b; Violette et al., 2000, 2001; Bohacek et al., 2001; Kawahata et al., 2001; Sundaramoorthi et al., 2003).

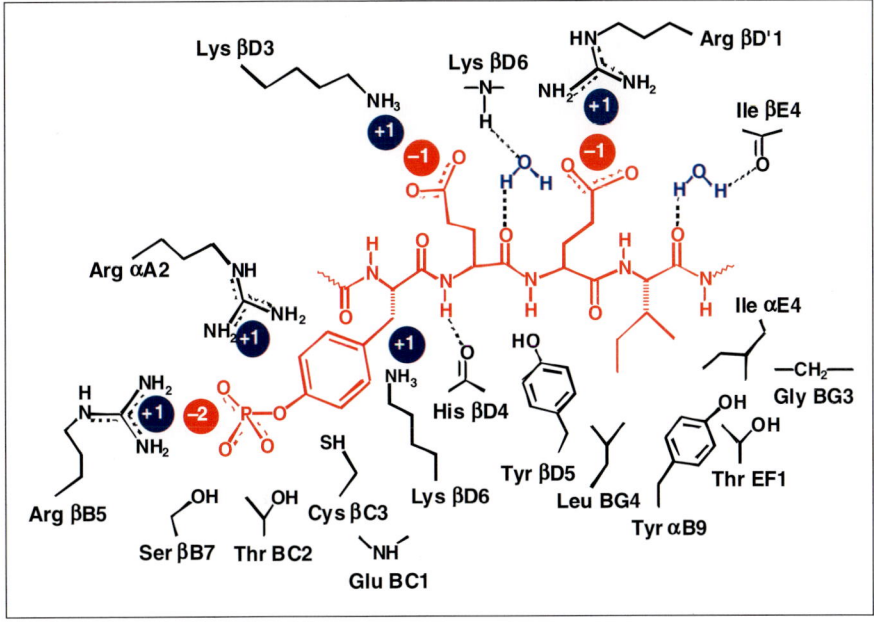

Figure 11.3. Schematic representation of the molecular interactions of pTyr-Glu-Glu-Ile binding to Src SH2. See text for detailed discussion and references.

With respect to small molecule SH2 inhibitor design, the X-ray structure of the complex between Src SH2 and a high-affinity, cognate phosphopeptide (e.g., containing pTyr-Glu-Glu-Ile) has provided important insights for a majority of the known peptidomimetic and nonpeptide lead compounds. Specifically, two essential binding pockets were revealed, including a positively charged, pocket for pTyr (i.e., the "pY site") and a hydrophobic pocket for Ile (i.e., the "pY + 3 site"). Relative to pTyr-Glu-Glu-Ile, essentially four key intermolecular contacts between the phosphopeptide ligand and the Src SH2 exist (Fig. 11.3). The first includes the paramount binding of pTyr to the pY site which contains several positively charged and/or H-bonding donor residues (see review by Waksman et al., 2004 for details and nomenclature): Arg αA2, Arg βB5, Lys βD6, Thr BC2, Glu BC1, and Ser βB7. The most critical of these residues is Arg βB5 (of the Phe-Leu-Val-Arg-Glu-Ser sequence) which forms two H-bonds with the phosphate oxygens of the pTyr sidechain. The second is a hydrophobic binding pocket for the Ile sidechain of the peptide ligand and it is comprised of several residues, including Tyr βD5, Ile βE4, Thr EF1, and Gly βG3. The third is that of a critical and direct intermolecular H-bond exists between the pY + 1 Glu backbone NH and the backbone C=O of His βD4. Finally, the fourth key contact is that of a water-mediated H-bond that exists between the pY + 1 Glu backbone C=O and the NH of Lys βD6.

In retrospect, the Src SH2 inhibitor drug discovery has focused on peptidomimetic modifications of cognate peptide sequences, structure-based de novo nonpeptide templates, and incorporation of various novel pTyr mimics (Gilmer et al.,

1994; Stankovic et al., 1997a,b; Metcalf et al., 1998, 2000, 2002; Rickles et al., 1998; Cody et al., 2000; Muller, 2000; Shakespeare et al., 2000a,b, 2003; Vu, 2000; Violette et al., 2000, 2001; Bohacek et al., 2001; Garcia-Echevarria, C., 2001; Kawahata et al., 2001; Sawyer et al., 2001, 2002; Sundaramoorthi et al., 2003). Such strategies are briefly described below.

2.2. Phosphotyrosine Mimicry and Drug Design

A major challenge of SH2 inhibitor drug discovery has been the pTyr moiety in terms of developing metabolically stable pTyr mimics that exhibit high affinity to a SH2 domain (Stankovic et al., 1997a; Burke et al., 2001; Shakespeare, 2001; Sawyer et al., 2002). Overall, the pTyr issue has been addressed by quite different approaches: (1) exploiting phosphonate groups to gain cellular and tissue selectivity as exemplified by the bone-targeted 3′,4′-diphosphonophenylalanine (Dpp) moiety of nonpeptide Src SH2 inhibitor AP22408 (Shakespeare et al., 2000b; Violette et al., 2001; Sawyer et al., 2002). (2) reduction of the charged nature of a phosphate or phosphonate group by replacement with carboxylate or phosphinate moieties as exemplified in Src, Lck and Grb2 SH2 inhibitors (Stankovic et al., 1997; Vu et al., 2000; Burke et al., 2001; Kawahata et al., 2001; Sawyer et al., 2002; Sundaramoorthi et al., 2003) (3) exploiting chemically reactive groups in the pTyr site as exemplified by the Cys residue in Src SH2 with pTyr mimics incorporating aldehyde or related electrophilic moieties (Alligood et al., 1998; Violette et al., 2000); and (4) masking the charged nature of a phosphonate or carboxylate group by prodrug moieties (Stankovic et al., 1997a,b; Rickles et al., 1998).

Recently, we have reported the X-ray crystallographic structure of Src SH2 complexed with citrate, a tricarboxylic acid, which forms multiple intermolecular contacts at the pTyr binding pocket (Bohacek et al., 2001). Similarly to the phosphate group of pTyr, the citrate forms several ionic and hydrogen bonds with the Src SH2 pTyr binding site in terms of both protein backbone and side-chain atoms (Fig. 11.4). Specifically, these include ionic interactions with the conserved Arg αA2 and Arg βB5

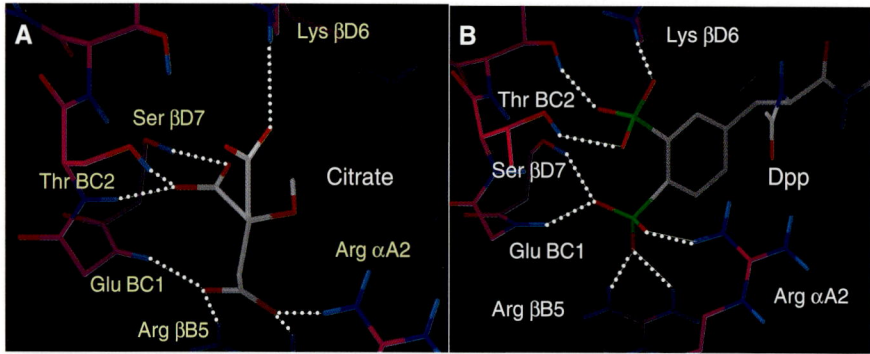

Figure 11.4. Comparative X-ray structures of citrate (**A**) and pTyr (**B**) complexed with Src SH2 phosphotyrosine binding pocket. See text for detailed discussion and references.

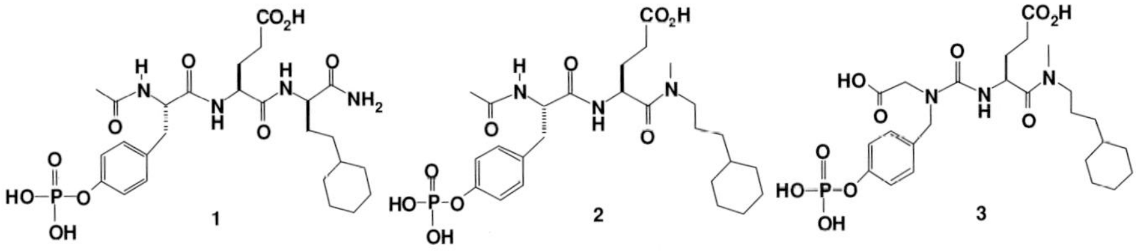

Figure 11.5. Chemical structures of peptidomimetic Src SH2 inhibitors **1–3**. See text for detailed discussion and references.

residues, as well as hydrogen bonds with Ser βB7, Thr BC2, and the backbone NH of Glu BC1. However, the citrate complex with Src SH2 revealed new molecular interactions not previously observed for pTyr. Specifically, these include an ionic bond between citrate and Lys βD6 as well as an intermolecular hydrogen bond with the backbone NH of Thr BC2. The unique interactions of citrate has provided impetus to the design of several novel pTyr mimetics (see below).

2.3. Novel Peptidomimetic Src SH2 Inhibitors

Iterative modification of the tetrapeptide sequence pTyr-Glu-Glu-Ile led to the series of peptidomimetic Src SH2 inhibitors **1–3** (Fig. 11.5) (Plummer et al., 1996, 1997a,b). The design of prototype peptidomimetics as exemplified by Src SH2 inhibitor **1** illustrated the use of D-amino acid replacement of the pY + 2 Glu of the cognate tetrapeptide ligand and elimination of the pY + 3 Ile (Plummer et al., 1996). Stepwise transformation of this series to dipeptide ligands such as compound **2** provided lead compounds for pTyr modifications (Plummer et al., 1997b). The ureido-modified, peptidomimetic **3** was designed to bind Src SH2 in such a manner that the single amino acid Glu at the pY + 1 position would access the pTyr binding site via a 4-phosphophenethyl-N-substituted Gly moiety as well as bind to the hydrophobic pY + 3 site via N,N-disubstituted C-terminal amide moiety (Plummer et al., 1997a). Indeed, the X-ray structure of peptidomimetic **3** complexed to Src SH2 revealed such molecular interactions. However, a somewhat surprising finding was that peptidomimetic **3** bound to the Src SH2 with a *cis*-conformation at the C-terminal amide linkage and that the structural water typically observed in Src SH2-phosphopeptide complexes (i.e., pY + 1 backbone carbonyl–H_2O–protein) was displaced by the *cis*-amide group (Plummer et al., 1997a). The latter finding indicated that compound **3** may actually not be a peptidomimetic (relative to its design), but rather be a prototypic nonpeptide (relative to its X-ray structure).

2.4. Novel Nonpeptide Src SH2 Inhibitors

In a related investigation that was focused on structure-based de novo designed nonpeptide Src SH2 inhibitors **4–6** (Fig. 11.5) using a benzamide template (Lunney et al., 1997; Para et al., 1999), X-ray studies revealed that the pY + 1 water molecule was effectively substituted (as predicted from molecular modeling) by the

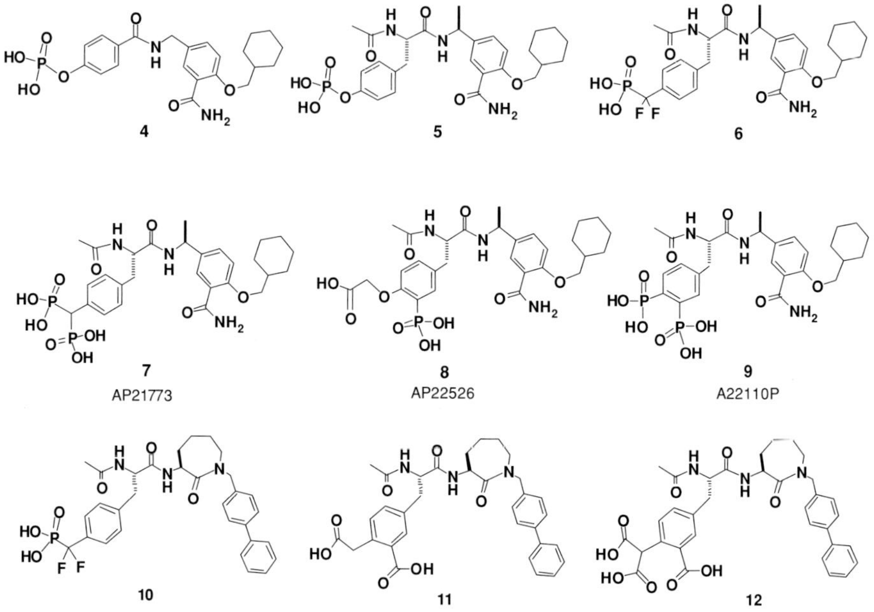

Figure 11.6. Chemical structures of nonpeptide Src SH2 inhibitors **4–12** incorporating pTyr mimetics. See text for detailed discussion and references.

carboxamide group of the template, whereas the phosphobenzoic acid and cyclohexyl moieties of the compound bound to the pY and pY + 3 sites, respectively. Compound **6** exemplified a benzamide template-based Src SH2 inhibitor that incorporated a non-hydrolyzable, 4-phosphodifluoromethyl-phenylalanine (F$_2$Pmp) mimic of pTyr that could be further modified by prodrug chemistry to provide efficacy in a Src-dependent cell assay (Rickles et al., 1998; Para et al., 1999). A series of second-generation, non-peptide inhibitors **7–12** of Src SH2 (Fig. 11.6) exemplify further exploitation of novel pTyr mimics and modifications of the benzamide template to confer bone-targeted and cellularly effective lead compounds (Shakespeare et al., 2000b; Bohacek et al., 2001; Kawahata et al., 2001; Violette et al., 2001; Sundaramoorthi, 2003). Conceptually significant to this work was the determination of an X-ray structure of Src SH2 complexed with citrate which showed multiple H-bonding and ionic interactions of citrate in the pY site (see above) and the ensuing structure-based design of a novel pTyr mimic 4-diphosphonomethyl-phenylalanine (Dmp) (Bohacek et al., 2001). In brief, the Dmp moiety was incorporated into nonpeptide **7** and was found to be significantly more potent and cellularly active than its monophosphonate analog (Bohacek et al., 2001). As predicted from molecular modeling, the Dmp moiety complexed with additional intermolecular H-bonding and electrostatic interactions relative to pTyr. Furthermore, the Dmp moiety endowed compound **7** with bone-targeting properties (hydroxyapatite binding affinity derived from the diphosphomethyl group) and antiresorptive activity in a cell-based assay utilizing osteoclasts and bone. A recently described (Kawahata et al., 2001) Src SH2 inhibitor **7** further exploits the pY site in terms of intermolecular H-bonding and ionic interactions, as previously observed in the Src SH2-citrate complex (Kawahata et al., 2001), by the design of a novel pTyr mimic 4′-carboxymethyloxy-3′-phosphono-phenylalanine (Cpp). The Cpp moiety of **8** was determined by X-ray structure studies to interact with the pY site such that its carboxy group is oriented similar to that of the phosphate group of pTyr. However, the phenyl ring of Cpp was found to be slightly tilted to accommodate binding of its phosphonate group with the ε-amino sidechain of Lys βD6 (Kawahata et al., 2001). These studies also showed that elimination of the 3′-phosphonate group of Cpp moiety results in an analog exhibiting markedly decreased Src SH2 binding affinity. Lastly, Src SH2 inhibitor **9** illustrates yet another pTyr mimic 3′,4′-diphosphono-phenylalanine (Dpp). The Dpp moiety confers bone-targeting properties and enhanced Src SH2 binding affinity relative to its *des*-3′-phosphonate analog as predicted by molecular modeling studies. This prototype series nonpeptide SH2 inhibitors provided proof-of-concept toward multiple functional group replacement of pTyr moiety through to structure-based design of pTyr mimics capable of molecular interactions at the pY site similar to that determined for citrate. In an independent approach, using structure-based design, a novel series of cyclic lactam-based, nonpeptide inhibitors of Src SH2 has been successfully advanced (Lesuisse et al., 2001a,b, 2002; Deprez et al., 2002a, b; Lange et al., 2002). Specifically, compounds **10–12** illustrate the use of a caprolactam template in which the carbonyl moiety was designed, and confirmed by X-ray structure, to displace the same structural water as the aforementioned ureido-type peptidomimetic/nonpeptide **3** and benzamide-based nonpeptide **4**. Functionalization of the caprolactam template by pTyr mimics and pY + 3 hydrophobic groups provided highly Src SH2 inhibitory potency Src SH2 inhibitors. Particularly noteworthy was the tricarboxy-modified pTyr mimic incorporated in the nonpeptide **12** relative to exhibiting high binding affinity to Src SH2.

2.5. AP22408 Series of Src SH2 Inhibitors

The first in vivo effective Src SH2 inhibitor AP22408 (**13**) (Shakespeare et al., 2000b) illustrates yet further structure-based design to optimize the nonpeptide template by virtue of both bicyclic benzamide modification and incorporation of Dpp as the pTyr mimic. The bicyclic benzamide moiety of AP22408 (Fig. 11.7) provides both increased hydrophobic interactions with Src SH2 as well as entropic advantage by virtue of locking the conformation of the template into a preferred arrangement with respect to its Src SH2 binding interactions (Shakespeare et al., 2000b). As predicted, Dpp moiety of AP22408 further endows bone-targeting properties (Violette et al., 2001) to selectivity inhibit osteoclast-mediated resorption of bone in both cell-based and in vivo assays. Relative to AP22408, elimination of the 3′-phosphonate group of the Dpp moiety resuled in an analog (**14**) that was significantly less potent in binding Src SH2 as well as osteoclast *in vitro* assays. Also, the simple benzylamide derivative (**15**) of Dpp was inactive in terms of Src SH2 binding, cell-based bone resorption, and

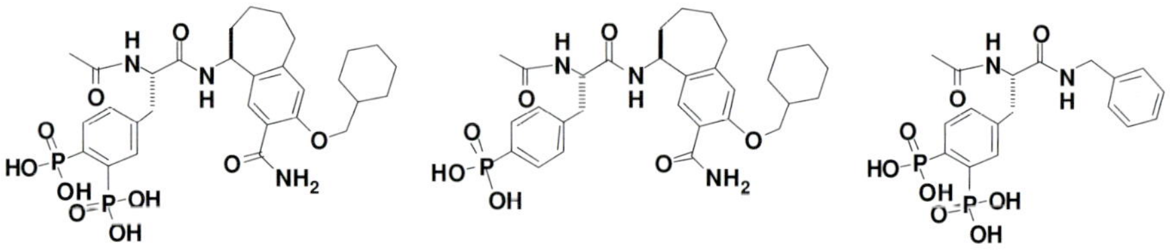

13
AP22408

14

15

Figure 11.7. Chemical structures of nonpeptide Src SH2 inhibitor **13** (AP22408) and analogs **14** and **15** which incorporate key modifications of the Dpp and bicyclic benzamide template, respectively. See text for detailed discussion and references.

in vivo bone resorption to further support the fact that the Dpp moiety did not possess intrinsic antiresorptive activity (Shakespeare et al., 2000b; Violette et al., 2001).

Molecular modeling studies comparing Ac-pTyr-Glu-Glu-Ile-NH$_2$ and AP22408 show (Shakespeare et al., 2000b) the unique 3D binding interactions of the functionalized, bicyclic benzamide template of the nonpeptide inhibitor (Fig. 11.8). Specifically, the central phenyl ring of AP22408 stacks perpendicular to the phenyl ring of Tyr-181 and the fused cycloheptyl ring provides extended hydrophobic interactions with the Src SH2 surface about the Tyr-181. Also, the cycloheptyl ring provides conformational constraint to orient the carboxamide group to H-bonding interactions directly

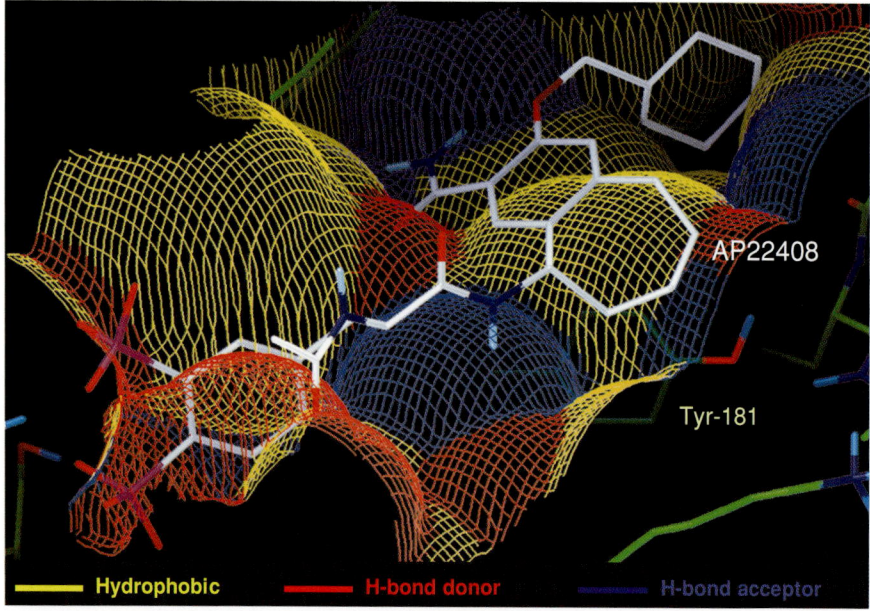

Figure 11.8. Comparative 3D molecular models of pTyr-Glu-Glu-Ile (**A**) and AP22408 (**B**) complexed with Src SH2. See text for detailed discussion and references.

with the Src SH2 domain (i.e., Lys-182) as well as guiding the cyclohexylmethy-loxy group into the Ile binding pocket. These predicted intermolecular interactions of AP22408 were supported by a 2.4 Å X-ray crystallographic structure of it com-plexed with Lck (Ser164Cys) SH2 domain (Shakespeare et al., 2000b). Relative to these molecular modeling and X-ray crystallographic studies, the Dpp moiety of AP22408 was also determined to interact with key residues of the pTyr binding site as predicted (Shakespeare et al., 2000b). Noteworthy, this AP22408 complex with Lck(Ser164Cys) SH2 revealed that the $3'$-PO_3H_2 group of Dpp formed ionic inter-actions with Lys BC6 and a hydrogen bond with Ser BC2 (Thr BC2 for Src SH2). Furthermore, both molecular modeling studies and X-ray crystallographic studies of AP22408 implicate that sp^3 hybridization of the $3'$-phosphonate group enable multi-ple intermolecular H-bonding and ionic interactions with the pTyr binding pocket. As predicted the bicyclic benzamide template provided complementarity in its contour over the hydrophobic surface and projecting the benzamide carbonyl group to form a hydrogen bond with the backbone NH of Lys $\beta D6$ displacing one of the two wa-ter molecules observed in the phosphopeptide complex. The second water molecule is not displaced and is hydrogen bonded to the benzamide NH group of AP22408 and the backbone carbonyl of Ile $\beta E4$, Finally, as predicted, the cyclohexyl group of AP22408 extends into the hydrophobic pY+3 pocket.

Recently reported (Sundaramoorthi et al., 2003) structure-activity studies have focused on AP22408 analogs **16–21** (Fig. 11.9) to explore multiple functional group modifications of the Dpp moiety aimed at incorporating carboxylate groups to de-crease anionic charge as well provide potential for prodrug modifications. The Tyr($4'$-CH_2CO_2H)-modified analog **16** was 100-fold less potent than the pTyr-containing par-ent inhibitor (Fig. 11.6). The $4'$-bis(carboxymethyl)-amino-Phe-modified analog **17**

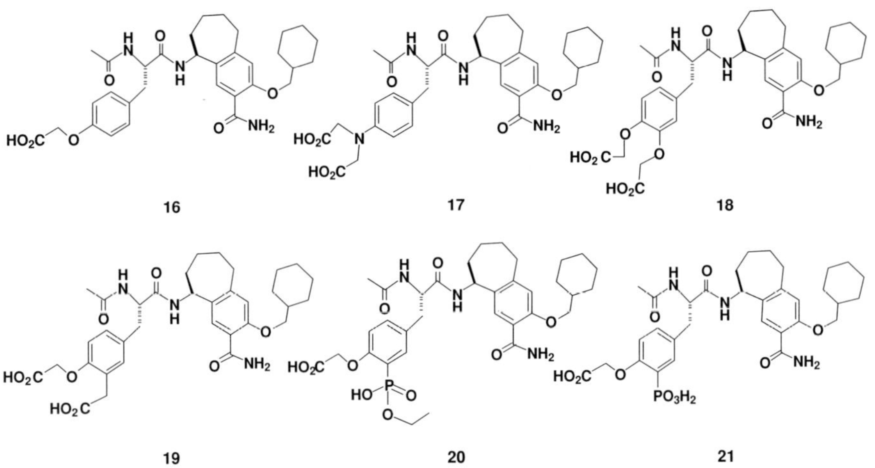

Figure 11.9. Chemical structures of AP22408 analogs **16–21** incorporating varying pTyr mimics. See text for detailed discussion and references.

was essentially equipotent to the compound **15**, and this result suggested that no significant additional binding was afforded by introduction of a second carboxymethyl functionality in this particular spatial configuration. However, the 4′,3′-dicarboxylic acid-containing analogs **18** and **19** revealed that 3′-carboxymethyloxy or 3′-carboxymethyl groups, respectively, significantly contributed to increased Src SH2 binding affinities. Importantly, these results confirmed that nonphosphonate groups at the 3′-position were capable of markedly enhancing binding affinities of pTyr mimetics having non-phosphonate groups at the 4′-position. Furthermore, compounds **20** and **21** showed that introduction of 3′-phosphonate modifications, as either monoethyl ester or the free acid, were also very effective in combination with a 4′-carboxymethyloxy group to achieve potent Src SH2 binding activity versus the monosubstituted parent compound **16**.

3. ATP MIMETIC-BASED SMALL MOLECULE SRC KINASE INHIBITORS

3.1. ATP/Peptide Substrate Binding to Src Kinase and Drug Design

As previously described, X-ray structures of Src tyrosine kinase and related SFKs (Yamaguchi and Hendrickson, 1996; Sicheri et al., 1997; Xu et al., 1997, 1999; Schindler et al., 1999) have provided detailed information on the catalytic domain, including complexes with the nonhydrolyzable ATP analog, AMP-PNP (where AMP refers to adenosine-5′-monophosphate and PNP refers to the N-linked, pyrophosphate derivative PO_2-NH-PO_3H). However, until most recently (see below). Src tyrosine kinase structures have not revealed the catalytically competent form of the enzyme, since key residues (e.g., Tyr-416 and Glu-310) and active site sequences or loops exist in the "inactive conformation" for which neither binding of substrate nor its phosphorylation by ATP was possible. Therefore, such structure-based design efforts have mostly exploited three-dimensional (3D) homology models of Src tyrosine kinase and have taken advantage of insights gained from X-ray structures of other protein tyrosine kinases and complexes thereof with inhibitors (Bridges, 1995; Hubbard, 1997; Mohammedi et al., 1998; Lamers et al., 1999; McTigue et al., 1999; Zhu et al., 1999; Schindler et al., 2000). The design of Src kinase inhibitors has focused on a number of strategies (for reviews see Levitzki and Gaztt 1995; Dalgarno et al., 1997; Klohs et al., 1997; Lawrence and Niu, 1998; Traxler, 1997; Traxler, 1998; Stover et al., 1999; Toledo et al., 1999; Sedlacek, 2000; Susa and Teti, 2000; Tsatsanis and Spandidos, 2000; Dumas, 2001; Sawyer et al., 2001, 2003; Boschelli, 2002; Metcalf et al., 2002; Shakespeare et al., 2003); including ATP mimetic- and peptide substrate-based compounds as well as molecules derived from natural products and combinatorial libraries (see below). A simplistic model of the Src kinase active site (Fig. 11.10) shows predicted ATP and peptide substrate binding sites and further illustrates the hydrophobic specificity pocket which differs to varying extents within the protein kinase superfamily in terms of size and molecular recognition properties to inhibitors interacting with it.

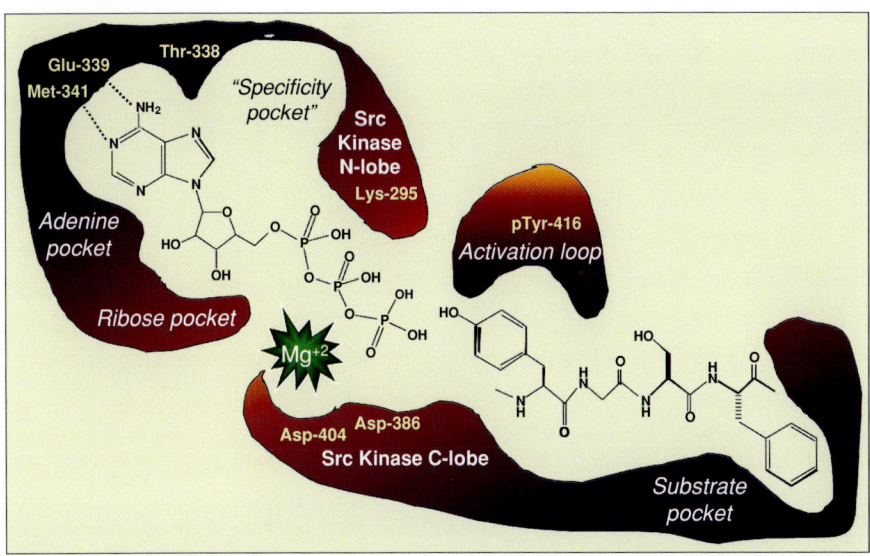

Figure 11.10. Schematic representation of Src kinase active site showing predicted binding interactions of ATP and peptide substrate. The Src kinase hydrophobic specificity pocket is further illustrated unoccupied in the case of ATP, but providing a promising binding site for Src kinase inhibitors with respect to drug design. See text for detailed discussion and references.

3.2. Purine Template-Based Src Kinase Inhibitors

The purine template has provided the basis for the development of promising Src kinase inhibitors as exemplified by **22–29** (Fig. 11.11). The purine-based Src inhibitor **22** (NVP-AAK980) has been described to be a potent inhibitor of Src kinase and effecting anti-resorptive activity in vivo (Missbach et al., 2000). In retrospect, purine template-based protein kinase inhibitors having 2,6,9-trisubstitutions have been advanced against cyclin-dependent kinases (CDKs) and X-ray structures of CDK2 complexed with such molecules have been determined (Gray et al., 1998; Legraverend et al., 2000; Dreyer et al., 2001) to further delineate their mode of binding at the ATP site. Albeit its Src inhibitory activity was not originally reported, Purvalanol B **(23)** was found to be a potent inhibitor of CDK2 kinase and an X-ray structure of it complexed with CDK2 kinase provide insight into its binding interactions (Gray et al., 1998). Relative to the development of a 3D molecular model of Src kinase (active conformation) a series of purine template-based Src kinase inhibitors has been advanced as exemplified by **24–29** (Sawyer et al., 2003). As highlighted by compounds AP23451 **(26)** and AP23464 **(29)**, this series of Src kinase inhibitors illustrate structure-based drug design strategies focused on bone-targeting and protein kinase selectivity, respectively, within the scope of a small molecule ARIAD therapeutics (dubbed "SMART") technology (see below) to advance proof-of-concept lead compounds to cancer and bone diseases (Dalgarno et al., 2003; Sawyer et al., 2003; Wang et al., 2003).

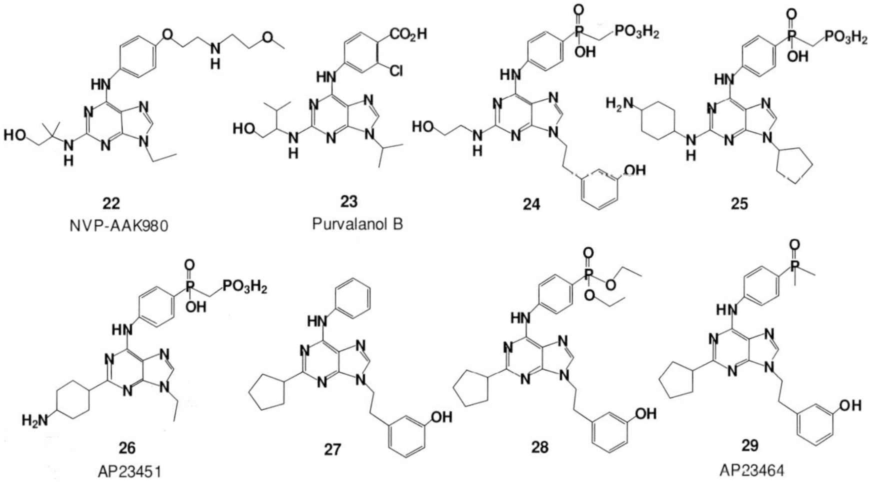

Figure 11.11. Examples of purine template-based Src kinase inhibitors **22–29**. Noteworthy are compounds **26** (AP23451) and **29** (AP23464) with respect to bone-targeting and protein kinase SMART drug design. See text for detailed discussion and references.

3.3. AP23451 and AP23464 Series of Src Kinase Inhibitors

Two purine template-based Src kinase inhibitors, namely AP23451 **(26)** and AP23464 **(29)** (Fig.11.11), illustrate the successful use a 3D molecular model of Src kinase to guide lead optimization. Specifically, a 3D molecular model of the catalytically active conformation of Src kinase was constructed using X-ray crystallographic structures of Src kinase in its inactive conformation and complexed with the nonhydrolyzable ATP mimic AMP-PNP (Xu et al., 1997), insulin receptor kinase (IRK) in its active conformation complexed with AMP-PNP and a peptide substrate (Hubbard, 1997), and Lck kinase apoprotein in its active conformation (Yamaguchi and Hendrickson, 1996).

Recently, we have described the structure-based design of a novel series of bone-targeted, Src kinase inhibitors **24–26** illustrating chemical diversity of the trisubstituted purine template and culminating in the X-ray structure of the AP23451 **(26)** complexed with Src kinase in its active conformation (von Stechow et al., 2001; Dalgarno et al., 2003; Boyce et al., 2003; Sawyer et al., 2003; Shakespeare et al., 2003; Wang et al., 2003). Noteworthy was AP23451 as a lead compound in terms of its in vivo efficacy in animal models of osteoporosis and osteolytic bone metastasis (Boyce et al., 2003; Shakespeare et al., 2003). The X-ray structure of AP23451 complexed with Src kinase (Fig. 11.12) revealed that is purine ring was oriented differently than that predicted for ATP. The ethyl group (R1 moiety in generic formula) of AP23451 projects partially into the hydrophobic specificity binding pocket and 4-amino-cyclohexyl group (R3 moiety) into the ribose binding site to make extensive contact with protein surface at the cleft between the N- and C-terminal lobes of Src

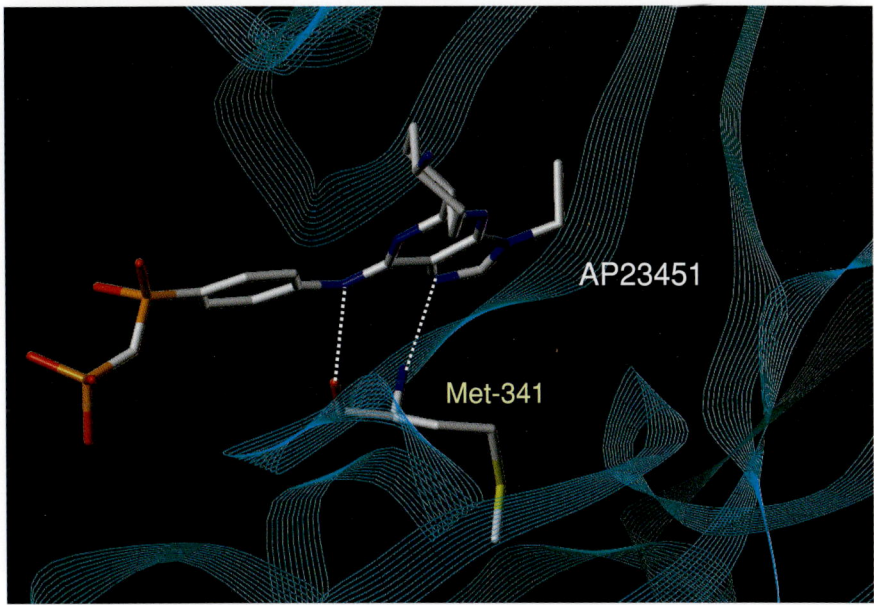

Figure 11.12. X-ray structure of AP23451 complexed with Src kinase illustrating its orientation in a manner to show H-bonding interactions between the protein (Met-341) and the purine template of the compound. See text for detailed discussion and references.

kinase. The overall mode of binding determined for AP23451 resemble that previously determined in CDK2 kinase complexes with purine template-based inhibitors (Gray et al., 1998; Legraverend et al., 2000; Dreyer et al., 2001). Key residues comprising the ATP binding site providing hydrophobic contacts with AP23451 included Leu-273, Val-281, Ala-293, Thr-338, Tyr-340, and Leu-393. Two key H-bonding interactions between Src kinase (Met-341 backbone amide and carbonyl) and AP23451 were identified. A single water-mediated hydrogen bond exists between the terminal phosphate group of the bone-targeting aniline group (R2 moiety) to the Lys-343 backbone carbonyl. A more complex, multiple water mediated H-bonding network involving the sidechains of Lys-295, Glu-310, and Asp-404 with the purine ring of AP23451 was also identified.

The aforementioned 3D Src kinase model also provided the framework for structure-based design of the novel Src kinase inhibitor series **27–29** (Fig. 11.11) which emphasized protein kinase selectivity and culminating in the X-ray structure of the AP23464 **(29)** complexed with Src kinase in its active conformation (Dalgarno et al., 2003; Metcalf et al., 2003a, 2004; O'Hare et al., 2003; Sawyer et al., 2003, 2004). By way of systematic modifications of the trisubstituted purine template guided by molecular modeling, the lead compound AP23464 **(29)** was advanced and was determined to possess highly potent Src and Abl kinase (including Bcr-Abl and mutants thereof) inhibitor potency (O'Hare et al., 2003; Metcalf et al., 2004) as well

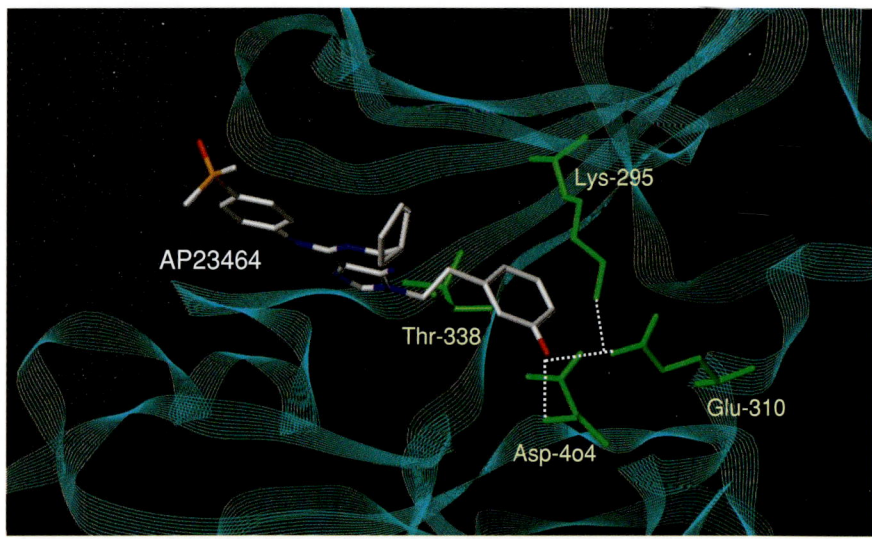

Figure 11.13. X-ray structure of AP23464 complexed with Src kinase illustrating its orientation in a manner to show H-bonding interactions between the protein (network of Lys-295, Glu-310, and Asp-494) and the 3-hydroxyphenethyl group of the compound. Also, the 3-hydroxyphenethyl group binds exceptionally well into the hydrophobic specificity pocket of the protein as indicated by the "gateway" residue, Thr-338. See text for detailed discussion and references.

as promising inhibitory activities against EGFR, Her2, PDGFR, FGFR3, Kit, and Raf kinases (Sawyer et al., 2004). Comparative structure–activity analysis of **27–29** show that incorporation of dialkyl esters of phosphonate and dialkyl phosphine oxide moieties at the 4-position of the aniline group (R2 site) resulted in significantly increased Src kinase inhibitory potency. The X-ray structure of AP23464 complexed with Src kinase (also in its active conformation) (Fig. 11.13) revealed a similar overall binding orientation at that for AP23451, yet with noteworthy differences to implicate an "induced fit" likely resulting from key H-bonding and hydrophobic interactions between 3-hydroxyphenethyl group of the AP23464 and the specificity pocket of Src kinase. Specifically, the 3-hydroxyphenethyl substituent penetrates deeply into the specificity binding pocket and forms an intricate network of hydrogen-bonding interactions involving residues Lys-295, Glu-310, and Asp-404. Most notably, the hydroxyl substituent of the 3-hydroxyphenethyl group forms hydrogen bonds with the sidechain carboxyl of Glu-310 and with the backbone amide of Asp-404. Furthermore, the aromatic ring of the 3-hydroxyphenethyl group is packed between the hydrophobic sidechains of Lys-295, Met-314, Ile-336, and Thr-338. The dimethylphosphine oxide group of AP23464 forms a weak water mediated H-bond with the side-chain phenolic oxygen atom of Tyr-340. Finally, the cyclopentyl group of AP23464 is oriented toward the ribose binding pocket of Src kinase in a similar manner to that of the 4-aminocyclohexyl group of AP23451.

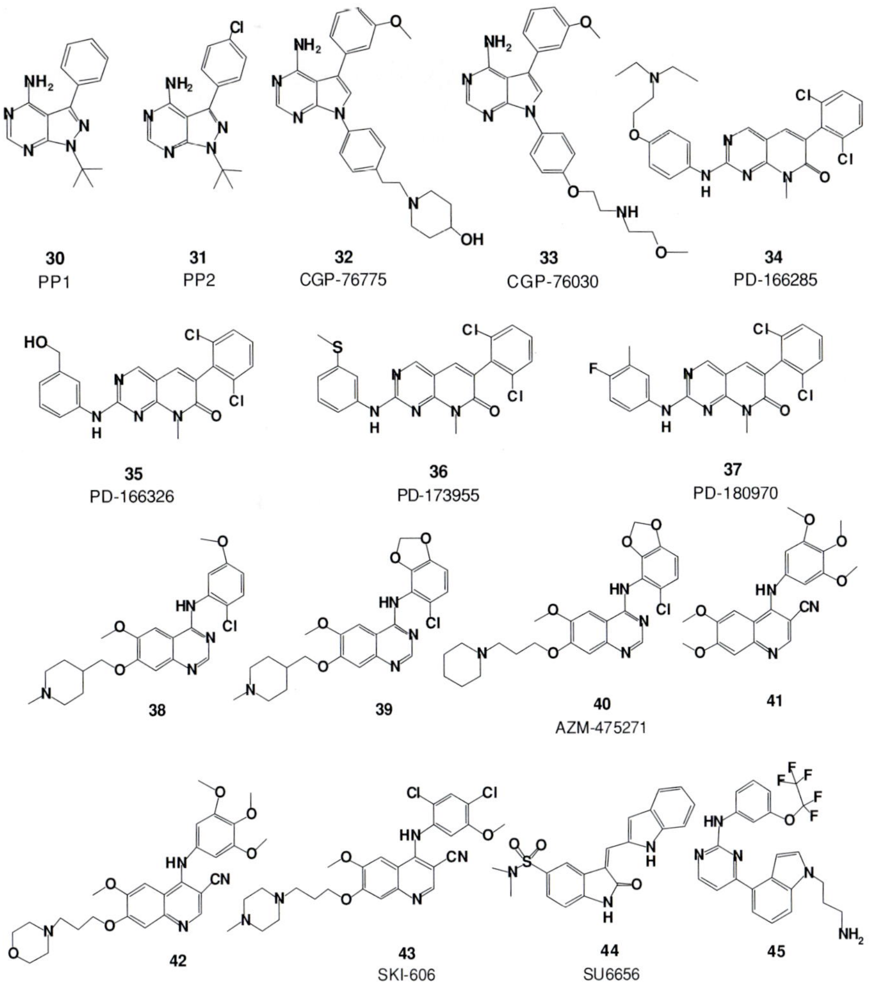

Figure 11.14. Examples of other ATP mimetic template-based Src kinase inhibitors. Representative classes include pyrazolopyrididine (**30** and **31**), pyrrolopyrimidine (**32** and **33**), pyridopyrimidinone (**34–37**), quinanzoline (**38–40**), quinolinecarbonitrile (**41–43**), indolinone (**44**), and phenylaminopyrimidine (**45**). See text for detailed discussion and references.

3.4. Other ATP Mimetic-Based Src Kinase Inhibitors

In addition to the above purine analogs, several other ATP mimetic-based Src kinase inhibitors **30–45** have been developed (Fig. 11.14). The pyrazolopyrimidine template-based compounds **30** (PP1) and **31** (PP2) have been described as potent inhibitors of Src-family kinases with marked selectivity versus ZAP-70, JAK2, EGF-R, and PKA kinases (Hanke et al., 1996). X-ray structures of Lck tyrosine kinase–PP2 and Hck tyrosine kinase–PP1 complexes are particularly noteworthy in that they provided the first 3D insight to understand SFK binding to ATP mimetic-based inhibitors, especially the hydrophobic specificity pockets of such SFKs (Schindler et al., 1999; Zhu et al., 1999). The role of Src in VEGF-mediated angiogenesis and vascular permeability has been determined using PP1 as a Src kinase inhibitor (Eliceiri et al., 1999; Paul et al., 2001). PP1 has been shown to inhibit human breast cancer cell lines with respect to both hergulin-dependent or independent growth as well as triggering apoptosis (Belsches-Jablonski et al., 2001). Furthermore, PP1 has been reported to inhibit collagen type-I-induced E-cadherin down-regulation and consequent effects on cell proliferation and metastatic properties (Menke et al., 2001). Very recently, PP1 has been determined to inhibit both Kit and Bcr-Abl kinases, including imatinib-resistant mutant Bcr-Abl kinases (Tatton et al., 2003; Warmuth et al., 2003b) as well as tumorogenesis induced by *RET* oncogenes (Carlomagno et al., 2002). Related to the pyrazolopyrimidine template has also been the development of allele-specific protein kinase inhibitors, including a series directed against Src kinase, to explore signal transduction pathways (Bishop et al., 1998; Bishop and Shokat, 1999; Kraybill et al., 2002).

The pyrrolopyrimidine template-based Src kinase inhibitors **32** (CGP-76775) and **33** (CGP-76030) have been described as potent inhibitors of Src tyrosine kinase with selectivity relative to a number of protein kinases (e.g., EGF-R, v-Abl, Cdc2, and Lck) as well as in vivo activity in animal models of osteoporosis (Missbach et al., 1999, 2000; Recchia et al., 2004). Furthermore, both CGP-76775 and CGP-76030 have been shown to reduce growth, adhesion, motility and invasion of prostrate cancer cells (Recchia et al., 2003). Also, CGP-76030 has been reported to inhibit imatinib-resistant mutant Bcr-Abl kinases (Warmuth et al., 2003b). The pyridopyrimidinone template-based compounds **34** (PD-166285), **35** (PD-166326), **35** (PD-166326), **36** (PD-173955), and **37**(PD-180970) have been advanced through systematic modifications of prototype lead compounds to provide a series of potent inhibitors of Src tyrosine kinase with varying selectivities to PDGF-R, fibroblast growth factor receptor (FGF-R), and EGF-R kinases (Hamby et al., 1997; Klutchko et al., 1998; Kraker et al., 2000; Wisniewski et al. 2002; Huron et al., 2003; von Bubnoff et al., 2003). Particularly noteworthy are PD-166326 and PD-180970 relative to their dual Src and Abl (Bcr-Abl) kinase inhibitory potency and cellular activities in Src- and Bcr-Abl-dependent cell lines (Kraker et al., 2000; Wisniewski et al., 2002; Huron et al., 2003; von Bubnoff et al., 2003). The quinazoline-based compounds **38–40** have been described as potent inhibitors of Src tyrosine kinase (Wang et al., 2000; Ple et al., 2004). Noteworthy, compound **40** (AZM- 475271) is effective in vivo to inhibit tumor growth in a Src-transfomed 3T3 xenograft model, hence providing proof-of-concept for the potential of a Src kinase inhibitor for cancer invasion and metastasis (Ple et al., 2004). Interestingly, the quinoline-based compounds **41–43** have been described as highly potent inhibitors of Src tyrosine kinase (Wang et al., 2000; Boschelli, 2002; Boschelli et al., 2001, 2004; Golas et al., 2003). Furthermore, compound **43** (SKI-606) has been shown to effect dual inhibition of both Src and Bcr-Abl kinases as well as antiproliferative efficacy in vitro and in vivo in Src-transformed fibroblast and Bcr-Abl-transformed leukemia cell xenograft models, respectively (Golas et al., 2003). The indolinone-based compound **44** (SU6656) has been described as a potent inhibitor of Src tyrosine kinase as well as Lck, Fyn, and, especially, Yes tyrosine kinases (Blake et al., 2001). Furthermore, compound SU6656 was found to inhibit PDGF-stimulated DNA synthesis and Myc induction in a fibroblast cell line. Finally, the phenylamino-pyrimidine-based compound **45** has been described as a potent inhibitor of Src tyrosine kinase, albeit is not selective versus Abl and EGF-R tyrosine kinases (Blake et al., 2001).

3.5. Peptide Substrate-Based and Other Small-Molecule Src Kinase Inhibitors

A number of Src tyrosine kinase inhibitors **46–51** of varying chemical structures have been described as derived from peptide substrate-, natural product-, and combinatorial library-based drug design strategies (Fig. 11.15) (Uehara et al., 1991; Yoneda et al., 1993; Hall et al., 1994; Slate et al., 1994; Alfaro-Lopez et al., 1998; Maly et al., 2000). Peptide substrate-based strategies have been advanced, as exemplified by the cyclic peptide **46** which is a potent inhibitor of Src tyrosine kinase (Alfaro-Lopez et al., 1998). Interestingly, compound **46** exhibits marked selectivity for Src versus Lyn and Lck tyrosine kinases to further support the opportunities of such peptide substrate-based inhibitor strategies. Several natural product-based inhibitors of Src or Src family tyrosine kinases have been reported, including herbimycin A **47** (Uehara et al., 1991; Yoneda et al., 1993; Hall et al., 1994), staurosporine **48** (Lamers et al., 1999), and halistanol trisulfate **47** (Slate et al., 1994). Herbimycin A

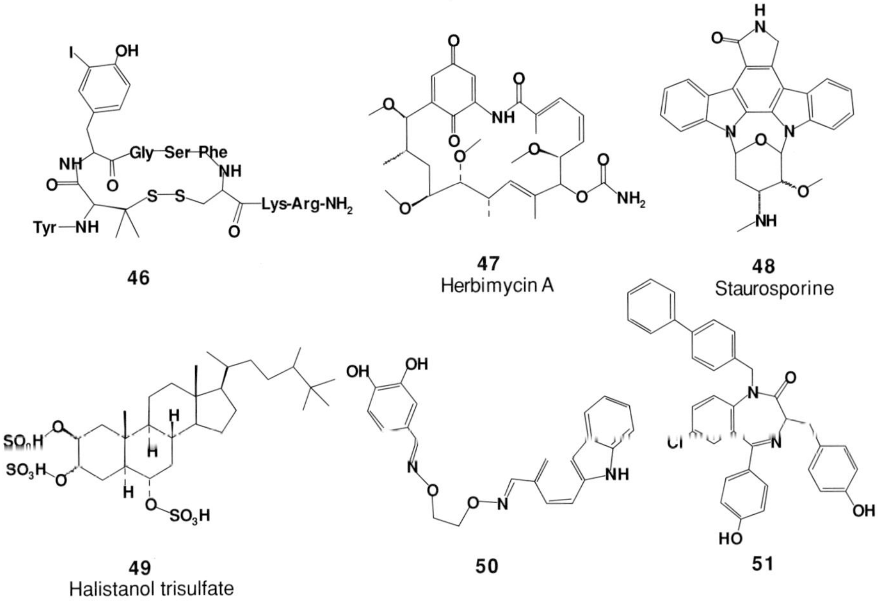

Figure 11.15. Examples of peptide substrate-, natural product-, and combinatorial library-based Src kinase inhibitors (**46–51**). See text for detailed discussion and references.

has been described as a potent inhibitor of Src tyrosine kinase (Uehara et al., 1991), and subsequent studies have shown it to effect inhibition of osteoclast-mediated bone resorption in vitro as well as hypercalcemia in vivo (Yoneda et al., 1993; Hall et al., 1994). Interestingly, it is an irreversible inhibitor of Src tyrosine kinase, but shows significant selectivity versus PKC and PKA (Uehara et al., 1991). Staurosporine **48** is a potent inhibitor of Lck tyrosine kinase and Lck-mediated substrate phosphorylation and signal transduction in T cells (Zhu et al., 1999). Furthermore, an X-ray structure of Lck tyrosine kinase–**48** complex has been determined (Zhu et al., 1999). Halistanol trisulfate **49** is an inhibitor of Src tyrosine kinase, and the sulfate moieties are critical for its inhibitory activity (Slate et al., 1994). These examples of natural product-based inhibitors of Src or Src family tyrosine kinases illustrate quite different chemical structures in comparison to the aforementioned ATP mimetic-based templates that have undergone a significant degree of lead optimization. Finally, combinatorial library-based Src kinase inhibitors **50** and **51** have been described (Ramdas et al., 1999; Maly et al., 2000) as novel inhibitors of Src tyrosine kinase. Noteworthy is compound **50** which was found to exhibit exceptional potency and selectivity to Src versus Fyn, Lyn, and Lck kinases (Maly et al., 2000).

4. SMART DRUG DESIGN TECHNOLOGIES, PROOF-OF-CONCEPT LEAD COMPOUNDS, AND TO BREAKTHROUGH MEDICINES

4.1. Bone-Targeting and Protein Kinase SMART Drug Design Technologies

From nearly a decade of drug discovery at ARIAD Pharmaceuticals that has been focused on therapeutic targets intimately involved in signal tranduction pathways critical for cancer and bone diseases, the development of SMART drug design technology has emerged (Sawyer et al., 2003). Specifically, two concepts have been advanced (Fig. 11.16): (1) bone-targeting SMART drug design to create novel small-molecule lead compounds exhibiting cellular (tissue) selectivity for potential application to bone diseases, including osteoporosis and osteolytic bone metastasis; and (2) protein kinase SMART drug design to create novel small-molecule lead compounds exhibiting molecular (singular or multiple) selectivity for potential application to cancer, including solid tumors and metastasis. As further described below, such SMART drug design technologies have integrated structural biology (X-ray/NMR), molecular modeling, virtual screening, synthetic chemistry, chemoinformatics, bioinformatics, and biological screening to advance proof-of-concept lead compounds. Importantly, a plethora of novel small molecules (peptidomimetics, nonpeptides, ATP mimetics, and natural products) have been created and provide the foundation of an ever-expanding intellectual property portfolio that harnesses both the bone-targeting and protein kinase SMART drug design technologies.

In the case of bone-targeting SMART drug design, the concept of incorporating unique chemical groups conferring binding to the mineral component of bone (i.e., hydroxyapatite) has been beautifully illustrated by bisphosphonate drugs such as zoledronate (**52**) and alendronate (**53**) (Fig. 11.17) in which the bisphosphonate moiety

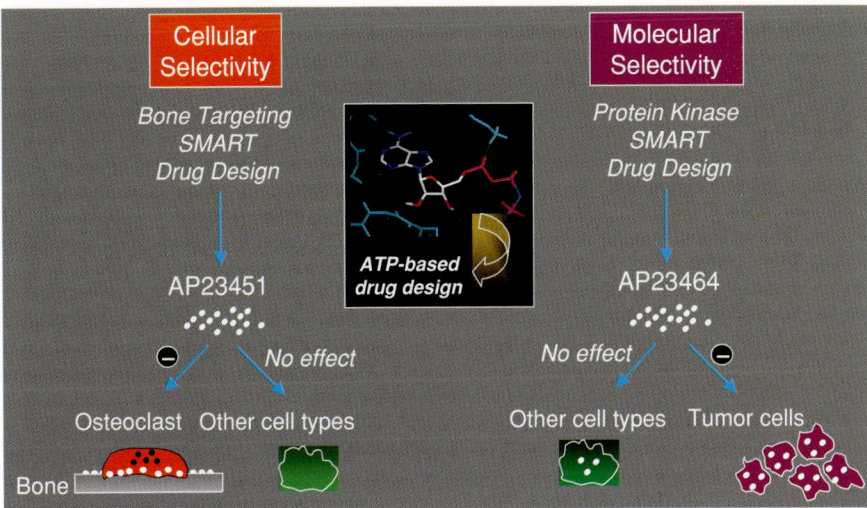

Figure 11.16. Schematic representation of SMART drug design technology within the scope of bone-targeting and protein kinase strategies to advance novel small molecule inhibitors of Src as exemplified in this review. See text for detailed discussion and references.

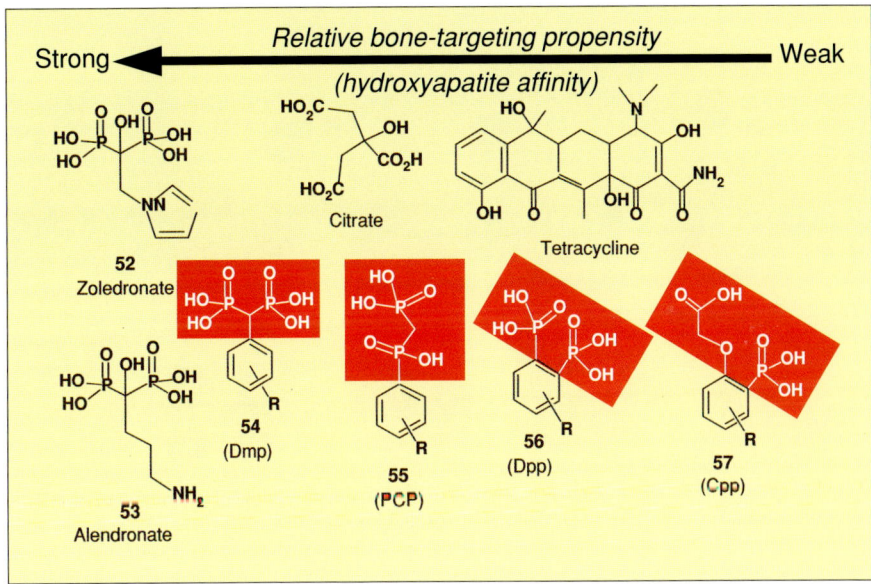

Figure 11.17. Examples of bone-targeting functionalities, illustrating chemical diversity within the scope of bisphosphonates **52** (zoledronate) and **53** (alendronate), citrate, tetracycline, and several phosphorus-modified phenyl groups (**54–57**) that may be incorporated into various small molecules relative to bone-targeting SMART drug design as exemplified by AP22408 and AP23451. See text for detailed discussion and references.

provides both bone-targeting and molecular recognition critical to inhibit farnesyl diphosphate synthase (FDS) in osteoclasts (Geddes et al., 1994; Rogers et al., 2000; Teitelbaum, 2000; Fleisch, 2001; Rogers, 2003; Uludag et al., 2002). This precedence provided impetus for drug discovery efforts at ARIAD Pharmaceuticals to create yet novel, chemically diverse bone-targeting functional groups for possible incorporation into any small molecule of interest. For example, a series of phosphonate-modified aryl groups **54–57** (Fig. 11.17), including both pTyr mimics and anilines, have been developed and tested within the scope of Src SH2 inhibitors and Src kinase inhibitors (see above). In contrast to bisphosphonates (in which modifications of the bone-targeting group most often compromise the FDS inhibitor potency and/or affinity to hydroxyapatite, the bone-targeting SMART drug design technology provides a wide spectrum of possibilities not chemically limited to the pharmacophoric constraints of FDS inhibition.

In the case of protein kinase SMART drug design, the concept of exploiting known ATP mimetic based templates and detailed 3D analysis of the protein kinase superfamily (i.e., X-ray structures and/or molecular models) to advance novel small molecules having high potency and desirable selectivity properties is the driving force which has successfully led to proof-of-concept lead compounds such as AP23464 (see above). To date, ATP mimetic-based templates such as **58–65** (Fig.11.18) have been exploited at ARIAD Pharmaceuticals using SMART drug design technologies to advance novel small molecule inhibitors of several protein kinases with varying molecular selectivities. Noteworthy has been the incorporation of chemically diverse

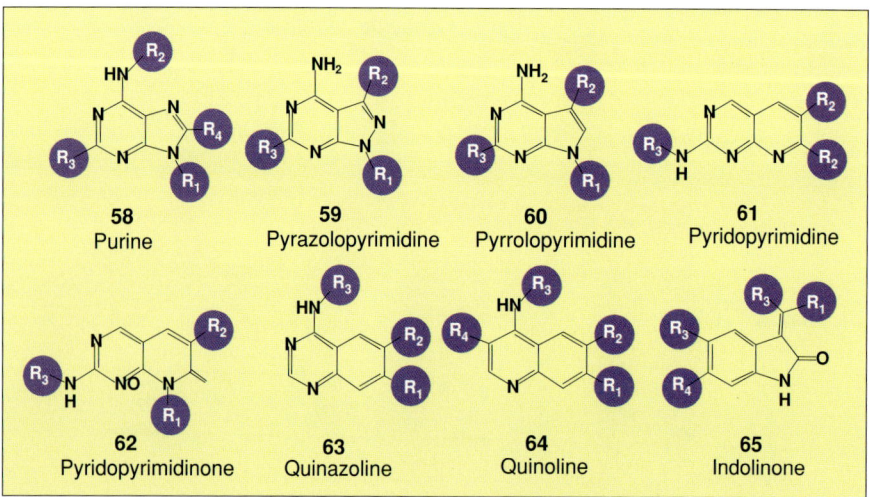

Figure 11.18. Examples of ATP mimetic-based templates **58–65** (generic structures) that have been optimized by functional group (e.g., R_1–R_4) elaboration relative to protein kinase SMART drug design. Representative templates include purines, pyrazolopyrimidines, pyrrolopyrimidines, pyridopyrimidines, pyridopyrimidinones, quinolinecarbonitriles, indolinones, and phenylaminopyrimidines. See text for detailed discussion and references.

phosphorus groups (e.g., dialkyl phosphonate esters and dialkyl phosphine oxides) to afford unique properties to lead compounds such as AP23464 relative to its high potency against both Src and Abl kinases as well as promising activity against several additional key oncogenic protein kinases (see above). Most importantly, such work requires state-of-the-art computational tools and programs to develop 3D molecular models for both lead compound generation and optimization (McMartin and Bohacek, 1997; Joseph-McCarthy, 1999; Gane and Dean 2000; Klebe, 2000; Scapin, 2000; Gould and Wong, 2002; Sotriffer and Klebe, 2002; Waszkowycz, 2002; Woolfrey and Weston, 2002).

4.2. Smart Proof-of-Concept Lead Compounds and Breakthrough Medicines

Key proof-of-concept lead compounds have included AP22408, AP23451, and AP23464 within the scope of both bone-targeted and protein kinase SMART drug design technologies focused on Src as a therapeutic target for cancer and bone diseases (see above).

Proof-of-concept studies for bone-targeted lead compounds AP22408 and AP23451 have included inhibition of Src-dependent cellular mechanisms relative to bisphosphonates, in vivo localization to bone tissue (using radiolabeled drug), and in vivo efficacy (osteoporosis and osteolytic bone metastasis models). Most recently, the proprietary chemistry methods have been successfully translated to natural product-based inhibitors of mTOR (i.e., novel bone-targeted SMART rapamycin analog) to identify a promising novel series of small molecules having yet different mechanisms of action for futher proof-of-concept in vivo studies, especially bone metastasis, bone cancer and related bone diseases.

Proof-of-concept studies for protein kinase lead compound AP23464 include selectivity (multiple inhibition of key therapeutic targets) in molecular and cellular assays as well as in vivo efficacy (e.g., Src-, Abl-, and other protein kinase-dependent cancer models). Noteworthy, the proprietary chemistry methods have been success-fully translated to the the first ARIAD small-molecule clinical candidate AP23573, a novel protein kinase SMART rapamycin analog (Clackson et al., 2002; Metcalf et al., 2004), which is currently in Phase I studies.

As a concluding remark, the integration of structural biology, molecular model-ing, computational chemistry, synthetic chemistry, chemoinformatics, bioinformatics, and biological screening in mechanism-based cellular and in vivo disease models has created a series of novel small molecules for cancer and bone diseases which will hopefully translate to future breakthrough medicines.

Special Acknowledgments

We would like to express our thanks to our colleagues at ARIAD Pharmaceuti-cals for their encouragement, support, and contributions relative to the work described in this review. We especially acknowledge Manfred Weigele, David Berstein, John Iuliucci, Camille Bedrosian, Roy Pollock, Victor Rivera, Tim Clackson, Maryann

Krane, Scott Wardwell, Joseph Snodgrass, Shuangying Liu, Kelly Kwan, Scott Lentini, Sonya Roeloffzen, Jennifer Saltmarsh, Jeff Keats, Hao Tang, Julie Ford, Karin Russian, Lawrence Andrade and Harvey Berger.

Furthermore, we would like to recognize our outstanding collaborations which have supported various structural biology, molecular and cell biology, and pharmacology aspects of the work described in this review. We especially acknowledge Thilo Stehle (Massachusetts General Hospital), Margaret Frame and Val Brunton (Beatson Cancer Institute, University of Glasgow), Brendan Boyce and Lianping Xing (University of Rochester Medical Center), and Brian Druker, Thomas O'Hare, Michael Deininger, and Amie Corbin (Oregon Health Sciences University).

REFERENCES

Abu-Amer, Y., Ross, F.P., Schlesinger, P., Tondravi, M.M., and Teitelbaum, S.L. (1997). Substrate recognition by osteoclast precursors induces c-Src/microtubule association. *J. Cell Biol.* 137:247–258.

Alexandropoulos, K., and Baltimore, D. (1996). Coordinate activation of c-Src by SH3- and SH2-binding sites on a novel p130Cas-related protein, Sin. *Genes Dev.* 10:1341–1355.

Alfaro-Lopez, J., Yuan, W., Phan, B.C., Kamath, J., Lou, Q., Lam, K.S., and Hruby, V.J. (1998). Discovery of a novel series of potent and selective substrate-based inhibitors of pp60^{c-src} protein tyrosine kinase: conformational and topographical constraints in peptide design. *J. Med. Chem.* 41:2252–2260.

Alligood, K.J., Charifson, P.S., Crosby, R., Consler, T.G., Feldman, P.L., Gampe, R.T. Jr., Gilmer, T.M., Jordan, S.R., Milstead, M.W., Mohr, C., Peel, M.R., Rocque, W., Rodriguez, M., Rusnak, D.W., Shewchuk, L.M., and Sternbach, D.D. (1998). The formation of a covalent complex between a dipeptide ligand and the SRC SH2 domain. *Bioorg. Med. Chem. Lett.* 8:1189–1194.

Avizienyte, E., Wyke, A.W., Jones, R.J., McLean, G.W., Westhoff, M.A., Brunton, V.G., and Frame, M.C. (2002). Src-induced de-regulation of E-cadherin in colon cancer cells requires integrin signalling. *Nat. Cell Biol.* 4:632–638.

Belsches-Jablonski, A.P., Biscardi, J.S., Peavy, D.R., Tice, D.A., Romney, D.A., and Parsons, S.J. (2001). Src family kinases and HER2 interactions in human breast cancer cell growth and survival. *Oncogene* 20:1464–1475.

Biscardi, J.S., Tice, D.A., and Parson, S.J. (1999). c-Src, receptor tyrosine kinases, and human cancer. *Adv. Cancer Res.* 76:61–119.

Bishop, A.C., and Shokat, K.M. (1999). Acquisition of inhibitor-senstive protein kinases through protein design. *Pharmacol. Ther.* 82:337–346.

Bishop, A.C., Shah, K., Liu, Y., Witucki, L, Kung, C., and Shokat, K.M. (1998). Design of allele-specific inhibitors to probe protein kinase signaling. *Curr. Biol.* 8:257–266.

Blake, R.A., Broome, M.A., Liu, X., Wu, J., Gishizky, M., Sun, L., and Courtneidge, S.A. (2000). SU6656, a selective Src family kinase inhibitor used to probe growth factor signaling. *Mol. Cell. Biol.* 20:9018–9027.

Bohacek, R.S., Dalgarno, D.C., Hatada, M., Jacobsen, V.A., Lynch, B.A., Macek, K.J., Merry, T., Metcalf, C.A. III, Narula, S.S., Sawyer, T.K., Shakespeare, W.C., Violette, S.M., and Weigele, M. (2001). X-ray structure of citrate bound to Src SH2 leads to high-affinity, bone-targeted Src SH2 inhibitor. *J. Med. Chem.* 44:660–663.

Boschelli, D.H. (2002). 4-Anilino-3-quinolinecarbonitriles: an emerging class of kinase inhibitors. *Curr. Top. Med. Chem.* 2:1051–1063.

Boschelli, D.H., Wang, Y.D., Ye, F., Wu, B., Zhang, N., Dutia, M., Powell, D.W., Wissner, A., Arndt, K., Weber, J.M., Boschelli, F. (2001). Synthesis and Src kinase inhibitory activity of a series of 4-phenylamino-3-quinolinecarbonitriles. *J. Med. Chem.* 44:822–833.

Boschelli, D.H., Wang, Y.D., Johnson, S., Wu, B., Ye, F., Barrios Sosa, A.C., Golas, J.M., and Boschelli, F. (2004). 7-Alkoxy-4-phenylamino-quinolinecarbonitriles as dual inhibitors of Src and Abl kinases. *J. Med. Chem.* 47:599–601.

Boyce, B.F., Yoneda, T., Lowe, C., Soriano, P., and Mundy, G.R. (1992). Requirement of pp60c-src expression for osteoclasts to form ruffled borders and resorb bone in mice. *J. Clin. Invest.* 90:1622–1627.

Boyce, B.F., Xing, L., Shakespeare, W., Wang, Y., Dalgarno, D., Iuliucci, J., and Sawyer, T. (2003). Regulation of bone remodeling and emerging breakthrough drugs for osteoporosis and osteolytic bone metastases. *Kidney Int.* 85:52–55.

Bradshaw, J.M., and Waksman, G. (2002). Molecular recognition by SH2 domains. *Adv. Protein Chem.* 61:161–210.

Bridges, A.J. (1995). The emerging role of protein phosphoryation and cell cycle control in tumor progression. *ChemTracts Org. Chem.* 8:73–107.

Bridges, A.J. (2001a). Chemical inhibitors of protein kinases. *Chem. Rev.* 101:2541–2572.

Bridges, A.J. (2001b). Current progress towards the development of tyrosine kinase inhibitors as anticancer agents. In: Bowman, W.C., Fitzgerald, J.D., Taylor, J.B. (eds.), *Emerging Drugs: The Prospect for Improved Medicines.* Ashley Publications, London, pp.:279–292.

Brugge, J.S., and Erikson, R.L. (1977). Identification of a transformation-specific antigen induced by an avian sarcoma virus. *Nature* 269:346–348.

Buchanan, J.L., Bohacek, R.S., Luke, G.P., Hatada, M., Lu, X., Dalgarno, D.C., Narula, S.S., Yuan, R., and Holt, D.A. (1999). Structure-based design and synthesis of a novel class of Src SH2 inhibitors. *Bioorg. Med. Chem. Lett.* 9:2353–2358.

Burke, T.R., Jr., and Lee, K. (2003). Phosphotyrosyl mimetics in the development of signal transduction inhibitors. *Acc. Chem. Res.* 36:426–400.

Burke, T.R., Jr., Yao, Z.-J., Liu, D.-G., Voigt, J., and Gao, Y. (2001). Phosphoryltyrosyl mimetics in the design of peptide-based signal transduction inhibitors. *Biopolymers (Peptide Sci.)* 60:32–44.

Carlomagno, F., Vitagliano, D., Guida, T., Napolitano, M., Veccio, G., Fusco, A., Gazit, A., Levitzki, A., and Santoro, M. (2002). The kinase inhibitor PP1 blocks tumorogenesis induced by RET oncogenes. *Cancer Res.* 62:1077–1082.

Catlett-Falcone, R., Dalton, W.S., and Jove, R. (1999). STAT proteins as novel targets for cancer therapy. Signal transducer and activatior of transcription. *Curr. Opin. Oncol.* 11:490–496.

Clackson, T., Metcalf C.A., III, Rozamus, L.W., Rivera, V.M., Knowles, H.L., Wardwell, S.D., Wang, X., Burns, K.D., Roses, J.B., Graytock, C., Pradeepan, S., Notari, S.D., Bohacek, R.S., Berstein, D.L., Weigele, M., Dalgarno, D.C., and Iuliucci, J.D. (2002). Regression of tumor xenografts in mice after oral administration of AP23573, a novel mTOR inhibitor that induces tumor starvation. *Proc. AACR* 43:ALB–95.

Cody, W.L., Lin, Z., Panek, R.L., Rose, D.W., and Rubin, J.R. (2000). Progress in the development of inhibitors of SH2 domains. *Curr. Pharm. Des.* 6:59–98.

Collet, M.S., and Erikson, R.L. (1978). Protein kinase activity associated with avian sarcoma virus src gene product. *Proc. Natl. Acad. Sci. USA* 75:2021–2024.

Dalgarno, D.C., Botfield, M.C., and Rickles, R.J. (1997). SH3 domains and drug design: ligands, structure, and biological function. *Biopolymers (Peptide Sci.)* 43:383–400.

Dalgarno, D.C., Metcalf, C.A., Shakespeare, W.C., and Sawyer, T.K. (2000). Signal transduction drug discovery: targets, mechanisms, and structure-based design. *Curr. Opin. Drug Dis. Dev.* 3:549–564.

Dalgarno, D., Bohacek, R., Stehle, T., Narula, S., Metcalf III, C., Shakespeare, W., Sundaramoorthi, R., Keenan, T., Wang, Y., Keats, J. Ram, M., Adams, S., van Schravendijk, M.R., Weigele, M. and Sawyer, T.K. (2003). Structure-based design and X-ray crystallographic analysis of a potent and selective Src tyrosine kinase inhibitor for the treatment of cancer. In: *AACR-NCI-EORTC Conference on Molecular Targets and Cancer Therapeutics,* November 17–23, Boston, MA, Abstract A5.

Deprez, P., Baholet, I., Burlet, S., Lange, G., Amengual, R., Schoot, B., Vermond, A., Mandine, E., Lesuisse, D. (2002a). Discovery of highly potent Src SH2 binders: structure-activity studies and X-ray structures. *Bioorg. Med. Chem. Lett.* 12:1291–1294

Deprez, P., Mandine, E., Gofflo, D., Meunier, S., and Lesuisse, D. (2002b). Small ligands interacting with the phosphotyrosine binding pocket of the Src SH2 protein. *Bioorg. Med. Chem. Lett.* 12:1295–1298.

Dreyer, M.K., Borcherding, D.R., Dumont, J.A., Peet, N.P., Tsay, J.T., Wright, P.S., Bitonti, A.J., Shen, J., and Kim, S.H. (2001). Crystal structure of human cyclin-dependent kinase 2 in complex with the adenine-derived inhibitor H717. *J. Med. Chem.* 44:524–530.

Dumas, J. (2001). Protein kinase inhibitors: emerging pharmacophores 1997–2000. *Expert Opin. Ther. Patents* 11:405–429.

Duong, L.T., Lakkakorpi, P.T., Nakamura, I., Machwate, M., Nagy, R.M., and Rodan, G.A. (1998). Pyk2 in osteoclasts is an adhesion kinase, localized in the sealing zone, activation by ligation of alpha(v)beta3 integrin and phosphorylated by Src kinase. *J. Clin. Invest.* 102:881–892.

Egan, C., Pang, A., Durda, D., Cheng, H.C., Wang, J.H., and Fujita, D.J. (1999). Activation of Src in human breast tumor cell lines: elevated levels of phosphotyrosine phosphatase activity that preferentially recognizes the Src carboxy terminal negative regulatory tyrosine 530. *Oncogene* 18:1227–1237.

Eliceiri, B.P., Paul, R., Schwartzbert, P.I., Hood, J.D., Leng, J., and Cheresch, D.A. (1999). Selective requirement for Src kinases during VEGF-induced angiogenesis and vascular permeability. *Mol. Cell* 4:915–924.

Ellis, L.M., Staley, C.A., Liu, W., Fleming, R.Y., Parikh, N.U., Bucana, C.D., and Gallick, G.E. (1998). Downregulation of vascular endothelial growth factor in a human colon carcinoma cell line transfected with an antisense expression vector specific for c-Src. *J. Biol. Chem.* 273:1052–1057.

Erpel, T., and Courtneidge, S.A. (1995). Src family protein tyrosine kinases and cellular signal transduction pathways. *Curr. Opin. Cell Biol.* 7:176–182.

Fleisch, H. (2001), Bisphosphonates: mechanisms of action. *Expert Opini. Ther. Patents* 11:1371–1381.

Frame, M.C. (2002). Src in cancer: deregulation and consequences for cell behavior. *Biochim. Biophys. Acta* 1602:114–130.

Gane, P.J., and Dean, P.M. (2000). Recent advances in structure-based rational drug design. *Curr. Opin. Struct. Biol.* 10:401–404.

Garcia-Echevarria, C. (2001). Antagonists of the Src homology 2 (SH2) domains of Grb2, Src, Lck and ZAP-70. *Curr. Med. Chem.* 8:589–604.

Geddes, A.D., D'Souaza, S.M., Ebetino, F.H., and Ibbotson, K.J. (1994). Bisphosphonates: structure activity relationships and therapeutic implications. *Bone Miner. Res.* 8:265–306.

Gilmer, T., Rodriguez, M., Jordan, S., Crosby, R., Alligood, K., Green, M., Kimery, M., Wagner, C., Kinder, D., Charifson, P., Hassell, A.M., Willard, D., Luther, M., Rusnak, D., Sternbach, D.D., Mehrotra, M., Peel, M., Shampine, L., Davis, R., Robbins, J., Patel, I.R., Kassel, D., Burkhart, W., Moyer, M., Bradshaw, T., and Berman, J. (1994). Peptide inhibitors of Src SH3-SH2-phosphoprotein interactions. *J. Biol. Chem.* 269:31711–31719.

Gmeiner, W.H., and Horita, D.A. (2001). Implications of SH3 domain structure and dynamics for protein regulation and drug design. *Cell Biochem. Biophys.* 35:127–140.

Gould, C. and Wong, C.F. (2002). Designing specific protein kinase inhibitors: insights from computer simulations and comparative sequence/structure analysis. *Pharmacol. Ther.* 93:169–178.

Gray, N.S., Wodicka, L., Thunnissen, A.M., Norman, T.C., Kwon, S., Espinoza, F.H., Morgan, D.O., Barnes, G., LeClerc, S., Meijer, L., Kim, S.H., Lockhart, D.J., and Schultz, P.G. (1998). Exploiting chemical libraries, structure, and genomics in the search for kinase inhibitors. *Science* 281:533–538.

Hall, T.J., Schaeriblin, M., and Missbach, M. (1994). Evidence that c-Src is involved in the process of osteoclastic bone resorption. *Biochem. Biophys. Res. Commun.* 199:1237–1244.

Hamby, J.M., Connolly, C.J., Schroeder, M.C., Winters, R.T., Showalter, H.D., Panek, R.L., Major, T.C., Olsewski, B., Ryan, M.J., Dahring, T., Lu, G.H., Keiser, J., Amar, A., Shen, C., Kraker, A.J., Slintak, V., Nelson, J.M., Fry, D.W., Bradford, L., Hallak, H., and Doherty, A.M. (1997). Structure-activity relationships for a novel series of pyrido[2,3-*d*]pyrimidine tyrosine kinase inhibitors. *J. Med. Chem.* 40:2296–2303.

Hanke, J.H., Gardner, J.P., Dow, R.L., Changelian, P.S., Brissette, W.H., Weringer, E.J., Pollok, B.A., and Connelly, P.A. (1996). Discovery of a novel, potent, and Src family-selective tyrosine kinase inhibitor. *J. Biol. Chem.* 271:695–701.

Hatada, M.H., Lu, X., Laird, E.R., Green, J., Morgenstern, J.P., Lou, M., Marr, C.S., Phillips, T.B., Ram, M.K., Theriault, K., Zoller, M.J., and Karas, J.L. (1995). Molecular basis for interaction of the protein tyrosine kinase ZAP-70 with the T-cell receptor. *Nature* 377:32–38.

Hubbard, S.R. (1997). Crystal structure of the activated insulin receptor tyrosine kinase in complex with peptide substrate and ATP analog. *EMBO J.* 16:5572–5581.

Hunter, T., and Sefton, B.M. (1980). Transforming gene product of Rous sarcoma virus phosphorylates tyrosine. *Proc. Natl. Acad. Sci. USA* 77:1311–1315.

Huron, D.R., Corre, M.E., Kraker, A.J., Sawyers, C.L., Rosen, N., and Moasser, M.M. (2003). A novel pyridopyrimidine inhibitor of Abl kinase is a picomolar inhibitor of Bcr-Abl-driven K562 cells and is effective against STI571-resistant Bcr-Abl mutants. *Clin. Cancer Res.* 9:1267–1273.

Irby, R.B., Mao, W., Coppola, D., Kang, J., Loubeau, J.M., Trudeau, W., Karl, R., Fujita, D.J., Jove, R., and Yeatman, T.J. (1999). Activating SRC mutation in a subset of advanced human colon cancers. *Nat. Genet.* 21:187–190.

Jeschke, M., Brandi, M.-L., and Susa, M. (1998). Expression of Src family kinases and their putative substrates in the human preosteoclastic cell line FLG 29.1. *J. Bone Miner. Res.* 13:1880–1889.

Joseph-McCarthy, D. (1999). Computational approaches to structure-based ligand design. *Pharmacol. Ther.* 84:179–191.

Karni, R., Jove, R., and Levitzki, A. (1999). Inhibition of pp60$^{c\text{-}Src}$ reduces Bcl-Xl expression and reverses the transformed phenotype of cells overexpressing EGF and HER-2 receptors. *Oncogene* 18:4654–462.

Kawahata, N., Yang, M.G., Luke, G.P., Shakespeare, W.C., Sundaramoorthi, R., Wang, Y., Johnson, D., Merry, T., Violette, S., Guan, W., Bartlett, C., Smith, J., Hatada, M., Lu, X., Dalgarno, D.C., Eyermann, C.J., Bohacek, R.S., and Sawyer, T.K. (2001). A novel phosphotyrosine mimetic 4′-carboxymethyloxy-3′-phosphophenylalanine (Cpp): exploitation in the design of nonpeptide inhibitors of pp60src SH2 domain. *Bioorg. Med. Chem. Lett.* 11:2319–2323.

Kauffman, E.C., Robinson, V.L., Stadler, W.M., Sokoloff, M.H., and Rinker-Schaeffer, C.W. (2003). Metastasis suppression: the evolving role of metastasis suppressor genes for regulating cancer cell growth at the secondary site. *J. Urol.* 169:1122–1133.

Klebe, G. (2000). Recent developments in structure-based drug design. *J. Mol. Med.* 78:269–281.

Klohs, W.D., Fry, D.W., and Kraker, A.J. (1997). Inhibitors of tyrosine kinase. *Curr. Opin. Oncol.* 9:562–268.

Klutchko, S.R., Hamby, J.M., Boschelli, D.H., Wu, Z., Kraker, A.J., Amar, A.M., Hartl, B.G., Shen, C., Klohs, W.D., Steinkampf, R.W., Driscoll, D.L., Nelson, J.M., Elliott, W.L., Roberts, B.J., Stoner, C.L., Vincent, P.W., Dykes, D.J., Panek, R.L., Lu, G.H., Major, T.C., Dahring, T.K., Hallak, H., Bradford, L.A., Showalter, H.D., and Doherty, A.M. (1998). 2-Substituted aminopyrido[2,3-*d*]pyrimidin-7(8*H*)-ones. Structure-activity relationships against selected tyrosine kinases and in vitro and in vivo anti-cancer activity. *J. Med. Chem.* 41:3276–3292.

Kraker, A.J., Hartl, B.G., Amar, A.M., Barvian, M.R., Showalter, H.D., and Moore, C.W. (2000). Biochemical and cellular effects of c-Src kinase-selective pyrido[2,3-*d*]pyrimidine tyrosine kinase inhibitors. *Biochem. Pharmacol.* 60:885–898.

Kraybill, B.C., Elkin, L.L. Blethrow, J.D., Morgan, D.O., and Shokat, K.M. (2002). Inhibitor scaffolds as new allele specific kinase substrates. *J. Am. Chem. Soc.* 124:12118–121128.

Lamers, M.B., Antson, A.A., Hukbbard, R.E., Scott, R.K., and Williams, D.H. (1999). Structure of C-terminal Src kinase (CSK) in complex with staurosporine. *J. Mol. Biol.* 28:713–725.

Lange, G., Lesuisse, D., Deprez, P., Schoot, B., Loenz, P., Marquette, J.-P., Broto, P., Sarubbi, E., and Mandinc, E. (2002). Principles governing the binding of a class of non-peptidic inhibitors to the SH2 domain of Src studied by X-ray analysis. *J. Med. Chem.* 45:2915–2922.

Lawrence, D.S., and Niu, J. (1998). Protein kinase inhibitors: the tyrosine-specific kinases. *Pharmacol. Ther.* 77:81–114.

Legraverend, M., Tunnah, P., Noble, M., Ducrot, P., Ludwig, O., Grierson, D.S., Leost, M., Meijer, L., and Endicott, J. (2000). Cyclin-dependent kinase inhibition by new C-2 alkynylated purine derivatives and molecular structure of a CDK2-inhibitor complex. *J. Med. Chem.* 43:1282–1292.

Lesuisse, D., Deprez, P., Albert, E., Duc, T.T., Sortais, B., Goffe, D., Jean-Baptiste, V., Marquette, J.-P., Schoot, B., Sarubbi, E., Lange, G., Broto, P., and Mandine, E. (2001a). Discovery of thioazepinone ligands for Src SH2: from non-specific to specific binding. *Bioorg. Med. Chem. Lett.* 11:2127–2131.

Lesuisse, D., Deprez, P., Bernard, D., Broto, P., Delettre, G., Jean-Baptiste, V., Marquette, J.P., Sarubti, E., Schoot, B., Mandine, E., and Lange, G. (2001b). SAR by crystallography: a new approach combining screening and rational drug design. Application to the discovery of nanomolar Src SH2 binders. *Chim. Nouv.* 19:3240–3241.

Lesuisse. D., Lange, G., Deprez, P., Benard, D., Schoot, B., Delettre, G., Marquette, J.P., Broto, P., Jean-Baptiste, V., Bichet, P., Sarubbi, E., and Mandine, E. (2002). SAR and X-ray. A new approach combining fragment-based screening and rational drug design: application to the discovery of nanomolar inhibitors of Src SH2. *J. Med. Chem.* 45:2379–2387.

Levinson, A.D., Oppermann, H., Levintow, L., Varmus, H.E., and Bishop, J.M. (1978). Evidence that the transforming gene of avian sarcoma virus encodes a protein kinase associated with a phosphoprotein. *Cell* 15:561–572.

Levitzki, A., and Gazit, A. (1995). Tyrosine kinase inhibition: an approach to drug development. *Science* 267:1782–1788.

Lowell, C.A., and Soriano, P. (1996). Knockouts of Src-family kinases: stiff bones, wimpy T cells, and bad memories. *Genes Dev.* 10:1845–1857.

Lunney, E.A., Para, K.S., Plummer, M.S., Shahripour, A., Holland, D., Rubin, J.R., Humblet, C., Fergus, J., Marks, J., Hubbell, S., Herrera, R., Saltiel, A.R., and Sawyer, T.K. (1997). Structure-based design of a novel series of nonpeptide ligands that bind to the pp60^{src} SH2 domain. *J. Am. Chem. Soc.* 119:12471–12476.

Lutz, M.P., Esser, I.B., Flossmann-Kast, B.B., Vogelmann, R., Luhrs, H., Friess, H., Buchler, M.W., and Adler, G. (1998). Overexpression and activation of the tyrosine kinase Src in human pancreatic carcinoma. *Biochem. Biophys. Res. Commun.* 243:503–508.

Maa, M.-C., Leu, T.H., McCarley, D.J., Schatzman, R.C., and Parsons, S.J. (1995). Potentiation of epidermal growth factor receptor-mediated oncogenesis by c-Src: implications for the etiology of multiple human cancers. *Proc. Natl. Acad. Sci. USA* 92:6981–6985.

Maly, D.J., Choong, I.C., and Ellman, J.A. (2000). Combinatorial target-guided ligand assembly: identification of potent subtype-selective c-Src inhibitors. *Proc. Natl. Acad. Sci. USA* 97:2419–2424.

Mao, W., Irby, R., Coppola, D., Fu, L., Wloch, M., Turner, J., Yu, H., Garcia, R., Jove, R., and Yeatman, T.J. (1997). Activation of c-Src by receptor tyrosine kinases in human colon cancer cells with high metastatic potential. *Oncogene* 15:3083–3090.

Martin, G.S. (2001). The hunting of the Src. *Nat. Rev. Mol. Cell Biol.* 2:467–475.

Marzia, M., Sims, N.A., Voit, S., Migliaccio, S., Taranta, A., Bernardini, S., Faraggiana, T., Yoneda, T., Mundy, G.R., Boyce, B.F., Baron, R., and Teti, A. (2000). Decreased c-Src expression enhances osteoblast differentiation and bone formation. *J. Cell Biol.* 151:311–320.

McMahon, G., Sukn, L., Liang, C., and Tang, C. (1998). Protein kinase inhibitors: structural determinants for target specificity. *Curr. Opin. Drug Disc. Dev.* 1:131–146.

McMartin, C., and Bohacek, R.S. (1997). QXP: powerful, rapid computer algorithms for structure-based drug design. *J. Comput. Aided Mol. Design* 11:333–344.

McTigue, M.A., Wickersham, J.A., Pinko, C., Showalter, R.E., Parast, C.V., Tempczyk-Russell, A., Gehring, M.R., Mroczkowski, B., Kan, C.C., Villafranca, J.E., and Appelt, K. (1999). Crystal structure of the kinase domain of human vascular endothelial growth factor receptor 2: a key enzyme in angiogenesis. *Struct. Fold. Des.* 7:319–330.

Menke, A., Philipp, C., Vogelmann, R., Seidel, B., Lutz, M.P., Adler, G., and Wedlich, D. (2001). Down-regulation of E-cadherin gene expression by collagen type I and type II in pancreatic cell lines. *Cancer Res.* 61:3508–3517.

Metcalf, C.A., and Sawyer, T.K. (2004). Src homology-2 domains and structure-based, small molecule library approaches to drug discovery. In: (A. Makriyannis, A., and Biegel, D. (eds.), *Drug Discovery Strategies and Methods*. Marcel Dekker, New York, pp. 23–59.

Metcalf III., C.A., Vu, C.B., Sundaramoorthi, R., Jacobsen, V.A., Laborde, E.A., Green, J., Green, Y., Macek, K.J., Merry, T.J., Pradeepan, S.G., Uesugi, M., Varkhedkar, V.M., and Holt, D.A., (1998). Novel phosphate ester-linked resins: the solid-phase generation of phenyl phosphate-containing compounds for SH2 inhibition. *Tetrahed. Lett.* 39:3435–3438.

Metcalf III, C.A., Eyermann, C.J., Bohacek, R.S., Haraldson, C.A., Varkhedkar, V.M., Lynch, B.A., Barlett, C., Violette, S.M., and Sawyer, T.K. (2000). Structure-based design and parallel synthesis of phosphorylated nonpeptides to explore hydrophobic binding at the Src SH2 domain. *J. Comb. Chem.* 2:305–313.

Metcalf, C., van Schravendijk, M.R., Dalgarno, D., and Sawyer, T.K. (2002). Targeting protein kinases for bone disease: discovery and development of Src inhibitors. *Curr. Pharm. Design* 8:2049–2075.

Metcalf, C., III, Wang, Y., Shakespeare, W., Sundaramoorthi, R., Keenan, T., Dalgarno, D., Bohacek, R., Burns, K., Roses, J., van Schravendijk, M.R., Ram, M., Keats, J., Liou, S., Adams, S., Snodgrass, J., Rivera, V., Weigele, M., Iuliucci, J., Clackson, T., Frame, M., Brunton, V., and Sawyer, T. (2003a). Structure-based methods to design potent and selective Src/Abl dual inhibitors and their development as antileukemic and antimetastatic agents. In: *ACS Abstracts* (226th ACS National Meeting). New York Sept 7–11, 2003.

Metcalf, C., III, Wang, Y., Shakespeare, W., Sundaramoorthi, R., Keenan, T., Dalgarno, D., Bohacek, R., Burns, K., Roses, J., van Schravendijk, M.R., Ram, M., Keats, J., Liou, S., Rivera, V., Weigele, M., Iuliucci, J., Clackson, T., Brunton, V., and Sawyer, T. (2003b). Discovery of potent and selective Src inhibitors and their development as antitumor and antimetastatic agents. *Proc. AACR* 44:A1716.

Metcalf III, C.A., Bohacek, R., Rozamus, L.W., Burns, K.D., Roses, J.B., Rivera, V.M., Tang, H., Keats, J.A., Dalgarno, D.C., Snodgrass, J., Berstein, D.L., Weigele, M., and Clackson, T. (2004). Structure-based design of AP23573, a phosphorus-containing analog of rapamycin for anti-tumor therapy. *Proc. AACR* 45:A2476.

Missbach, M., Jeschke, M., Feyen, J., Muller, K., Glatt, M., Green, J., Susa, M. (1999). A novel inhibitor of the tyrosine kinase Src suppresses phosphorylation of its major cellular substrates and reduces bone resorption in vitro and in rodent models in vivo. *Bone* 24:437–449.

Missbach, M., Attmann, E., and Susa, M. (2000). Tyrosine kinase inhibition in bone metabolism. *Curr. Opin. Drug Disc.* 3:541–548.

Mohammadi, M., McMahon, G., Sun, L., Tang, C., Hirth, P., Yeh, B.K., Hubbard, S.R., and Schlessinger, J. (1997). Structures of the tyrosine kinase domain of fibroblast growth factor receptor in complex with inhibitors. *Science* 276:955–960.

Mohammadi, M., Froum, S., Hamby, J.M., Schroeder, M.C., Panek, R.L., Lu, G.H., Eliseenkova, A.V., Green, D., Schlessinger, J., and Hubbard, S.R. (1998). Crystal structure of an angiogenesis inhibitor bound to the FGF receptor tyrosine kinase. *EMBO J.* 17:5896–5904.

Muller, G. (2000). Peptidomimetic SH2 domain antagonists for targeting signal transduction. *Top. Curr. Chem.* 211:17–59.

Musacchio, A. (2002). How SH3 domains recognize proline. *Adv. Protein Chem.* 61:211–268.

O'Hare, T., Pollock, R., Stoffreger, E.P., Keats, J.A., Abdullah, O.M., Moseson, E.M., Rivera, V.M., Tang, H., Metcalf III, C.A., Bohacek, R.S., Wang, Y., Sundaramoorthi, R., Shakespeare, W.C., Dalgarno, D.C., Clackson, T., Sawyer, T.K., Deininger, M.W., Drucker, B.J. (2004). Inhibition of wild-type and mutant Bcr-Abl by Ap23464, a potent ATP-bised oncogenic protein kinas: inhibitor: Implications for CML. *Blood* 104:2532–2539.

Para, K., Lunney, E.A., Plummer, M., Stankovic, C.J., Shahripour, A., Holland, D., Rubin, J.R., Humblet, C., Fergus, J., Marks, J., Hubbell, S., Herrera, R., Saltiel, A.R., and Sawyer, T.K. (1999). Structure-based de novo design and discovery of nonpeptide antagonists of the pp60^{src} SH2 domain. In: Tam, J., and Kaumaya, P. (eds.), *Proceedings of the Fifteenth American Peptide Symposium*. Kluwer, Dordrecht, The Netherlands, pp. 173–175.

Paul, R., Zhang, Z.G., Eliceiri, B.P., Jiang, Q., Boccia, A.D., Zhang, R.L., Chopp, M., and Cheresh, D.A. (2001). Src deficiency or blockade of Src activity in mice provides cerebral protection during stroke. *Nat. Med.* 7:222–227.

Pawson, T. (2004). Specificity in signal transduction: from phosphotyrosine-SH2 domain interactions to complex cellular systems. *Cell* 116:191–203.

Ple, P.A., Green, T.P., Hennequin, L.F., Curwen, J., Fennell, M., Allen, J., Lambert-van der Brempt, C., and Costello, G. (2004). Discovery of a new class of anilinoquinazoline inhibitors with high affinity and specificity for the tyrosine kinase domain of c-Src. *J. Med. Chem.* 47:871–887.

Plummer, M.S., Lunney, E.A., Para, K.S., Prasad, J.V., Shahripour, A., Singh, J., Stankovic, C.J., Humblet, C., Fergus, J.H., Marks, J.S., and Sawyer, T.K. (1996). Hydrophobic D-amino acids in the design of peptide ligands for the pp60src SH2 domain. *Drug Des. Discov.* 13:75–81.

Plummer, M.S., Holland, D.R., Shahripour, A., Lunney, E.A., Fergus, J.H., Marks, J.S., McConnell, P., Mueller, W.T., and Sawyer, T.K. (1997a). Design, synthesis, and cocrystal structure of a nonpeptide Src SH2 domain ligand *J. Med. Chem.* 40:3710–3725.

Plummer, M.S., Lunney, E.A., Para, K.S., Shahripour, A., Stankovic, C.J., Humblet, C., Fergus, J.H., Marks, J.S., Herrera, R., Hubbell, S., Saltiel, A., and Sawyer, T.K. (1997b). Design of peptidomimetic ligands for the pp60src SH2 domain. *Bioorg. Med. Chem.* 5:41–47.

Rahuel, J., Gay, B., Erdmann, D., Strauss, A., Garcia-Echeverria, C., Furet, P., Caravatti, G., Fretz, H., Schoepfer, J., and Grutter, M.G. (1996). Structural basis for specificity of Grb2-SH2 revealed by a novel ligand binding mode. *Nat. Struct. Biol.* 3:586–589.

Ramdas, L., Bunnin, B.A., Plunkett, M.J., Sun, G., Ellman, J., Gallick, G., and Budde, R.J.A. (1999). Benzodiazepine compounds as inhibitors of the Src protein tyrosine kinase: screening of a combinatorial library of 1,4-benzodiazepines. *Arch. Biochem. Biophys.* 368:394–300.

Recchia, I., Rucci, N., Festuccia, C., Bologna, M., MacKay, A.R., Migliaccio, S., Longo, M., Susa, M., Fabbro, D., and Teti, A. (2003). Pyrrolopyrimidine c-Src inhibitors reduce growth, adhesion, motility and invasion of prostate cancer cells in vitro. *Eur. J. Cancer* 39:1927–1935.

Recchia, I., Rucci, N., Funari, A., Migliaccio, S., Taranta, A., Longo, M., Kneissel, M., Susa, M., Fabbro, D., Teti, A. (2004). Reduction of c-Src activity by substituted 5,7-diphenyl-pyrrolo[2,3-d]-pyrimidines induces osteoclast apoptosis in vivo and in vitro. Involvement of ERK1/2 pathway. *Bone* 34: 65–79.

Rickles, R.J., Henry, P.A., Guan, W., Azimioara, M., Shakespeare, W.C., Violette, S., Zoller, M.J. (1998). A novel mechanism-based mammalian cell assay for the identification of SH2-domain-specific protein-protein inhibitors. *Chem. Biol.* 5:529–538.

Rogers, M.J. (2003). New insights into the molecular mechanisms of action of bisphosphonates. *Curr. Pharm. Des.* 9:2643–2658.

Rogers, M.J., Gordon, S., Benford, H.L., Coxon, FP., Luckman, SP., Monkkonen, J, and Frith JC: (2000). Cellular and molecular mechanisms of action of bisphosphonates. *Cancer 88(12, Suppl.)*: 2961–2978.

Russello, S.V., and Shore, S.K. (2003). Src in human carcinogenesis. *Front. Biosci.* 8:1068–1073.

Sawyer, T.K. (2004). Novel dual Src/Abl kinase inhibitors for cancer therapy. *Keystone Symposium on Protein Kinases and Cancer: The Promise of Molecular-Based Therapies*, February 24–29 2004, Tahoe City. CA.

Sawyer, T.K., Boyce, B., Dalgarno, D., and Iuliucci, J. (2001). Src inhibitors: genomics to therapeutics. *Expert Opin. Invest. Drugs* 10:1327–1344.

Sawyer, T.K., Bohacek, R.S., Dalgarno, D.C., Eyermann, C. J., Kawahata, N., Metcalf, C., Shakespeare, W.C., Sundaramoorthi, R., Wang, Y., and Yang, M. (2002). *Mini Rev. Med. Chem.* 2:475–489.

Sawyer, T.K., Bohacek, R.S., Metcalf, C.A., III, Shakespeare, W. C., Wang, Y., Sundaramoorthi, R., Keenan, T., Narula, S., Weigele, M., Dalgarno, D.C. (2003). Novel protein kinase inhibitors: SMART drug design technology. *BioTechniques* (Suppl) June:2–15.

Scapin, G. (2000). Structural biology in drug design: selective protein kinase inhibitors. *Drug Discov Today* 7:601–611.

Schindler, T., Sicheri, F., Pico, A., Gazit, A., Levitzki, A., and Kuriyan, J. (1999). Crystal structure of Hck in complex with a Src family-selective tyrosine kinase inhibitor. *Mol. Cell* 3:639–648.

Schindler, T., Bornmann, W., Pellicena, P., Miller, W.T., Clarkson, B., and Kuriyan, J. (2000). Structural mechanism for STI-571 inhibition of Abelson tyrosine kinase. *Science* 289:1938–1942.

Sedlacek, H.H. (2000). Kinase inhibitors in cancer therapy. *Drugs* 59:435–476.

Shakespeare, W.C. (2001). SH2 domain inhibition: a problem solved? *Curr. Opin. Chem. Biol.* 5:409–415.

Shakespeare, W.C., Bohacek, R.S., Azimioara, M.D., Macek, K.J., Luke, G.P., Dalgarno, D.C., Hatada, M.H., Lu, X., Violette, S.M., Bartlett, C., and Sawyer, T.K. (2000a). Structure-based design of novel bicyclic inhibitors for the Src SH2 domain. *J. Med. Chem.* 43:3815–3819.

Shakespeare, W., Yang, M., Bohacek, R., Cerasoli, F., Stebbins, K., Sundaramoorthi, R., Azimioara, M., Vu, C., Pradeepan, S., Metcalf, C. III, Haraldson, C., Merry, T., Dalgarno, D., Narula, S., Hatada, M., Lu, X., van Schravendijk, M.R., Adams, S., Violette, S., Smith, J., Guan, W., Bartlett, C., Herson, J., Iuliucci, J., Weigele, M., and Sawyer, T. (2000b). Structure-based design of an osteoclast-selective, nonpeptide Src homology 2 inhibitor with in vivo antiresorptive activity. *Proc. Natl. Acad. Sci. USA* 97:9373–9378.

Shakespeare, W.C., Metcalf, C.A. III, Wang, Y., Sundaramoorthi, R., Keenan, T., Weigele, M., Bohacek, R.S., Dalgarno, D.C. and Sawyer, T.K. (2003a). Novel bone-targeted Src tyrosine kinase inhibitor drug discovery. *Curr. Opin. Drug Disc. Devel.* 6:729–741.

Shakespeare, W., Wang, Y., Metcalf, C., Sundaramoorthi, R., Keenan, T., Bohacek, R., vanSchravendijk, M.R., Snodgrass, J., Dilauro, A., Roeloffzen, S., Liu, S., Saltmarsh, J., Paramanathan, G., Pradeepan, S., Naugle, J., Wardwell, S., Bogus, J., Keats, J., Ram, M., Andrade, L., Liou, S., Narula, S., Adams, S., Boyce, B., Xing, L., Weigele, M., Dalgarno, D., Iuliucci, J., and Sawyer, T. (2003b). Development of a novel bone-targeted Src tyrosine kinase inhibitor AP23451 having potent activity in an animal model of osteolytic bone metastasis. *Proc. AACR* 44:A3871.

Sicheri, F., Moarefi, I., and Kuriyan, J. (1997). Crystal structure of the Src family tyrosine kinase Hck. *Nature* 385:602–609.

Slate, D.L, Lee, R.H., Rodriguez, J., and Crews, P. (1994). The marine nature product, halistanol trisulfate, inhibits pp60v-src protein tyrosine kinase activity. *Biochem. Biophys. Res. Commun.* 203:260–264.

Songyang, Z., Shoelson, S.E., Chaudhuri, M., Gish, G., Pawson, T., Haser, W.G., King, F., Roberts, T., Ratnofsky, S., Lechleider, R.J., Neel, B.G., Rirge, R.B., Fajardo, J.E., Chou, M.M., Hanafusa, H., Schaffhausen, B., and Cantley, C.L. (1993). SH2 domains recognize specific phosphopeptide sequences. *Cell* 72:767–778.

Soriano, P., Montgomery, C., Geske, R., and Bradley, A. (1991). Targeted disruption of the c-src proto-oncogene leads to osteoporosis in mice. *Cell* 64:693–702.

Sotriffer, C., and Klebe, G. (2002). Identification and mapping of small-molecule binding sites in proteins: computational tools for structure-based drug design. *Farmaco* 57:243–251.

Staley, C.A., Parikh, N.U., and Gallick, G.E. (1997). Decreased tumorigenicity of a human colon carcinoma cell line by an antisense expression vector specific for c-Src. *Cell Growth Differ.* 8:269–274.

Stankovic, C.J., Plummer, M.S., and Sawyer, T.K. (1997a). Peptidomimetic ligands for Src homology-2 domains. *Adv. Amino Acid Mimet. Peptidomimet.* 1:127–163.

Stankovic, C.J., Surendran, N., Lunney, E., Plummer, M.S., Para, K., Shahipour, A., Fergus, J.H., Marks, J.S., Herrera, R., Hubbell, S.E., Humblet, C., Saltiel, A.R., Stewart, B.H., and Sawyer, T.K. (1997b). The role of 4-phosphonodifluoromethyl- and 4-phosphono-phenylalanine in the selectivity and cellular uptake of SH2 domain ligands. *Bioorg. Med. Chem. Lett.* 7:1909–1914.

Stover, D.R., Lydon, N.B., and Nunes, J.J. (1999). Recent advances in protein kinase inhibition: current molecular scaffolds used for inhibitor synthesis. *Curr. Opin. Drug Discov. Dev.* 2:274–285.

Strawn, L.M., and Shawver, L.K. (1998). Tyrosine kinases in diseases. Overview of kinase inhibitors as therapeutic agents and current drugs in clinical trials. *Expert Opin. Invest. Drugs* 7:533–573.

Summy, J.M., and Gallick, G.E. (2003). Src family kinases in tumor progression and metastasis. *Cancer Metastas. Rev.* 22:337–358.

Sundaramoorthi, R., Kawahata, N., Yang, M.G., Shakespeare, W.C., Metcalf, C.A. III, Wang, Y., Merry, T., Eyermann, C.J., Bohacek, R.S., Narula S., Dalgarno, D.C., Sawyer, T.K. (2003). Structure-based design of novel nonpeptide inhibitors of the Src SH2 domain:phosphotyrosine mimetics exploiting multifunctional group replacement chemistry. *Biopolymers (Pept. Sci.)* 71:717–729.

Susa, M., and Teti, A. (2000). Tyrosine kinase Src inhibitors: potential therapeutic applications. *Drug News Perspect.* 13:169–175.

Susa, M., Missbach, M., and Green J. (2000). Src inhibitors: drugs for the treatment of osteoporosis, cancer or both? *Trends Pharmacol. Sci.* 21:489–495.

Talamonti, M.S., Roh, M.S., Curley, S.A., and Gallick, G.E. (1993). Increase in activity and level of pp60^c-src in progressive stages of human colorectal cancer. *J. Clin. Invest.* 91:3–60.

Tanaka, S., Amling, M., Neff, L., Peyman, A., Uhlmann, E., Levy, J.B., and Baron, R. (1996). c-Cbl is downstream of c-Src in a signalling pathway necessary for bone resorption. *Nature* 383:528–531.

Tatton, L., Morley, G.M., Chopras, R. and Khwaja, A. (2003). The Src-selective kinase inhibitor PP1 also inhibits Kit and BcrAbl tyrosine kinases. *J. Biol. Chem.* 278:4847–4853.

Teitelbaum, S.L. (2000). Bone resorption by osteoclasts. *Science* 289:1504–1508.

Thomas, J.W., Ellis, B., Boerner, R.J., Knight, W.B., White, G.C., II, and Schaller, M.D. (1998). SH2- and SH3-mediated interactions between focal adhesion kinase and Src. *J. Biol. Chem.* 273:577–583.

Thomas, S.M., and Brugge, J.S. (1997) Cellular functions regulated by Src family kinases. *Annu. Rev. Cell Dev. Biol.* 13:513–609.

Toledo, L.M., Lydon, N.B., and Elbaum, D. (1999). The structure-based design of ATP-site directed protein kinases. *Curr. Med. Chem.* 6:775–805.

Traxler, P. (1997). Protein tyrosine kinase inhibitors in cancer treatment. *Expert Opin. Ther. Patents* 7:571–587.

Traxler, P. (1998). Tyrosine kinase inhibitors in cancer treatment (part II). *Expert Opin. Ther. Patents* 8:1599–1628.

Tsai, Y.T., Su, Y.H., Fang, S.S., Huang, T.N., Qiu, Y., Jou, Y.S., Shih, H.M., Kung, H.J., and Chen, R.H. (2000). Etk, a Btk family tyrosine kinase, mediates cellular transformation by linking Src to STAT3 activation. *Mol. Cell. Biol.* 20:2043–2054.

Tsatsanis, C., and Spandidos, D.A. (2000). The role of oncogenic kinases in human cancer. *Int. J. Mol. Med.* 5:583–590.

Turkson, J., Bowman, T., Garcia, R., Caldenhoven, E., De Groot, R.P., and Jove, R. (1998). STAT3 activation by Src induces specific gene regulation and is required for cell transformation. *Mol. Cell. Biol.* 18:2545–2552.

Uehara, Y., and Fukuzawa, H. (1991). Use and selectivity of herbimycin A as inhibitor of protein-tyrosine kinases. *Methods Enzymol.* 201:370–379.

Uludag, H. (2002). Bisphosphonates as a foundation of drug delivery to bone. *Curr. Pharm. Des.* 8:929–1944.

Van Oijen, M.G., Rijkseng, G., Ten Broek, F.W., and Slootweg, P.J. (1998). Overexpression of c-Src in areas of hyperproliferation in head and neck cancer, premalignant lesions and benign mucosal disorders. *J. Oral Pathol. Med.* 27:147–152.

Verbeek, B.S., Vroom, T.M., Adriaansen-Slot, S.S., Ottenhoff-Kalff, A.E., Geertzema, J.G., Hennipman, A., and Rijksen, G. (1996). c-Src protein expression is increased in human breast cancer. An immuno-histochemical and biochemical analysis. *J. Pathol.* 180:383–388.

Vetter, S.W., and Zhang, Z.Y. (2002). Probing the phosphopeptide specificities of protein tyrosine phosphatases, SH2 and PTB domains with combinatorial library methods. Curr. *Protein Pept. Sci.* 3:265–397.

Vidal, M., Gigoux, V., and Garbay, C. (2001). SH2 and SH3 domains as targets for antiproliferative agents. *Crit. Rev. Oncol. Hematol.* 40:175–186.

Violette, S.M., Shakespeare, W.C., Bartlett, C., Guan, W., Smith, J.A., Rickles, R.J.; Bohacek, R.S., Holt, D.A., Baron, R., and Sawyer, T.K. (2000). A Src SH2 selective binding compound inhibits osteoclast-mediated resorption. *Chem. Biol.* 7:225–235.

Violette, S.M., Guan, W., Bartlett, C., Smith, J.A., Bardelay, C., Antoine, E., Rickles, R.J., Mandine, E., van Schravendijk, M.R., Adams, S. E., Lynch, B.A., Shakespeare, W.C., Yang, M., Jacobsen, V.A., Takeuchi, C.S., Macek, K.J., Bohacek, R.S., Dalgarno, D.C., Weigele, M., Lesuisse, D., Sawyer, T.K., and Baron, R. (2001). Bone-targeted Src SH2 inhibitors block Src cellular activity and osteoclast-mediated resorption. *Bone* 28:54–64.

von Bubnoff, N., Veach, D.R., Miller, T., Li, W., Sanger, J., Peschel, C., Bornmann, W.G., Clarkson, B., and Duyster, J. (2003). Inhibition of wild-type and mutant Bcr-Abl by pyrido-pyrimidine-type small molecule kinase inhibitors. *Cancer Res.* 63:6395–6404,

von Stechow, D., Alexander, J., Chorev, M., Fish, S., Iuliucci, J., Wang, Y., Metcalf, C., Keenan, T., Sundaramoorthi, R., Shakespeare, W., van Schravendijk, M.R., Dalgarno, D., and Sawyer, T. (2001).

Novel Src tyrosine kinase inhibitors prevent ovariectomy-induced bone loss in a Swiss-Webster mouse model of post-menopausal osteoporosis. *J. Bone Miner. Res.* 16 (*Suppl. 1*), Abstract F425.

Vu, C.B. (2000). Recent advances in the design and synthesis of SH2 inhibitors of Src, Grb2, and ZAP-70. *Curr. Med. Chem.* 7:1081–1100.

Waksman, G. Kominos, D., Robertson, S.C., Pant, N., Baltimore, D., Birge, R.B., Cowburn, D., Hanafusa, H., Mayer, B.J., Overduin, M., Resh, M.D., Rios, C.B., Silverman, L., and Kuriyan, J. (1992). Crystal structure of the phosphotyrosine recognition domain SH2 of v-Src complexed with tyrosine-phosphorylated peptides. *Nature* 358:646–653.

Waksman, G., Shoelson, S.E., Pant, N., Cowburn, D., and Kuriyan, J. (1993). Binding of a high affinity phosphotyrosyl peptide to the Src SH2 domain: crystal structures of the complexed and peptide-free forms. *Cell* 72:779–790.

Waksman, G., Kumaran, S., and Lubman, O. (2004). SH2 domains: role, strucure and implications for molecular medicine. *Exp. Rev. Molec. Med.* 6:1–21.

Wang, Y.D., Miller, K., Boschelli, D.H., Ye, F., Wu, B., Floyd, M.B., Powell, D.W., Wissner, A., Weber, J.M., Boschelli, F. (2000). Inhibitors of Src tyrosine kinase: the preparation and structure-activity relationship of 4-anilino-3-cyanoquinolines and 4-anilinoquinazolines. *Bioorg. Med. Chem. Lett.* 10:2477–2480.

Wang, Y., Metcalf, C.A., III, Shakespeare, W.C., Sundaramoorthi, R., Keenan, T.P., Bohacek, R.S., van Schravendijk, M.R., Violette, S.M., Narula, S.S., Dalgarno, D.C., Haraldson, C., Keats, J., Liou, S., Mani, U., Pradeepan, S., Ram, M., Adams, S., Weigele, M., and Sawyer TK. (2003). Bone-targeted 2,6,9-trisubstituted purines: novel inhibitors of Src tyrosine kinase for the treatment of bone diseases. *Bioorg. Med. Chem. Lett.* 13:3067–3070.

Warmuth, M., Damoiseaux, R., Liu, Y., Fabbro, D., and Gray, N. (2003a). Src family kinases: potential targets for the treatment of human cancer and leukemia. *Curr. Pharm. Design* 9:2043–2059.

Warmuth, M., Simon, N., Mitina, O., Mathes, R., Fabbro, D., Manley, P. W., Buchdunger, E., Forster, K., Moarefi, I., and Hallek, M. (2003b). Dual-specific Src and Abl kinase inhibitors, PP1 and CGP76030, inhibit growth and survival of cells expressing imatinib mesylate-resistant Bcr-Abl kinases. *Blood* 101:664–672.

Waszkowycz, B. (2002). Structure-based approaches to drug design and virtual screening. *Curr. Opin. Drug Discov. Devel.* 5:407–413.

Wisniewski, D., Lambek, C.I., Liu, C., Strife, A., Veach, D.R., Nagar, B., Young, M.A., Schindler, T., Bornmann, W.G., Bertino, J.R., Kuriyan, J., and Clarkson, B. (2002). Characterization of potent inhibitors of the Bcr-Abl and the c-Kit receptor tyrosine kinases. *Cancer Res.* 62:4244–4255.

Woolfrey, J.R., and Weston, G.S. (2002). The use of computational methods in the discovery and design of kinase inhibitors. *Curr. Pharm. Des.* 8:1527–1545.

Wong, B.R., Besser, D., Kim, N., Arron, J.R., Vologodskaia, M., Hanafusa, H., and Choi, Y. (1999). TRANCE, a TNF family member, activates Atk/PKB through a signalling complex involving TRAF6 and c-Src. *Mol. Cell* 4:1041–1049.

Xu, W., Doshi, A., Lei, M., Eck, M.J., and Harrison S.C. (1999). Crystal structures of c-Src reveal features of its autoinhibitory mechanism. *Mol. Cell* 3:629–638.

Xu, W., Harrison, S.C., and Eck, M.J. (1997). Three-dimensional structure of the tyrosine kinase c-Src. *Nature* 385:595–602.

Yamaguchi, H., and Hendrickson, W.A. (1996). Structural basis for activation of human lymphocyte kinase Lck upon tyrosine phosphorylation. *Nature* 384:484–489.

Yoneda, T., Lowe, C., Lee, C.-H., Gutierrez, G. Niewolna, M., Williams, P.J., Izbicka, E., Uehara, Y., Mundy, G. R. (1993). Herbimycin A, a pp60[c-src] tyrosine kinase inhibitor, inhibits osteoclastic bone resorption in vitro and hypercalcemia in vivo. *J. Clin. Invest.* 91:2791–2795.

Zhu, X., Kim, J.L., Newcomb, J.R., Rose, P.E., Stover, D.R., Toledo, L.M., Zhao, H., Morgenstern, K.A. (1999). Structural analysis of the lymphocyte-specific Lck in complex with non-selective and Src family-selective kinase inhibitors. *Struct. Fold. Des.* 7:651–661.

Zimmerman, J., Buchdunger, E., Mett, H., Meyer, T., and Lydon, N.B. (1997). Potent and selective inhibitors of the Abl-kinase: phenylaminopyrimidine (PAP) derivatives. *Bioorg. Med. Chem. Lett.* 7:187–192.

12

Disrupting Protein–Protein Interaction: Therapeutic Tools Against Brain Damage

Michelle Aarts and Michael Tymianski

ABSTRACT

Brain damage caused by stroke, epilepsy, head and spinal trauma, and degenerative neurological conditions represents a significant source of morbidity and mortality in Westernized
society. Neuronal death occurs as a result of a complex combination of excitotoxicity, necrosis,
apoptosis, edema, and inflammatory reactions. Neuroprotection via glutamate receptor blockade, antioxidant, or anti-inflammatory therapy have not proven effective in clinical treatment
of brain damage because of narrow therapeutic windows, poor pharmacokinetics or blockade
of signaling essential for normal excitatory neurotransmission and neuronal survival. Recent
work in neuronal biochemistry, genomics, and proteomics has increased understanding of the
molecular organization of the excitatory synapse and the neuronal postsynaptic density. This
understanding in kind has led to the dissection of the intracellular signaling cascades responsible for excitotoxicity. In addition, we have increased our understanding of the intracellular
and extracellular signaling, networks responsible for apoptosis and inflammation. It has thus
become possible to uncouple toxic second messenger pathways from their membrane receptors by targeting the interactions between the receptor and downstream proteins. In addition
the use of cell-permeable protein transduction domains now allows for the design of fusion
peptides and small proteins for in vivo therapies while overcoming limitations of poor cell

MICHELLE AARTS • Toronto Western Research Institute, Toronto, Ontario MST 258, Canada.
MICHAEL TYMIANSKI • Department of Neurosurgery, Toronto Western Hospital and Department
of Physiology, University of Toronto. Toronto, Ontario MST 258, Canada.

Proteomics and Protein–Protein Interactions: Biology, Chemistry, Bioinformatics, and Drug Design,
edited by Waksman. Springer, New York, 2005.

membrane permeability and in the instance of neurological disorders, poor delivery across the blood–brain barrier. This technology, together with an intricate knowledge of the enzymes, protein networks, and signaling mechanisms related to evolution of brain pathology, will aid in the design of future effective therapeutics.

1. INTRODUCTION

Injuries to the human central nervous system (CNS) occur frequently and can cause devastating, irreversible damage to patients. Brain damage caused by stroke, epilepsy, head and spinal trauma, and degenerative neurological conditions have considerable impact on health and economics in Westernized society. For example, ischemic stroke, a consequence of transient or permanent reduction in cerebral blood flow, represents a leading cause of death and long-term disability in Western society (for current statistics see American Stroke Association, http://www.strokeassociation. org; Heart and Stroke Foundation, ww2.heartandstroke.ca; The Stroke Association, http://www.stroke.org.uk). In addition to the significant morbidity, the direct and indirect costs of stroke care are a substantial burden on health care systems (Lewandowski and Barsan, 2001; McGovern and Rudd, 2003). Likewise head and spinal cord injuries are common and occur in the prime of life, often leaving victims with permanent damage and clinical disability (Thurman et al., 1999; Adekoya et al., 2002). Despite the enormous impact of acute neurological injuries on society, there exist few clinical treatments capable of altering the natural progression of brain damage caused by stroke, trauma, or neurological disorders. Current therapies consist mainly of providing symptomatic relief and care to minimize secondary complications.

Considerable research efforts have been directed toward finding methods for treating acute injury to CNS neurons. The slow progress of results may reflect the complex nature of the mammalian CNS and the limited regenerative capacity of central neurons and their processes. The inability of CNS tissue to regenerate imposes serious restrictions on the ability of treatment to restore function to damaged tissue. The significant advances in the understanding and promotion of nerve regeneration have not yet progressed to the point of clinical application. Therefore the most promising strategies for reducing damage to neurons involve interrupting the injury process before the damage becomes irreversible. As most of these strategies involve pharmacological therapies, understanding the normal and abnormal physiology and function of the nervous tissue at the subcellular level is critical to the development of practical treatments. The pathophysiology of acute neuronal damage after stroke or trauma involves a complex combination of processes including excitotoxicity, inflammation, necrosis, and apoptosis. In the last two decades, a better understanding of this pathophysiology, particularly of the molecular mechanisms of brain injury, has dramatically changed the approach to treatment. Rather than the "wait and see" approach, it is now recognized that patients presenting with stroke require rapid assessment and therapeutic action to restore blood flow and protect the brain from further injury. However, although progress has been made with blood flow restoration using thrombolytic agents, no neuroprotective compounds have been shown to be effective in stroke patients. To date, compounds that block or decrease excitatory neurotransmission, that inhibit reactive oxygen species, or that block the inflammatory response have been studied. Despite promising preclinical data, all of these promising therapies have failed to show benefit in extensive human clinical trials. Although the reasons for this remain controversial, a main hypothesis is that the negative consequences of administering agents that inhibit excitatory neurotransmission in the CNS had outweighed their utility as neuroprotectants. Thus, a more sophisticated approach to treating acute neuronal death is needed. This level of sophistication has been made possible through recent advances in genomics and proteomics. These have allowed the identification of intracellular proteins that form the structural basis of signal transmission at central synapses. New knowledge has permitted the identification of specific postsynaptic protein–protein interactions as therapeutic targets for stroke drugs that block neurotoxic signals with high specificity. This can be achieved by procedures that result in the uncoupling of toxic signal pathways from the normal neurotransmission that is essential to neuronal survival. Here we will address the use of cell permeable peptides as a novel treatment for excitotoxic neurodegeneration.

1.1. Pathophysiology of Neuronal Death Following Injury

Our understanding of cellular and molecular brain injury mechanisms and the approach to treatment has changed dramatically in the last two decades. Acute pathophysiology involves a complex combination of processes including excitotoxicity, inflammation, necrosis, and apoptosis (Fig. 12.1). Following an acute insult neurons can die almost immediately by excitotoxic mechanisms (Chan, 1994) or as many as 10 days later (Saunders et al., 1995) via a programmed cell death mechanism (apoptosis) (Tominaga et al., 1993; Kihara et al., 1994; Bartus et al., 1995;

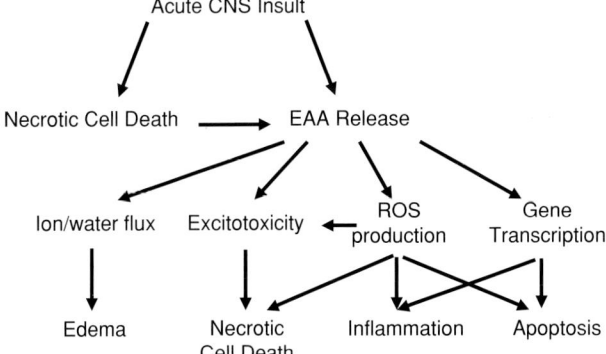

Figure 12.1. Flowchart denoting the multiple interrelated stages that encompass the pathophysiologic processes that lead to brain damage following acute CNS injury.

Hill et al., 1995). The brain has a relatively high demand for oxygen and glucose and depends heavily on oxidative phosphorylation for energy production. Focal impairment of cerebral blood flow, by embolus, hemorrhage, or trauma, restricts the delivery of substrates and impairs the maintenance of ionic gradients. There is an ensuing loss of membrane potential and neurons depolarize. Consequently, dendritic as well as presynaptic voltage-dependent Ca^{2+} channels become activated and excitatory amino acids (EAAs) are released into the extracellular space. EAA receptors subsequently mediate the critical excitotoxic events that result in neuronal death. Receptor-mediated entry of Na^+ and Cl^- leads to passive influx of water, and the ensuing edema can compound initial perfusion injury as well as increase intracranial pressure. Extensive research has shown that Ca^{2+} influx via EAA receptors is key in mediating ischemic and posttraumatic neurodegeneration. Ca^{2+} is a ubiquitous intracellular messenger capable of triggering signal cascades as well activating enzymes responsible for protein, lipid, and DNA cleavage. Increasing knowledge of the synaptic organization of EAA receptors and their associated proteins has allowed identification of specific, neurotoxic signaling pathways that transduce Ca^{2+} events during ischemia. For instance, nitric oxide (NO) synthesized by the Ca^{2+}-dependent enzyme, neuronal nitric oxide synthase (nNOS) reacts with a superoxide anion to form the highly reactive species peroxynitrite that promotes tissue damage. The important role played by reactive oxygen species (ROS) in cell damage during stroke is emphasized by the fact that treatment with free radical scavengers can be effective in experimental focal cerebral ischemia (Chan, 1994; Dugan and Choi, 1994; Dawson et al., 1996; Iadecola, 1997). In addition to extensive cellular damage, EAA-triggered production of oxygen free radicals can initiate inflammation and apoptosis. Activation of intracellular second messengers and ROS can elicit a proinflammatory response via transcription factors, nuclear factor-κB, hypoxia inducible factor, interferons, and STAT proteins (Planas et al., 1996; O'Neill and Kaltschmidt, 1997; Iadecola et al., 1999b); and proinflammatory genes such as tumor necrosis factor-α and interleukin-1β (Rothwell and Hopkins, 1995). This process attracts inflammatory cells to the injured brain so that monocytes and macrophages become the predominant cells within several days (Lees, 1993; Tomita and Fukuuchi, 1996; Hallenbeck, 1997; Stoll and Jander, 1999; Schwab et al., 2000; Beschorner et al., 2002; Schilling et al., 2003). This postinjury inflammatory response can contribute to further brain damage by the production of toxic mediators and microvascular obstruction (del Zoppo et al., 1991). An apoptotic cascade may be initiated by a similar sequence of intracellular messenger activation, ROS production, and proapoptotic gene expression. Injured neurons are particularly susceptible to apoptotic death, and both inhibition of caspases or augmentation of Bcl-2 expression has been shown to reduce brain damage (Yakovlev et al., 1997; Adams and Cory, 1998; Thornberry and Lazebnik, 1998). Thus there is a complex interplay of excitotoxicity, inflammation, and apoptosis that ultimately results in the permanent brain damage initiated by an acute neuronal injury. Each phase in turn represents a target for neuroprotection and must be fully investigated so that appropriate therapeutic interventions can be designed.

2. EXCITOTOXICITY

It is now widely accepted that the predominant mechanism of cell death in diseases such as stroke, CNS trauma, epilepsy, and chronic neurodegenerative disorders is "excitotoxicity" (Olney, 1969). Excitotoxicity results from an excessive release and inadequate reuptake of synaptic glutamate, the predominant excitatory amino acid neurotransmitter in the CNS. The role of glutamate in hypoxic neurodegeneration was established in the 1980s by studies that showed reduced neuronal sensitivity to hypoxia when postsynaptic glutamate receptors were blocked (Kass and Lipton, 1982; Rothman, 1983). The majority of glutamate receptor (GluR) subtypes have since been implicated in mediating neurotoxicity and there is general agreement that the mechanism is largely calcium dependent (Choi, 1987, 1995). Glutamate receptor overactivation and the resultant failure in Ca^{2+} homeostasis activate intracellular signalling events leading to free radical production and cell death. Postsynaptic responses to glutamate are mediated via pharmacologically and functionally distinct metabotropic (G-protein–coupled) or ionotropic (ligand-gated ion channel) glutamate receptor families. During ischemia, glutamate receptors are believed to mediate an imbalance in neuronal ion homeostasis.

2.1. Metabotropic GluRs

Metabotropic receptors (mGluRs) are G-protein–coupled receptors belonging to the same family as the Ca^{2+}-sensing and γ-aminobutyric acid B (GABA-B) receptors. They exist as a family of pharmacologically discernible subtypes (mGluR1-8) that act through the inositol phosphate–stimulated release of intracellular Ca^{2+}. Although not directly implicated in excitotoxicity during stroke, mGluRs play an important modulatory role in the neuronal response to glutamate signaling. Group-I mGluRs have been shown to downregulate K^+ channels and upregulate nonselective cation channels, inhibit GABA receptor activity, and potentiate iGluR function, resulting in enhanced neuronal excitability (Gerber et al., 1992; Crepel et al., 1994; Hoffpauir and Gleason, 2002; Chong et al., 2003). Group II and III mGlu receptors located in the presynaptic terminals modulate release of glutamate and the inhibitory neurotransmitter GABA (Pin and Duvoisin, 1995). Thus the mGlu receptors are important mediators of neuronal plasticity, nociception, pain, and in some instances, neurodegeneration. Agonists of group I mGluRs (mGluR1 and mGluR5) have been shown to amplify N-methyl-D-aspartate (NMDA)-mediated excitotoxicity in vitro (Nicoletti et al., 1999) and both competitive and noncompetitive group I antagonists reduce neuronal toxicity to NMDA and are neuroprotective in experimental stroke (Nicoletti et al., 1999; Pellegrini-Giampietro et al., 1999).

2.2. Ionotropic Glutamate Receptors (iGLuRs)

iGluRs are activated during ischemia, resulting in increased permeability to sodium, potassium, and calcium. The ionotropic glutamate receptors are divided into

Table 12.1. Ionotropic glutamate receptor family

Receptor subunit (Rat)	Ligand selectivity
GluR1, GlurR2, GluR3, GluR4	AMPA-selective
GluR5, GluR6, GluR7	Kainate-selective
KA-1, KA-2	Kainate-selective
NR2A, NR2B, NR2C, NR2D	NMDA
NR1	NMDA
NR3A, NR3B	NMDA

distinct subfamilies based on their affinity for NMDA: α-amino-3-hydroxyl-5-methyl-4-isoxazolepropionic acid (AMPA) or kainate (for review of iGluRs see Cull-Candy et al., 2001; Madden, 2002). IGluRs are heteromeric structures assembled from a combination of subunits (Table 12.1) that form a selective ion channel pore. Each subunit has four transmembrane domains with an extracellular N-terminus and an intracellular C-terminal tail (Hollmann and Heinemann, 1994). iGluR signaling plays an important role in mediating the synaptic plasticity that is implicated in our ability to learn and form memories (Kind and Neumann, 2001; Sheng and Kim, 2002).

2.2.1. AMPA/Kainate Receptors

AMPA and kainate receptors mediate the fast excitatory component of glutamate neurotransmission. These channels open and close more quickly than NMDA receptors and are responsible for propagating membrane depolarization by opening postsynaptic voltage-sensitive channels (Bettler and Mulle, 1995). AMPARs are comprised of combinations of GluR1 to GluR4 (GluR-A to GluR-D) subunits. Each subunit exists as two alternatively spliced isoforms whose expression is regulated in a developmental manner (Sommer et al., 1990; Schoepfer et al., 1994). In addition, RNA editing of the mRNA transcripts can result in a variety of functionally distinct receptor isoforms. The receptors subunits are permeable to Na^+ and K^+. They may also be permeable to Ca^{2+} and Zn^{2+} ions unless the receptor contains at least one GluR2 (Hollmann et al., 1991). The GluR2 transcripts undergo RNA editing within the second transmembrane domain so that a glutamine in the pore-lining region is replaced with a positively charged arginine. This makes GluR2-containing AMPARs impermeable to Ca^{2+} (Hume et al., 1991; Sommer et al., 1991). Though the majority of AMPARs in the CNS contain GluR2, downregulation of this subunit during ischemia and subsequent increased Ca^{2+} permeability has been implicated in delayed neuronal death. (Pellegrini-Giampietro et al., 1997). AMPA receptors (AMPARs) undergo trafficking from intracellular pools to synaptic sites, a property that is implicated in the activity-dependent modulation of synaptic strength (Rose and Konnerth, 2000). AMPAR activity can also be modulated by subunit phosphorylation. Receptor activation is potentiated S/T phosphorylation by protein kinase A (PKA), PKC, and calmodulin-dependent kinase type II (CaMKII) (Blackstone et al., 1994; Wang et al., 1994a; Barria et al., 1997; Derkach et al., 1999; Banke et al., 2000; McDonald et al., 2001).

Kainate receptors are comprised of subunits that are homologous in transmembrane topology and stoichiometry to AMPA and NMDA receptor subunits. Kainate receptor subunits can be further divided into two groups based on their high affinity (KA1 and KA2) or lower affinity (GluR5-7) binding to kainic acid (Bettler et al., 1990; Werner et al., 1991; Herb et al., 1992). Like AMPARs, kainate subunits undergo RNA editing so that some subunits (GluR5 and GluR6) exist as Q to R substituted isoforms that are Ca^{2+} impermeable (Hume et al., 1991). Compared with the other iGluR subtypes there has been little specific data or understanding of kainate receptor physiology. Until the mid-1990s, research was limited by a lack of highly specific agonists or antagonists and kainate receptors have been lumped together with AMPARs as non-NMDA iGluRs. Although kainic acid has a higher affinity for kainate subunits, it is also an effective agonist for AMPAR subunits, and non-NMDA antagonists do not distinguish between kainate and AMPAR subunits (for review see Lerma et al., 2001). With the discovery of more selective agonists and antagonists research can now discern specific roles for kainate receptors in neuronal physiology and pathophysiology (Paternain et al., 1995; Wilding and Huettner, 1995). It is now known that in the absence of AMPAR function, kainate elicits a small, slowly deactivating but rapidly desensitizing current in hippocampal neurons (Paternain et al., 1995; Castillo et al., 1997). Kainate receptor subunits are found localized at both the pre- and postsynaptic surfaces of neurons (Petralia et al., 1994; Charara et al., 1999; Kieval et al., 2001). Although not primarily implicated in postsynaptic communication, kainate receptors have recently been shown to have synaptic activation in hippocampal neurons (Castillo et al., 1997; Vignes et al., 1997; Frerking et al., 1998) and may be important for synaptic plasticity (Bortolotto et al., 1999; Contractor et al., 2001). Most notably, kainate receptors are important modulators of the presynaptic release of neurotransmitter at both excitatory and inhibitory synapses (Rodriguez-Moreno and Lerma, 1998; Lauri et al., 2001; Schmitz et al., 2000, 2001).

2.2.2. NMDA Receptors

NMDA receptor (NMDAR) channels are made up of various subunits with distinct temporal and spacial distribution, pharmacological properties, and signal transduction. NMDAR subunits share amino acid sequence and structural homology with AMPA/kainate receptors and similarly form tetrameric structures (Mori and Mishina, 1995). Three types of NMDAR subunits have been identified: the ubiquitously expressed NR1 subunit, four distinct NR2 subunits (A to D), and two NR3 members (A, B) (Madden, 2002). The NR1 subunit exists as eight possible isoforms owing to splicing variations (Zukin and Bennett, 1995). In addition, NR2C, 2D, and 3A have been shown to exist as alternatively spliced isoforms (Rafiki et al., 2000; Cull-Candy et al., 2001). NR1 subunits contain an asparagine residue within the second transmembrane domain, analogous to the edited Q/R site of AMPAR subunits, which governs ion permeability of the receptor channel (Sakurada et al., 1993). The NR1 subunit is essential for receptor formation; however, highly active, functional NMDARs are produced only when they contain both NR1 and NR2 subunits and in some cases

NR3 subunits (Ciabarra et al., 1995; Sucher et al., 1995). Studies of recombinant receptors have shown that subunit composition dictates the functional and pharmacological properties of NMDAR channels. Both the affinity for glutamate and the time course for channel opening are unique to the NR2 subunit, with further contributions to receptor properties made by the various NR1 isoforms. NMDARs are slow gating channels that are highly permeable to Ca^{2+} and Na^+ (Dale and Roberts, 1985). While Na^+ contributes to membrane depolarization the influx of exogenous Ca^{2+} that generates the intracellular calcium transients that are responsible for the physiological effects of NMDAR signaling. NMDARs are unique in that they require binding to both L-glutamate and glycine (D-serine is the endogenous modulator) for efficient channel opening (Johnson and Ascher, 1987). NMDA receptors also differ from their AMPA/kainate cousins in that they require membrane depolarization for full activation. Membrane depolarization, supplied in many cases by colocalized AMPARs, releases Mg^{2+} block of the receptor, allowing glutamate to bind and open the ion pore. Thus NMDAR channels are coincidence detectors, opening only when neurotransmitter activation is paired temporally with depolarization of the postsynaptic membrane.

At least a portion of the signaling properties of NMDARs is governed by the interactions of its intracellular domains with cytoskeletal and signal transduction molecules. The intracellular tail of NR2 subunits is critical for the proper function of NMDARs as noted by transgenic mouse studies. Mouse mutants with targeted deletion of the NR2 cytoplasmic tails are phenotypically indistinguishable from the corresponding NR2 subunit knockouts (Sprengel et al., 1998). For instance, both knockout of NR2B and deletion of its C-terminus are neonatal lethal whereas both knockout of NR2A and deletion of its tail result in mice with impaired synaptic plasticity and memory (Sakimura et al., 1995; Kadotani et al., 1996; Sprengel et al., 1998). The C-terminal tails possess multiple motifs for phosphorylation and protein–protein interactions. The regulatory protein CaMKII, activated on glutamate stimulation, phosphorylates serine residues within the C-terminal tail of NR2A and 2B and may play a role in regulating synaptic plasticity (Bayer et al., 2001; Fong et al., 2002). NMDAR signaling may be upregulated by tyrosine phosphorylation by Src kinase family members and inhibited by the actions of the phosphatase calcineurin (Lieberman and Mody, 1994; Wang et al., 1994b; Yu et al., 1997). NMDAR assembly can also be modulated by phosphorylation. For instance, PKC phosphorylation of NR1 plays a role in subunit clustering (Ehlers et al., 1995). Interaction of NR1 and NR2B with α-actinin-2, an actin binding protein enriched in the postsynaptic density (PSD), may be important for the clustering of NMDARs at the synapse (Wyszynski et al., 1997; Allison et al., 1998). Ca^{2+}/calmodulin binds multiple high- and low-affinity sites on the tail of NR1, where it inhibits channel opening and decreases open time in response to Ca^{2+} (Ehlers et al., 1996; Zhang et al., 1998).

2.3. Excitotoxicity and Calcium Signaling

Calcium is a ubiquitous intracellular messenger, governing a wide range of cellular functions including cell growth, membrane excitability, synaptic activity, and cell death. Its intracellular concentrations are tightly regulated to efficiently control signaling. Localized increases in intracellular Ca^{2+} concentration ($[Ca^{2+}]_i$) are used to trigger physiological events such as the activation of enzymes or ion channels. It is believed that excessive Ca^{2+} loading can exceed the capacity of Ca^{2+}-regulatory mechanisms and inappropriately activate processes (i.e., activation of proteases and endonucleases) that lead to cell death (for review see Rahn et al., 1991; Bindokas and Miller, 1995). The majority of glutamate receptor subtypes have been implicated in mediating neurotoxicity, and there is general agreement that the mechanism is largely calcium dependent (Choi, 1987, 1995). It is also generally accepted that the NMDA receptor family plays a key role in glutamate toxicity owing to their high Ca^{2+} permeability (Choi, 1988; Choi et al., 1988; Tymianski, 1996). IGluR activation causes influx of Na^+, which mediates toxic swelling of dendritic spines, and Ca^{2+}, which is responsible for delayed neuronal degeneration. Choi et al. (Choi, 1985, 1987) demonstrated that removal of extracellular Ca^{2+} but not Na^+ could reduce cell death in response to glutamate challenge. This and later works have established the critical role for Ca^{2+} influx in neurodegeneration. There are differing opinions, however about the mechanism by which Ca^{2+} influx triggers neuronal cell death during excitotoxicity.

The "calcium load hypothesis" suggests that neurodegeneration is simply a function of the quantity of Ca^{2+} entering the cell. This theory was based on experiments showing that cultured neurons experience delayed Ca^{2+} accumulation and studies demonstrating that Ca^{2+} uptake and cell death were correlated with glutamate exposure (Choi et al., 1989; Manev et al., 1989; Marcoux et al., 1990). However, there is a compelling body of evidence that shows dissociation between Ca^{2+} accumulation and neuronal death. Several studies have shown that calcium channel blockers can prevent Ca^{2+} accumulation but not neurotoxicity during anoxia (Marcoux et al., 1989, 1992; Madden et al., 1990; Dubinsky and Rothman, 1991). Thus a general elevation in calcium does not necessarily predict neuronal death, and additional factors may influence the outcome of Ca^{2+} influx. We know that the various calcium-dependent processes are regulated via distinct signal pathways linked to specific routes of Ca^{2+} influx (Bading et al., 1993; Ghosh and Greenberg, 1995). The "source specificity hypothesis" reasons that Ca^{2+} toxicity occurs not simply as a function of increased Ca^{2+} concentration, but is instead linked to the route of Ca^{2+} entry and the distinct second messenger pathways activated as a result. Source specificity was originally based on experiments performed with free Ca^{2+} indicators showing that Ca^{2+} loads produced by voltage sensitive Ca^{2+} channels were not harmful whereas similar $[Ca^{2+}]_i$ increases via NMDARs were toxic (Tymianski et al., 1993). Thus, distinct influx pathways, rather than calcium load, determine neuronal vulnerability to glutamate and calcium (Sattler et al., 1998). These findings directed research in our laboratory toward identifying the Ca^{2+}-activated processes that are associated with glutamate toxicity. The source specificity hypothesis proposes that molecular targets, such as rate-limiting enzymes, are physically linked or colocalized with glutamate receptors and can be manipulated to block Ca^{2+}-dependent neurotoxicity.

2.4. Excitotoxic Glutamate Signaling as a Target for Therapeutic Intervention

"Neuroprotection" is an important concept in the approach to finding new stroke therapies. Because glutamate plays such a critical role in excitotoxicity, blocking glutamate neurotransmission and ion flow through glutamate receptors has become an important aspect of the neuroprotective strategy. In theory, controlling excitotoxic glutamate signaling in neurodegenerative states should be as simple as blocking glutamate receptors. Indeed, a vast array of competitive and noncompetitive NMDAR and AMPAR antagonists have been developed in addition to compounds that selectively block GluR subunits, such as the noncompetitive blocker of NR2B, ifenprodil (Williams, 1993). Despite benefits in animal models of stroke, exhaustive studies using glutamate receptor antagonists for the treatment of stroke have not resulted in clinically useful compounds. These failures represent a combination of problems with clinical trial design, narrow therapeutic windows, poor pharmacokinetic properties of the drugs, and adverse side effects (Ikonomidou and Turski, 2002)(see http://www.stroketrials.org for clinical trials with glutamate antagonists). NMDAR signaling is essential to neuronal survival, and blockade of NMDARs triggers apoptosis in the developing brain (Ikonomidou et al., 1999). Normal NMDAR signaling may in fact be neuroprotective after brain injury (Ikonomidou et al., 2000). These disappointments have caused researchers to turn away from neuroprotection as a stroke treatment. Rather than cast aside the importance of GluR mediated toxicity, we have proposed a more detailed and sophisticated methodology. An alternative approach to blocking GluRs focuses on knowledge gained from studies of postsynaptic signaling pathways and mechanisms of cell death. Recent evidence suggests that clustering of glutamate receptors at the PSD allows coupling of glutamate receptor activity to second messengers capable of mediating neurotoxicity. Understanding this arrangement has allowed us to separate neurotoxic signaling from signals essential to neuronal survival and to propose novel strategies for the treatment of excitotoxic damage in stroke.

3. MOLECULAR ORGANIZATION OF GluR SIGNALING AT THE PSD

Signal molecules with the potential to act as neurotoxic triggers most likely exist within the PSD. The PSD is a specialized structure located beneath the postsynaptic membrane aligned with active zones of presynaptic terminals within the CNS. It is an electron-dense region comprised of multiple membrane-bound, scaffolding and cytoskeletal proteins (Kennedy, 1993). Several functions including cell-to-cell adhesion, regulation of receptor clustering, and modulation of receptor function have been attributed to the PSD. Excitatory CNS synapses are especially enriched in PSDs, which play a critical role in the clustering and function of glutamate receptors (Landis and Reese, 1974). PSD morphology can be influenced by synaptic activity and events triggered by GluR activation affect not only signal transmission but also structural remodeling of the PSD (Kennedy, 1993).

3.1. Components of the PSD

Four major types of molecules constitute the PSD: membrane bound, cytoskeletal and scaffolding proteins, and modulatory enzymes. Knowledge about these is increasing as new PSD components continue to be identified. Cell junction proteins play a key role in the colocalization of the PSD with the appropriate presynaptic region; however, NMDA and AMPA receptors make up the major membrane-bound component of the PSD (Moon et al., 1994; Baude et al., 1995). On the intracellular surface of the PSD are found a variety of enzymes that regulate phosphorylation of PSD components or are key effectors in the glutamate receptor signal pathway (Kennedy, 1998; Allison et al., 2000; Shin et al., 2000). The PSD is enriched in a number of cytoskeletal elements including actin and neurofilaments, and scaffolding proteins including spectrin, α-actinin-2, and PDZ containing proteins and their associated binding partners (Sheng, 2001). Cytoskeletal proteins are considered important to the localization and clustering of PSD receptor complexes whereas scaffolding proteins are the "glue" that function in bringing the various PSD components into association. For instance, α-actinin-2 serves as the intermediary that links actin to NMDAR subunits, thereby influencing clustering of the receptors (Wyszynski et al., 1997).

The PSD is particularly enriched in specialized scaffolding proteins that contain PDZ domains. The PDZ proteins of excitatory synapses fall into two main families related to either PSD-95 or GRIP1 (Table 12.2) (Kornau et al., 1997; Hung and Sheng, 2002). Four members of the PSD-95 family are found in mammalian synapses: PSD-95/SAP90, SAP97, PSD-93/chapsyn-110, and SAP102. These are membrane-associated guanylate kinases (MAGUKs), each characterized by three N-terminal PDZ domains followed by a Src-homology 3 (SH3) and guanylate kinase-like (GK) domain. Each domain is capable of mediating protein–protein interactions with multiple binding partners (Table 12.2). The PDZ domains are roughly 100 amino acids long and bind to small consensus motifs (tSXV) at the C-terminus of associated proteins such as NMDAR NR2 subunits. In addition, PDZ domains can self-associate, both within PSD-95 family members and with the PDZ domains of other proteins. This

Table 12.2. Postsynaptic density PDZ proteins and binding partners

PDZ protein	Type of interaction	Binding partners
PSD 95 Family (PSD-95/SAP90,	Class I PDZ (T/SxV)	NR2A-D, GluR1, GluR6, neuroligins, SynGAP, citron, CRIPT
SAP97, SAP102, PSD-93/chapsyn-110)	PDZ-PDZ	nNOS, PSD-95 members
	Other-PDZ	Src-family kinases
	SH3	Pyk2, KA2
	GK-like	GKAP, KA2, SPAR
GRIP Family (GRIP1, ABP)	Class II PDZ (SVKI)	GluR2, GluR3
	PDZ-PDZ	GRIP1/ABP
PICK1	Class II PDZ (SVKI)	PKC, GluR2/3

feature furthers their ability to cluster PSD proteins into functional complexes. For instance, PSD-95 forms PDZ–PDZ interactions with neuronal nitric oxide synthase (nNOS), cysteine-rich interactor of PDZ3 (CRIPT)— a microtubule-binding protein— and citron (Brenman et al., 1996a; Niethammer et al., 1998; Passafaro et al., 1999; Zhang et al., 1999). The SH3 domains of PSD-95 and SAP102 bind Pyk2, which in turn localizes and activates Src-kinase within the PSD (Seabold et al., 2003). The GK domain of PSD-95 binds to SPAR and guanylate kinase-associated protein (GKAP). SPAR links PSD-95 to both the actin cytoskeleton and the Rap signalling pathway (Pak et al., 2001). GKAP may functionally link PSD-95 to metabotropic glutamate receptors via its interaction with Shank and Homer (Ehlers, 1999; Naisbitt et al., 1999; Tu et al., 1999).

The second family of PDZ proteins includes GRIP and ABP, with seven and six PDZ domains respectively (Dong et al., 1997; Srivastava et al., 1998). These proteins do not contain SH3 or GK domains. The PDZ interactions of ABP/GRIP with AMPAR GluR2/3 subunits occur via a class II hydrophobic interaction, distinct from the class I PSD-95–NMDAR interaction. GRIP and ABP are also capable of forming homo- and heteromultimers through PDZ–PDZ interactions (Srivastava et al., 1998). Protein interacting with C kinase (PICK1) is yet another PSD scaffolding molecule initially identified by its ability to bind via its single PDZ domain to the catalytic domain fragment of PKCα (Staudinger et al., 1995). PICK1 binds in vivo to a variety of transmembrane proteins, including the GluR2 AMPA receptor subunit, the eph receptor tyrosine kinases, and to mGluR7 (Xia et al., 1999; Perez et al., 2001).

3.2. AMPA Receptors in the PSD

Yeast two-hybrid screening has been instrumental in identifying binding partners for both AMPAR and NMDAR channels including the variety of PDZ-containing proteins found associated with the receptors (Fig. 12.2) (Niethammer and Sheng, 1998). It is now established that the wide array of proteins found in the PSD interact to form the glutamatergic signal transduction machinery. Such interactions have been proposed to govern the activity-dependent and independent receptor targeting and trafficking that is important for long-term potentiation (LTP) and long-term depression (LTD) (Contractor and Heinemann, 2002; Malinow and Malenka, 2002; Sheng and Kim, 2002). Indeed, AMPAR function in synapses is partly regulated by receptor trafficking from intracellular pools to synaptic sites, a property that controls their number and is implicated in the activity-dependent modulation of synaptic strength (Rose and Konnerth, 2000). Preventing the interaction of GluR2 with GRIP1/ABP proteins impairs LTD in both the hippocampus and cerebellum (Osten et al., 2000). In addition, AP2, an adaptor protein for clathrin-mediated endocytosis, and N-ethylmaleimide-sensitive fusion protein (NSF) bind the AMPAR GluR2 subunit and are implicated in AMPAR membrane trafficking (Osten et al., 1998; Lee et al., 2002). AMPAR activity can also be potentiated by subunit phosphorylation by serine/threonine phosphorylation by PKA, PKC, and CaMKII (Wang et al., 1994a; Barria et al., 1997; Hayashi et al., 2000).

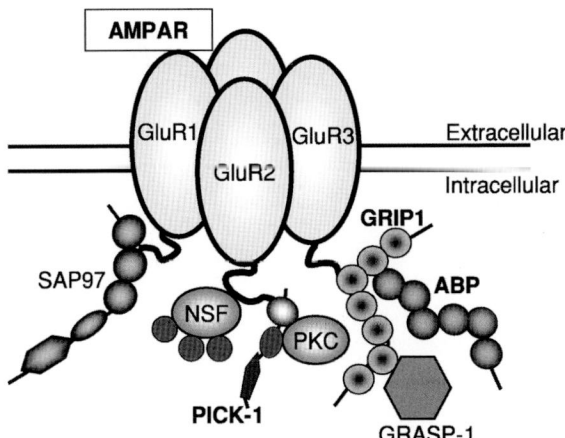

Figure 12.2. Schematic representation of AMPA receptor-associated proteins at the postsynaptic density. AMPA receptors are linked to a variety of intracellular regulators via PDZ scaffolding proteins including members of the PSD-95 family, PICK-1 and ABP/GRIP. It is important to note that protein interactions within the PSD are dynamic and may be altered by glutamatergic signalling. *NSF*, N-ethylmaleimide-sensitive fusion protein; *PKC*, protein kinase C.

Synaptic localization of Ca^{2+}-impermeable GluR2 subunits is thought to be important in modulating the neurotoxic effects of AMPAR signaling. Several studies suggest that CA1 hippocampal neurons, the cells that exhibit the greatest vulnerability to delayed death following global ischemia, also downregulate GluR2 AMPAR subunits (Bennett et al., 1996; Pellegrini-Giampietro et al., 1997; Takuma et al., 1999). This may induce a delayed increase in Ca^{2+} permeability in these cells that correlates with their degeneration. However, although increased Ca^{2+} permeability may play a role in the vulnerability of CA1 neurons to degeneration, there are also studies that indicate that additional processes could be involved, as the GluR2 subunit mediates many of the interactions of AMPARs with intracellular proteins, and regulates the affinity of AMPARS for extracellular ligands (Iihara et al., 2001). To date, none of the AMPAR-associated proteins have been ascribed a role in excitotoxic signaling. These proteins have the potential for mediating excitotoxic insults via interactions with other synaptic molecules, however. For example, GRIP1 binding to GRASP-1 may couple AMPAR activation to the Ras signaling pathway (Ye et al., 2000). GRASP-1 is cleaved in apoptotic neurons during ischemia, disrupting its regulation of Ras signaling. In addition, GRASP-1 has been shown to downregulate synaptic targeting of AMPARs (Ye et al., 2002). Thus its cleavage during ischemia may result in increased synaptic AMPAR activity and vulnerability to glutamate overactivity.

3.3. NMDA Receptors in the PSD

At least a portion of the signaling properties of NMDARs is governed by the interactions of their intracellular domains with intracellular cytoskeletal and postsynaptic

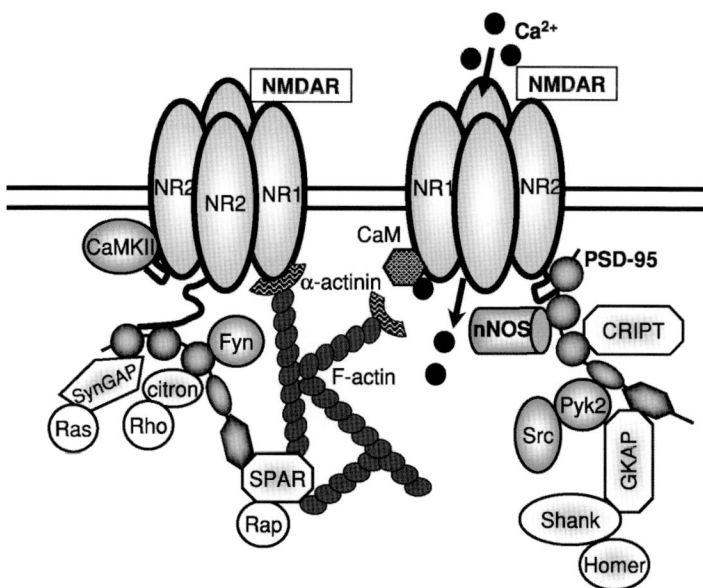

Figure 12.3. NMDA receptors are linked to a variety of enzymatic signalling pathways and regulators via protein–protein interactions. The associated postsynaptic signal pathways represent potential targets for therapeutic intervention such as coupling of Ca^{2+}-dependent activation of nNOS to the NMDAR via PSD-95. *CaM*, calmodulin; *CaMKII*, Ca^{2+}/calmodulin-dependent kinase II; *CRIPT*, cysteine-rich interactor of PDZ3; *GKAP*, guanylate kinase-associated protein; *MT*, microtubule; *nNOS*, neuronal nitric oxide synthase; *SPAR*, a GTPase protein for Rap.

signal transduction molecules (Fig. 12.3). The C-terminal tail of NR2 subunits is critical for the proper function of NMDARs, as noted by transgenic mouse studies. Mouse mutants with targeted deletion of the NR2 cytoplasmic tails are phenotypically indistinguishable from the corresponding NR2 subunit knockouts (Sprengel et al., 1998). The C-terminal tails possess multiple motifs for phosphorylation and protein–protein interactions including binding to members of the PSD-95 protein family. PSD-95 in turn binds multiple intracellular signaling and scaffold molecules and links NMDARs to signaling partners, regulatory enzymes, and adaptor molecules. Interaction of NMDAR subunits with intracellular signal molecules via scaffolding proteins represents an important mechanism whereby NMDARs could mediate excitotoxicity in addition to normal signal transduction. The Ca^{2+}-dependent activation of synaptic signal molecules such as nNOS represents a key pathway whereby NMDARs can produce neurotoxic signals in response to glutamate overstimulation.

Close coupling of receptor channels to downstream signal machinery allows for efficient, local Ca^{2+}-dependent activation of signaling cascades. Unlike the situation for AMPA receptors, compelling evidence exists demonstrating that Ca^{2+}-dependent neuronal death is triggered most efficiently through NMDARs (Tymianski et al., 1993; Sattler et al., 1998; Sattler and Tymianski, 2001). These studies suggest that the synaptic organization of NMDARs brings them into close association with intracellular pathways capable of causing neurotoxicity. NR2A and NR2B subunits bind to the first two PDZ domains of PSD-95. The third PDZ domain has been shown to bind neuroligins who in turn bind the cell–cell adhesion molecule β-neurexin. This interaction indicates that PSD-95 binding is important for synaptic NMDAR clustering. However, PSD-95 knockout mice demonstrate normal synaptic localization of NMDARs, suggesting interaction with PSD-95 is not sufficient for clustering (Migaud et al., 1998). Instead, synaptic localization of NMDA receptors is governed by the interaction of the NR1 subunit α-actinin-2 and the actin cytoskeleton. Depolymerizing F-actin results in redistribution of synaptic NMDAR clusters to extrasynaptic sites (Allison et al., 1998; Sattler et al., 2000). Interestingly calmodulin, a negative regulator of NMDAR signaling, has been shown to directly antagonize α-actinin-2 binding to NR1 (Wyszynski et al., 1997; Krupp et al., 1999). Our laboratory has shown that depolymerizing F-actin reduces signaling and excitotoxicity evoked by synaptic glutamate release but not by exogenous glutamate application (Sattler et al., 2000). These studies indicate that synaptic and extrasynaptic NMDARs are equally capable of mediating neurotoxicity. Thus, although actinin and calmodulin interactions with NR1 are optimized for synaptic signaling, some other interaction may be critical in regulating excitotoxic signal cascades.

It is notable that the hippocampal neurons of PSD-95 mutant mice exhibit a dramatic increase in LTP, indicating an important role for NR2–PSD-95 interactions in NMDAR signal transduction. Indeed, through its various protein binding domains PSD-95 brings an array of signal molecules into close approximation with Ca^{2+} influx through the NMDAR channel (Fig. 12.3). The discovery of distinctive signal molecules as binding partners of PSD-95 lead to the hypothesis that PSD-95 acts as the scaffolding link between incoming calcium ions and intracellular signal molecules such as nNOS (Brenman et al., 1996a,b; Brenman and Bredt, 1996; Stricker et al., 1997). Suppression of PSD-95 expression selectively attenuated excitotoxicity triggered by NMDARs but not other GluR or Ca^{2+} channels (Sattler et al., 2000). Further study of NMDAR signaling mechanisms revealed that suppressing PSD-95 selectively reduced Ca^{2+}-activated production of NO, a second messenger implicated in excitotoxic mechanisms, without affecting NOS expression (Sattler et al., 1999). Thus PSD-95 was required for efficient coupling of NMDAR activation to NO signaling and toxicity (Fig. 12.4).

4. TARGETING INTRACELLULAR SIGNAL EVENTS AS A THERAPEUTIC STRATEGY

In theory, a CNS disorder in which neuronal loss is caused by glutamate overactivity has the potential to be treated by blocking the receptors that transduce the neurotransmitter signal. To this end numerous specific NMDAR and AMPAR channel blockers have been developed and studied in the treatment of brain damage during stroke. However, antagonism of glutamate receptor activity has not proven clinically useful, in part owing to the pharmacokinetic difficulties and adverse side effects.

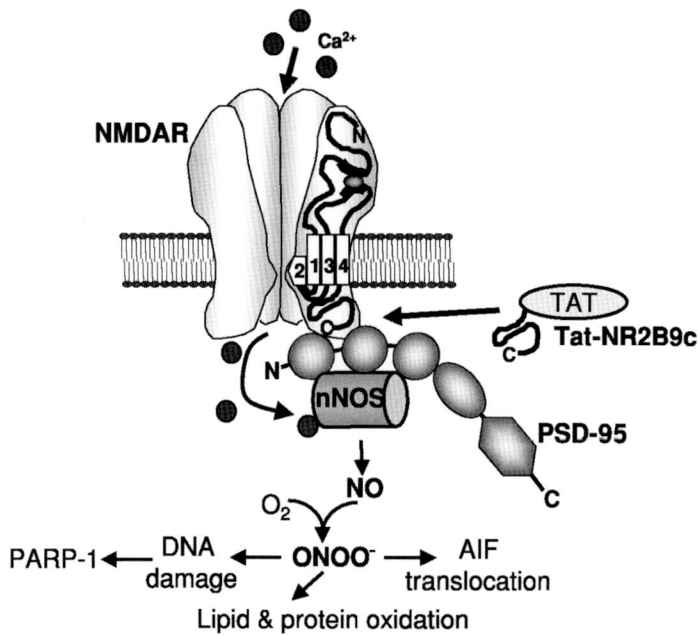

Figure 12.4. The NMDAR/PSD-95 complex represents a critical starting point for multiple downstream neurotoxic signal events. This signal cascade may be averted by dissociation of NMDAR/PSD-95 interaction using HIV-1 TAT peptide fused to the C-terminus of NR2B (Tat-NR2B9c). Thus excitotoxic signaling via Ca^{2+}-dependent signaling molecules such as nNOS can be reduced. Ca^{2+}, calcium; $ONOO^-$, peroxynitrite; *NMDAR*, NMDA receptor; *nNOS*, neuronal nitric oxide synthase; *NO*, nitric oxide; *PARP-1*, poly(ADP-ribose) polymerase.

The high levels of NMDAR antagonists needed to treat excitotoxic damage produce undesirable effects in the healthy brain and can induce hallucinations, centrally mediated hypo- or hypertension, catatonia, and sometimes anaesthesia. AMPAR blockers have shown more potential for therapeutic use, especially in the protection of CA1 hippocampal neurons during global ischemia, yet their clinical applications have been limited by poor solubility, poor CNS penetration, and nephrotoxicity (Madsen et al., 2001; Nikam and Kornberg, 2001).

Excitatory synaptic activity through glutamatergic signaling, and the resulting synaptic Ca^{2+} influx, are vital to neuronal function and survival. Blocking NMDARs has been shown to cause extensive apoptosis in perinatal rats (Ikonomidou et al., 1999; Ishimaru et al., 1999), suggesting that GluR antagonists have direct neurotoxicity. To correct for this, alternative strategies for treating excitotoxic damage have been proposed to prevent the complete block of the receptor. These strategies include the use of antagonists selective for particular receptor subunits (ifenprodil for NMDAR NR2B), partial block of receptor activity (such as glycine antagonists), and the use of low-affinity blockers whose binding is more easily displaced, allowing for some glutamatergic signaling. Yet another approach in the search for a therapeutic agent in excitotoxicity targets the specific intracellular signal pathways that propagate excito-toxic signals downstream from glutamate receptors. Given current knowledge of the molecular organization of the PSD, it is possible to derive a strategy to target specific protein–protein interactions to uncouple glutamate receptors from their potentially neurotoxic downstream effectors.

Based on our previous findings linking NMDAR activation to downstream NO toxicity through the scaffolding protein PSD-95 (Sattler et al., 1999), we investigated the idea that the NMDAR NR2–PSD-95 interaction might constitute a therapeutic target for diseases involving excitotoxicity. Targeting the PDZ interaction with the C-terminal tSXV motif of NR2B represents a therapeutic strategy that may circumvent the negative consequences of blocking NMDAR function. We questioned whether in-terfering with NR2B–PSD-95 interactions could suppress excitotoxicity in a manner similar to knockdown of PSD-95 expression. We designed a targeted peptide com-prised of the nine C-terminal residues of NR2B (NR2B9c), including the tSXV motif, which is anticipated to bind the second PDZ domain of PSD-95 (Fig. 12.4). To allow for efficient delivery of the peptide into cells in vitro and across the blood–brain barrier in vivo, we conjugated NR2B9c to the cell membrane transduction domain of HIV-1 TAT protein (Tat-NR2B9c). The TAT-transduction domain is able to transport proteins of variable size across membranes in a rapid, dose-dependent manner independent of receptors or transporters (Frankel and Pabo, 1988; Mann and Frankel, 1991; Becker-Hapak et al., 2001). We predicted that our Tat-NR2B9c peptide would disrupt the interaction between NMDAR NR2B and PSD-95, protecting treated neurons from NMDAR-activated NO production and excitotoxic death (Fig. 12.4).

We found that fluorescent-labeled Tat-NR2B9c rapidly crossed the plasma mem-brane of cultured mouse neurons and accumulated in the brains of mice after intraperi-toneal injection. Tat-NR2B9c but not control peptides protected cultured neurons from NMDA-mediated excitotoxicity without affecting NMDAR Ca^{2+} signaling or elec-trophysiology (Aarts et al., 2002). Tat-NR2B9c disrupted the interaction of NR2B and PSD-95 in coimmunoprecipitates and significantly depressed NMDAR-evoked stim-ulation of NO-cGMP signaling. Since Tat-peptides that target the NMDAR–PSD-95 interaction protect against NMDA toxicity without blocking NMDARs, we reasoned that treatment with Tat-NR2B9c in vivo could serve as an improvement on NMDA blockers in the treatment of ischemic brain damage. We found that Tat-NR2B9c but not control peptide dramatically reduced cerebral infarction and improved neuro-logical function in rats subjected to transient focal cerebral ischemia. The peptide treatment was effective 1 h before, and most importantly, 1 h after the onset of ex-citotoxicity and cerebral ischemia (Aarts et al., 2002). In addition, the peptide dose used to dramatically reduce infarction had no effect on blood pressure, respiratory rate, or cognitive function in experimental animals (unpublished results). Together these results indicate that the strategy of treating neurons with Tat-fusion peptides is effective in reducing vulnerability to excitotoxicity in vitro and stroke damage in vivo. As this occurs without affecting NMDAR activity, the adverse consequences of blocking NMDARs are not expected. Research currently underway indicates that Tat-NR2B9c treatment is highly effective 3 h after MCAo onset and significantly improves long-term functional recovery. This longer time window for efficacy after the insult onset suggests that targeting the NMDAR–PSD-95 interaction is a practical future strategy for treating stroke.

Further support for the use of this cell-permeable peptide approach to treat excitotoxic damage can be seen in the recent work by Arundine et al. (Arundine et al., 2003) regarding secondary traumatic injury. In this study, neurons were subjected to a sublethal stretch trauma followed by various secondary insults including treatment with calcium ionophore, kainic acid, and NMDA. Treatment with physiological concentrations of NMDA resulted in a dramatic increase in neuronal death while other secondary insults produced no significant difference from background stretch. Thus they showed that secondary traumatic damage occurred by an excitotoxic mechanism mediated specifically by NMDARs. Most importantly, the authors demonstrated that pre- and posttreatment with Tat-NR2B9c could protect the stretched neurons from the secondary excitotoxic insult. Therefore specifically targeting protein–protein interactions with the NMDA receptor using cell-permeable peptides represents a feasible approach to treating excitotoxic brain damage produced by various conditions. It is likely that targeting other intracellular proteins using the same approach could be used to modulate additional signaling mechanisms to further aid the treatment of brain damage and the protein–protein interactions that lead to other diseases.

4.1. Other Targets for Therapeutic Intervention

4.1.1. Production of Reactive Oxygen Species

Following acute neuronal injury and activation of excitotoxic signaling, a number of enzymatic processes are inhibited or activated that result in the generation of toxic free radicals. Mitochondrial function is critical for cell survival as they produce the energy needed for normal cell processes, help regulate Ca^{2+} levels and free radical production, and control release of proapoptotic factors such as cytochrome c. Deficits in mitochondrial electron chain functioning after injury can result in excessive reactive oxygen species production and lead to neurotoxicity (Chan, 1994; Dugan et al., 1995; Schinder et al., 1996; Fiskum, 2000). Both ischemia and reperfusion have been shown to cause increases in superoxide anion production (O_2^-) (Kumar et al., 1990). Excess intracellular Ca^{2+} can result in futile mitochondrial Ca^{2+} cycling that in turn increases ROS production (Richter and Kass, 1991). ROS production also occurs in the cytoplasm following intracellular Ca^{2+} elevations. For instance, excess cytoplasmic Ca^{2+} can initiate production of xanthine oxidase (XOD), which uses molecular oxygen as an electron acceptor, thus resulting in elevations of O_2^- (Sussman and Bulkley, 1990). Elevations in intracellular Ca^{2+} can also activate phospholipase A_2, which releases arachadonic acid whose metabolism by oxidases results in the production of oxygen free radicals (Chanock et al., 1994; Shami et al., 1998).

Neuronal NO production is one ROS pathway specifically associated with the NMDA receptor subtype via PSD-95 (Sattler et al., 1999)(see above), and it has been shown that disrupting the NMDAR–PSD-95 interaction can protect neurons from the consequences of ischemic and traumatic injury (Aarts et al., 2002; Arundine et al., 2003). It has been argued, however, that targeting signal mechanisms further downstream of the receptor may extend the time window for therapeutic intervention. For example, glutamate-mediated neurotoxicity results from the overproduction of both NO and O_2^-. The reaction product of these two species, $ONOO^-$, is chemically complex, as it has the activity of both the hydroxyl radical and the nitrogen dioxide radical (Koppenol et al., 1992). $ONOO^-$ is a potent oxidant that reacts with sulfhydryls (Radi et al., 1991a) and zinc-thiolate moieties (Crow et al., 1995). It can also nitrate and hydroxylate aromatic rings on amino acid residues (Beckman et al., 1992) and oxidize lipids (Radi et al., 1991b), protein, and DNA (Darley-Usmar et al., 1992). NO and $ONOO^-$ have been show to inhibit mitochondrial respiratory chain enzymes, compounding the problems of mitochondrial damage in neuronal injury (Bolanos et al., 1997). Both NO and $ONOO^-$ can DNA strand breaks, leading to the activation of the nuclear DNA repair enzyme, poly(ADP ribose) polymerase 1 (PARP-1). Although PARP-1 is a DNA repair agent it has been shown to be a key regulator of NMDA toxicity via the NO signal pathway. Once activated, PARP-1 catalyzes an energy-dependent reaction that attaches ADP ribose units to nuclear proteins. Excessive activation leads to depletion of ATP and NAD, however, and is implicated as a cell death mechanism in ischemia, chemically induced Parkinsonism, and CNS trauma (Mandir et al., 2000). Research suggests that inhibition of PARP-1 may provide long-term protection after ischemia (Zhang et al., 1994; Mandir et al., 2000) and thus provides a therapeutic target for stroke therapy. Thus the accumulation of intracellular ROS can mediate cellular destruction and represents critical pathways that may be modulated to treat brain damage. Therapeutics that mimic free radical scavengers or inhibit ROS production have shown some success in treating ischemic and traumatic brain damage (Smith et al., 1980; Faden and Salzman, 1992; Lu et al.; Muizelaar, 1993). The active sites of enzymes that generate ROS with brain-permeable peptides may therefore represent another target for peptide therapeutics.

4.1.2. Inflammatory Processes

Activated by excitotoxic signaling and the production of ROS, proinflammatory genes and mediators represent another potentially critical target to treat brain damage. There is increasing evidence that inflammation can exacerbate brain injury. Inhibiting the activity of transcription factors such as nuclear factor-kappa B (NF-κB) and STATs may preclude the expression of inflammatory mediators such platelet-activating factor, tumor necrosis factor-α (TNF-α), and interleukin (IL)-1β that are produced by injured neurons (Feuerstein et al., 1997; Berti et al., 2002). NF-κB is inhibited by its interaction with the IκB whereas STAT proteins require dimerization for transcription factor activity. Mimicking these protein–protein interactions may help to prevent or reduce the immune response in the injured brain tissue. Migratory immune cells are attracted to the damaged brain by the expression of adhesion factors, cytokines, and chemokines (Yamasaki et al., 1995a,b; Iadecola, 1997; Feuerstein et al., 1998). Infiltrating neutrophils can contribute to further brain damage by microvascular obstruction as well as production of toxic amounts of NO via inducible NOS (del Zoppo et al., 1991; Forster et al., 1999). NO produced by iNOS may represent a significant delayed response to brain injury as pharmacological inhibition of iNOS reduces ischemic brain damage (Iadecola, 1997). Blocking the receptors for the proteins that attract immune cells may help to prevent the accumulation of blood-borne cells in the damaged brain tissue and alleviate further complications. Damaged neurons may also express cyclooxygenase 2, an enzyme that mediates injury by producing superoxide and toxic prostanoids (Nogawa et al., 1997; Iadecola et al., 1999a). Inhibition of this enzyme as well as iNOS may result in significant reduction of delayed neuronal injury.

4.1.3. Apoptotic Processes

Classical apoptotic cell death involves the mitochondrial release of cytochrome c mediated through both free radical–dependent mechanisms and/or oligomerization of Bax proteins (for review see Wang, 2001). Cytoplasmic cytochrome c binds to apoptosis protease-activating factor-1 (Apaf-1), which subsequently recruits multiple procaspase-9 molecules to facilitate their autoactivation. Once activated, caspase-9 cleaves procaspase-3 into active caspase-3, the primary effector enzyme in neuronal apoptosis (Liu et al., 1996; Rodriguez and Lazebnik, 1999). The hallmarks of apoptosis include internucleosmal DNA cleavage, somal shrinkage and neuronal condensation, nuclear membrane breakdown, externalization of phosphatidylserine, and the formation of apoptotic bodies (Raghupathi et al., 2000)(for review see (Hengartner, 2000)).

Although the full contribution of programmed cell death to brain damage in human CNS injury is of some debate, research has shown that it cannot be overlooked in considering therapeutic targets. A recent study demonstrated that anti-apoptotic Bcl-xL can be transduced and protect against ischemic insult and apoptosis (Cao et al., 2002). Mice injected with recombinant Bcl-xL protein fused to the TAT transduction domain had decreased cerebral infarction (up to 40%) after focal ischemia and the TAT-Bcl-xL protein attenuated ischemia-induced caspase-3 activation in neurons. A later, independent study demonstrated that a similar TAT-Bcl-xL protein also reduced infarct volume, inhibited caspase-3 activation and reduced the number of neurons with DNA fragmentation (Kilic et al., 2002). The Bcl family members represent but one possible target in preventing apoptotic death. The promiscuous cleavage enzymes known as caspases play a critical role in apoptosis, and are considered to be key targets for the design of cytoprotective drugs. Administering short peptides that bind and irreversibly deactivate the enzyme catalytic site can block caspase activity (Nicholson and Thornberry, 1997). In recent years studies have shown that caspase inhibition is neuroprotective following in vitro excitotoxicity and experimental stroke (Le et al., 2002; Yang et al., 2003) and in experimental head trauma (Yakovlev et al., 1997). Several lines of evidence also suggest that the tumor suppressor gene $p53$ is a salient upstream initiator of apoptosis following neuronal injury in response to excitotoxins, hypoxia, and ischemia (Banasiak and Haddad, 1998; Xiang et al., 1998; Gilman et al., 2003). In addition to classic apoptotic death there are now known to exist several caspase-independent programmed cell death mechanisms. Apoptosis-inducing factor (AIF) and endonuclease G are mitochondrial-released proteins that cleave DNA and trigger apoptosis in a caspase-independent manner. It has been recently shown that AIF is important in mediating some aspects of delayed cell death following brain injury (Zhu et al., 2003).

Although not as specific to neurons as the glutamate receptors, both the caspase-dependent and independent delayed cell death pathways represent important targets for peptide intervention. In each case activation and propagation of the signal cascades are dependent on specific protein–protein interactions. In addition to providing multiple peptide therapeutic targets, apoptotic protein interactions may add desperately needed time to the therapeutic window for treatment. For instance, injection of a specific caspase-3 inhibitor was shown to be effective up to 9 h after reversible ischemia (Fink et al., 1998). Thus the delayed nature of neuronal suicide after acute injury may be instrumental in the design of pharmacological agents that could be used alone or in combination with upstream, antiexcitotoxic peptides.

4.1.4. Aquaporins

Water homeostasis in the CNS is critically important to normal neuronal activity and brain physiology. The movement of K^+ from areas of high neuronal activity is coupled to water flux. Cerebral edema is a common complication of numerous neurological diseases that may rapidly become life threatening because of the rigid encasement of the brain (Papadopoulos et al., 2002). Aquaporins are a family of water channels that facilitate water transport through the plasma membrane of many cell types. Within the rodent brain aquaporin 1 (AQP1) is expressed on epithelial cells in the choroid plexus whereas AQP4, AQP5, and AQP9 are found on astrocytes and ependymal cells (Venero et al., 1999, 2001; Papadopoulos et al., 2002). Under physiological conditions, AQP4 and AQP9 are thought to regulate brain homeostasis and central plasma osmolarity. Aquaporins may play a role in pathologic edema, as AQP4-knockout mice have reduced edema after water intoxication and focal cerebral ischemia (Manley et al., 2000). In addition AQP4 and AQP9 expression was shown to be upregulated following ischemia and traumatic injuries (Vizuete et al., 1999; Badaut et al., 2001; Aoki et al., 2003; Sun et al., 2003). Thus aquaporin expression and distribution may be important not only for normal water homeostasis but also for the development of edema after acute cerebral insults, and regulation of these water flux channels represents a potential target for controlling neurological complications and brain damage.

4.1.5. TRP Channels

In addition to GluR proteins neurons express members of the transient receptor protein (TRP) family of cation channels. The first TRP member was identified as the Ca^{2+}-permeable channel responsible for depolarization of *Drosophila* photoreceptor cells in response to light (Montell et al., 2002a). As many as 20 mammalian homologs of the *Drosophila* TRP have now been characterized in this ion channel superfamily that is united by a common primary structure and permeability to monovalent cations and Ca^{2+} (reviewed in Montell et al., 2002a). TRP channels are expressed in various tissues including brain and some have be implicated in mediating physiological processes that require Ca^{2+}-dependent signaling ranging from sensory transduction to behavior (Liman et al., 1999; de Bono et al., 2002). Each member contains six membrane-spanning domains with intracellular amino and carboxyl tails. The transmembrane portions contribute to a cation channel that is formed by oligomerization of possibly four subunits of one or more TRP member. Based on structural homology the TRP family is subdivided in three main subfamilies: the TRPC (canonical) group, the TRPV (vanilloid) group, and the TRPM (melastatin) group (Montell et al., 2002b). TRPM members have exceptionally long intracellular tails and two members of the TRPM group, TRPM2 and TRPM7, possess the unique characteristic of being both ion channels and enzymes. TRPM7 is a non–voltage-dependent cation channel that also exhibits kinase activity and is gated by intracellular Mg^{2+} (Nadler et al., 2001; Runnels et al., 2001, 2002). TRPM7 currents are rapidly activated at low Mg-ATP levels in stably transfected cell lines (Nadler et al., 2001; Monteilh-Zoller et al., 2003), suggesting that they could be activated during anoxia. In addition, TRPM7 requires proper regulation for cell survival; either overexpression or knockout of the channel is lethal (Nadler et al., 2001). TRPM2 is gated by ADP-ribose and contains ADP-ribose pyrophosphatase activity, which may serve to self-regulate channel activity (Perraud et al., 2001). Also of particular not is the fact that H_2O_2 and agents that produce reactive oxygen/nitrogen species gate TRPM2. In heterologous cell lines, TRPM2 expression enhances and TRPM2 suppression reduces, vulnerability to H_2O_2 toxicity (Hara et al., 2002; Kraft et al., 2003). Thus during ischemic conditions decreased availability of Mg-ATP and increased production of oxygen radicals could lead to improper activation of these channels and unregulated cation flux that has toxic consequences. Perhaps some neurotoxicity could be prevented by inhibiting the function of these TRP channels such as blocking the ion channel pore or controlling the intracellular regulatory site. Further study of this interesting channel family may reveal unique therapeutic targets for neuroprotection.

5. CONCLUSIONS

The use of peptides and small proteins in in vivo therapies has met the limitations of poor permeability of the cell membrane and in the instance of neurological disorders, poor delivery across the blood–brain barrier. However, research over the last decade has revealed that a series of small protein domains, termed protein transduction domains (PTDs) can cross cell membranes independent of specific receptors or transporters. Peptide PTDs may simply be synthetic polycation sequences (arginine or lysine rich) or they may be derived from proteins such as *Drosophila* Antennapaedia protein, Herpes virus VP22 and HIV-1 TAT (Shen and Ryser, 1978; Bergmann et al., 1984; Frankel and Pabo, 1988; Derossi et al., 1996; Elliott and O'Hare, 1997;

Lindgren et al., 2000; Prochiantz, 2000). The use of PTDs can ensure efficient delivery of attached proteins into cells and across the blood–brain barrier (Schwarze et al., 1999). PTD conjugation to other macromolecules such as DNA also represents an effective, nonviral approach to gene therapy. Such a mechanism could be used in the delivery of PSD-95 antisense DNA for treatment of long-term neurodegenerative conditions such as epilepsy. However, gene delivery is not a suitable treatment strategy for acute neurological damage owing to the expedient nature with which treatment is needed and fact that protein synthesis in the cerebrum may be compromised (Hata et al., 2000; Hermann et al., 2001). Rather, the use of peptides to target specific protein–protein interactions allows for rapid treatment and short-term modulation of signaling cascades, a strategy particularly suited to the narrow therapeutic window offered during stroke or neurotrauma. Intricate knowledge of the enzymes, protein networks and signaling mechanisms related to evolution of brain pathology will aid in the design of these effective therapeutics.

The strategy outlined above, targeted disruption of protein–protein interactions based on a molecular understanding of excitotoxic mechanisms, may amount to practical future treatments for human neurological disorders. As our understanding of protein-binding domains has grown, so has the potential for therapeutic intervention. SH2, SH3, and ligand binding domains; enzyme active sites; and protein dimerization sites have all been investigated as targets for therapeutic intervention (Pawson, 1995; Gadek and Nicholas, 2003). The coupling of Ca^{2+}-dependent nNOS signaling to NMDAR activation is but one possible pathway in glutamate-mediated excitotoxicity. As we expand our knowledge of the signaling machinery attached to glutamate receptors and the complex pathophysiology initiated by neuronal injury new potential therapeutic targets may also arise. Since TAT proteins rapidly enter into cells after administration (Lee and Pardridge, 2001), local administration may allow for some CNS specificity in targeting apoptotic pathways. Little is yet known about the role of GRIP1/ABP multimers in clustering synaptic proteins with AMPARs at the synapse and the role this might play in mediating ischemic injury. It may be that specific enzymes or signal molecules, clustered with AMPARs, are deregulated on glutamate overactivity or cell stress and mediate a toxic second messenger cascade. In addition, nNOS may not be the only signal pathway influenced by the Tat-NR2B9c peptide that disrupts the NR2B–PSD-95 interaction. As discussed above, PSD-95 binds and clusters a wide variety of enzymes and modulators that may be important in NMDAR signaling and/or neurotoxicity. Further investigation into the organization of the excitatory PSD and of the molecular mechanisms of glutamate-mediated excitotoxicity should reveal additional targets for pharmacological intervention in the treatment of acute neuronal injury and other neurodegenerative conditions.

REFERENCES

Aarts, M., Liu, Y., Liu, L., Besshoh, S., Arundine, M., Gurd, J.W., Wang, Y.T., Salter, M.W., and Tymianski, M. (2002). Treatment of ischemic brain damage by perturbing NMDA receptor- PSD-95 protein interactions. *Science* 2985594:846–850.

Adams, J.M., and Cory, S. (1998). The Bcl-2 protein family: arbiters of cell survival. *Science* 2815381:1322–1326.

Adekoya, N., Thurman, D.J., White, D.D., and Webb, K.W. (2002). Surveillance for traumatic brain injury deaths—United States, 1989–1998. *MMWR Surveill. Summ.* 5110:1–14.

Allison, D.W., Gelfand, V.I., Spector, I., and Craig, A.M. (1998). Role of actin in anchoring postsynaptic receptors in cultured hippocampal neurons: differential attachment of NMDA versus AMPA receptors. *J. Neurosci.* 187:2423–2436.

Allison, D.W., Chervin, A.S., Gelfand, V.I., and Craig, A.M. (2000). Postsynaptic scaffolds of excitatory and inhibitory synapses in hippocampal neurons: maintenance of core components independent of actin filaments and microtubules. *J. Neurosci.* 2012:4545–4554.

Aoki, K., Uchihara, T., Tsuchiya, K., Nakamura, A., Ikeda, K., and Wakayama, Y. (2003). Enhanced expression of aquaporin 4 in human brain with infarction. *Acta Neuropathol. (Berl).* 1062:121–124.

Arundine, M., Chopra, G.K., Wrong, A.W., Lei, S., Aarts, M., MacDonald, J.F., and Tymianski, M. (2003). Enhanced vulnerability to NMDA toxicity in sublethal traumatic neuronal injury in-vitro. *J. Neurotrauma* (in press).

Badaut, J., Hirt, L., Granziera, C., Bogousslavsky, J., Magistretti, P.J., and Regli, L. (2001). Astrocyte-specific expression of aquaporin-9 in mouse brain is increased after transient focal cerebral ischemia. *J. Cereb. Blood Flow Metab.* 215:477–482.

Bading, H., Ginty, D.D., and Greenberg, M.E. (1993). Regulation of gene expression in hippocampal neurons by distinct calcium signaling pathways. *Science* 2605105:181–186.

Banasiak, K.J., and Haddad, G.G. (1998). Hypoxia-induced apoptosis: effect of hypoxic severity and role of p53 in neuronal cell death. *Brain Res.* 7972:295–304.

Banke, T.G., Bowie, D., Lee, H., Huganir, R.L., Schousboe, A., and Traynelis, S.F. (2000). Control of GluR1 AMPA receptor function by cAMP-dependent protein kinase. *J. Neurosci.* 201:89–102.

Barria, A., Muller, D., Derkach, V., Griffith, L.C., and Soderling, T. R. (1997). Regulatory phosphorylation of AMPA-type glutamate receptors by CaM-KII during long-term potentiation. *Science* 2765321:2042–2045.

Bartus, R.T., Elliott, P.J., Hayward, N.J., Dean, R.L., Harbeson, S., Straub, J.A., Li, Z., and Powers, J.C. (1995). Calpain as a novel target for treating acute neurodegenerative disorders. *Neurol. Res.* 174:249–258.

Baude, A., Nusser, Z., Molnar, E., McIlhinney, R.A., and Somogyi, P. (1995). High-resolution immunogold localization of AMPA type glutamate receptor subunits at synaptic and non-synaptic sites in rat hippocampus. *Neuroscience* 694:1031–1055.

Bayer, K.U., De Koninck, P., Leonard, A.S., Hell, J.W., and Schulman, H. (2001). Interaction with the NMDA receptor locks CaMKII in an active conformation. *Nature* 4116839:801–805.

Becker-Hapak, M., McAllister, S.S., and Dowdy, S.F. (2001). TAT-mediated protein transduction into mammalian Cells. *Methods* 243:247–256.

Beckman, J.S., Ischiropoulos, H., Zhu, L., van der, W.M., Smith, C., Chen, J., Harrison, J., Martin, J.C., and Tsai, M. (1992). Kinetics of superoxide dismutase- and iron-catalyzed nitration of phenolics by peroxynitrite. *Arch. Biochem. Biophys.* 2982:438–445.

Bennett, M.V., Pellegrini-Giampietro, D.E., Gorter, J.A., Aronica, E., Connor, J.A., and Zukin, R.S. (1996). The GluR2 hypothesis: Ca(++)-permeable AMPA receptors in delayed neurodegeneration. *Cold Spring Harb. Symp. Quant. Biol.* 61:373–384.

Bergmann, P., Kacenelenbogen, R., and Vizet, A. (1984). Plasma clearance, tissue distribution and catabolism of cationized albumins with increasing isoelectric points in the rat. *Clin. Sci. (Lond.)* 671:35–43.

Berti, R., Williams, A.J., Moffett, J.R., Hale, S.L., Velarde, L.C., Elliott, P.J., Yao, C., Dave, J.R., and Tortella, F.C. (2002). Quantitative real-time RT-PCR analysis of inflammatory gene expression associated with ischemia-reperfusion brain injury. *J. Cereb. Blood Flow Metab.* 229:1068–1079.

Beschorner, R., Schluesener, H.J., Gozalan, F., Meyermann, R., and Schwab, J.M. (2002). Infiltrating CD14+ monocytes and expression of CD14 by activated parenchymal microglia/macrophages contribute to the pool of CD14+ cells in ischemic brain lesions. *J. Neuroimmunol.* 1261-2:107–115.

Bettler, B., and Mulle, C. (1995). Review: neurotransmitter receptors. II. AMPA and kainate receptors. *Neuropharmacology* 342:123–139.

Bettler, B., Boulter, J., Hermans-Borgmeyer, I., O'Shea-Greenfield, A., Deneris, E.S., Moll, C., Borgmeyer, U., Hollmann, M., and Heinemann, S. (1990). Cloning of a novel glutamate receptor subunit, GluR5: expression in the nervous system during development. *Neuron* 55:583–595.

Bettler, B., Egebjerg, J., Sharma, G., Pecht, G., Hermans-Borgmeyer, I., Moll, C., Stevens, C.F., and Heinemann, S. (1992). Cloning of a putative glutamate receptor: a low affinity kainate-binding subunit. *Neuron* 82:257–265.

Bindokas, V.P., and Miller, R.J. (1995). Excitotoxic degeneration is initiated at non-random sites in cultured rat cerebellar neurons. *J. Neurosci* 1511:6999–7011.

Blackstone, C., Murphy, T.H., Moss, S.J., Baraban, J.M., and Huganir, R.L. (1994). Cyclic AMP and synaptic activity-dependent phosphorylation of AMPA-preferring glutamate receptors. *J. Neurosci.* 1412:7585–7593.

Bolanos, J.P., Almeida, A., Stewart, V., Peuchen, S., Land, J.M., Clark, J.B., and Heales, S.J. (1997). Nitric oxide-mediated mitochondrial damage in the brain: mechanisms and implications for neurodegenerative diseases. *J. Neurochem.* 686:2227–2240.

Bortolotto, Z.A., Clarke, V.R., Delany, C.M., Parry, M.C., Smolders, I., Vignes, M., Ho, K.H., Miu, P., Brinton, B.T., Fantaske, R., Ogden, A., Gates, M., Ornstein, P.L., Lodge, D., Bleakman, D., and Collingridge, G.L. (1999). Kainate receptors are involved in synaptic plasticity. *Nature* 4026759:297–301.

Brenman, J.E., and Bredt, D.S. (1996). Nitric oxide signaling in the nervous system. *Methods Enzymol.* 269:119–129.

Brenman, J.E., Chao, D.S., Gee, S.H., McGee, A.W., Craven, S.E., Santillano, D.R., Wu, Z., Huang, F., Xia, H., Peters, M.F., Froehner, S.C., and Bredt, D.S. (1996a). Interaction of nitric oxide synthase with the postsynaptic density protein PSD-95 and alpha1-syntrophin mediated by PDZ domains. *Cell* 845:757–767.

Brenman, J.E., Christopherson, K.S., Craven, S.E., McGee, A.W., and Bredt, D.S. (1996b). Cloning and characterization of postsynaptic density 93, a nitric oxide synthase interacting protein. *J. Neurosci.* 1623:7407–7415.

Cao, G., Pei, W., Ge, H., Liang, Q., Luo, Y., Sharp, F.R., Lu, A., Ran, R., Graham, S.H., and Chen, J. (2002). In vivo delivery of Bcl-XL fusion protein containing the Tat protein transduction domain protects against ischemic brain injury and neuronal apoptosis. *J. Neurosci.* 2213:5423–5431.

Castillo, P.E., Malenka, R.C., and Nicoll, R.A. (1997). Kainate receptors mediate a slow postsynaptic current in hippocampal CA3 neurons. *Nature* 3886638:182–186.

Chan, P.H. (1994). Oxygen radicals in focal cerebral ischemia. *Brain Pathol.* 41:59–65.

Chanock, S.J., el Benna, J., Smith, R.M., and Babior, B.M. (1994). The respiratory burst oxidase. *J. Biol. Chem.* 26940:24519–24522.

Charara, A., Blankstein, E., and Smith, Y. (1999). Presynaptic kainate receptors in the monkey striatum. *Neuroscience* 914:1195–1200.

Choi, D.W. (1985). Glutamate neurotoxicity in cortical cell culture is calcium dependent. *Neurosci. Lett.* 583:293–297.

Choi, D.W. (1987). Ionic dependence of glutamate neurotoxicity. *J. Neurosci.* 72:369–379.

Choi, D.W. (1988). Calcium-mediated neurotoxicity: relationship to specific channel types and role in ischemic damage. *Trends Neurosci.* 1110:465–469.

Choi, D.W. (1995). Calcium: still center-stage in hypoxic-ischemic neuronal death. *Trends Neurosci.* 182:58–60.

Choi, D.W., Koh, J.Y., and Peters, S. (1988). Pharmacology of glutamate neurotoxicity in cortical cell culture: attenuation by NMDA antagonists. *J. Neurosci.* 81:185–196.

Choi, D.W., Weiss, J.H., Koh, J.Y., Christine, C.W., and Kurth, M.C. (1989). Glutamate neurotoxicity, calcium, and zinc. *Ann. NY Acad. Sci.* 568:219–224.

Chong, Z.Z., Kang, J.Q., and Maiese, K. (2003). Metabotropic glutamate receptors promote neuronal and vascular plasticity through novel intracellular pathways. *Histol. Histopathol.* 181:173–189.

Ciabarra, A.M., Sullivan, J.M., Gahn, L.G., Pecht, G., Heinemann, S., and Sevarino, K.A. (1995). Cloning and characterization of chi-1: a developmentally regulated member of a novel class of the ionotropic glutamate receptor family. *J. Neurosci.* 1510:6498–6508.

Contractor, A., and Heinemann, S.F. (2002). Glutamate receptor trafficking in synaptic plasticity. *Sci. STKE* 2002156:RE14–

Contractor, A., Swanson, G., and Heinemann, S.F. (2001). Kainate receptors are involved in short- and long-term plasticity at mossy fiber synapses in the hippocampus. *Neuron* 291:209–216.

Crepel, V., Aniksztejn, L., Ben Ari, Y., and Hammond, C. (1994). Glutamate metabotropic receptors increase a Ca(2+)-activated nonspecific cationic current in CA1 hippocampal neurons. *J. Neurophysiol.* 724:1561–1569.

Crow, J.P., Beckman, J.S., and McCord, J.M. (1995). Sensitivity of the essential zinc-thiolate moiety of yeast alcohol dehydrogenase to hypochlorite and peroxynitrite. *Biochemistry* 3411:3544–3552.

Cull-Candy, S., Brickley, S., and Farrant, M. (2001). NMDA receptor subunits: diversity, development and disease. *Curr. Opin. Neurobiol.* 113:327–335.

Dale, N., and Roberts, A. (1985). Dual-component amino-acid-mediated synaptic potentials: excitatory drive for swimming in *Xenopus* embryos. *J. Physiol* 363:35–59.

Darley-Usmar, V.M., Hogg, N., O'Leary, V.J., Wilson, M.T., and Moncada, S. (1992). The simultaneous generation of superoxide and nitric oxide can initiate lipid peroxidation in human low density lipoprotein. *Free Radic. Res. Commun.* 171:9–20.

Dawson, V.L., Kizushi, V.M., Huang, P.L., Snyder, S.H., and Dawson, T.M. (1996). Resistance to neurotoxicity in cortical cultures from neuronal nitric oxide synthase-deficient mice. *J. Neurosci.* 168:2479–2487.

de Bono, M., Tobin, D.M., Davis, M.W., Avery, L., and Bargmann, C.I. (2002). Social feeding in Caenorhabditis elegans is induced by neurons that detect aversive stimuli. *Nature* 4196910:899–903.

del Zoppo, G.J., Schmid-Schonbein, G.W., Mori, E., Copeland, B.R., and Chang, C.M. (1991). Polymorphonuclear leukocytes occlude capillaries following middle cerebral artery occlusion and reperfusion in baboons. *Stroke*. 2210:1276–1283.

Derkach, V., Barria, A., and Soderling, T.R. (1999). Ca2+/calmodulin-kinase II enhances channel conductance of alpha-amino-3- hydroxy-5-methyl-4-isoxazolepropionate type glutamate receptors. *Proc. Natl. Acad. Sci. USA* 966:3269–3274.

Derossi, D., Calvet, S., Trembleau, A., Brunissen, A., Chassaing, G., and Prochiantz, A. (1996). Cell internalization of the third helix of the Antennapedia homeodomain is receptor-independent. *J. Biol. Chem.* 27130:18188–18193.

Dong, H., O'Brien, R.J., Fung, E.T., Lanahan, A.A., Worley, P.F., and Huganir, R.L. (1997). GRIP: a synaptic PDZ domain-containing protein that interacts with AMPA receptors. *Nature* 3866622:279–284.

Dubinsky, J.M., and Rothman, S.M. (1991). Intracellular calcium concentrations during "chemical hypoxia" and excitotoxic neuronal injury. *J. Neurosci.* 118:2545–2551.

Dugan, L.L., and Choi, D.W. (1994). Excitotoxicity, free radicals, and cell membrane changes. *Ann. Neurol.* 35(Suppl):S17–S21.

Dugan, L.L., Sensi, S.L., Canzoniero, L.M., Handran, S.D., Rothman, S.M., Lin, T.S., Goldberg, M.P., and Choi, D.W. (1995). Mitochondrial production of reactive oxygen species in cortical neurons following exposure to *N*-methyl-D-aspartate. *J. Neurosci.* 1510:6377–6388.

Ehlers, M.D. (1999). Synapse structure: glutamate receptors connected by the shanks. *Curr. Biol.* 922:R848–R850.

Ehlers, M.D., Tingley, W.G., and Huganir, R.L. (1995). Regulated subcellular distribution of the NR1 subunit of the NMDA receptor. *Science* 2695231:1734–1737.

Ehlers, M.D., Zhang, S., Bernhadt, J.P., and Huganir, R.L. (1996). Inactivation of NMDA receptors by direct interaction of calmodulin with the NR1 subunit. *Cell* 845:745–755.

Elliott, G., and O'Hare, P. (1997). Intercellular trafficking and protein delivery by a herpesvirus structural protein. *Cell* 882:223–233.

Faden, A.I., and Salzman, S. (1992). Pharmacological strategies in CNS trauma. *TIPS* 13:29–35.

Feuerstein, G.Z., Wang, X., and Barone, F.C. (1997). Inflammatory gene expression in cerebral ischemia and trauma. Potential new therapeutic targets. *Ann. NY Acad. Sci.* 825:179–193.

Feuerstein, G.Z., Wang, X., and Barone, F.C. (1998). The role of cytokines in the neuropathology of stroke and neurotrauma. *Neuroimmunomodulation* 53–4:143–159.

Fink, K., Zhu, J., Namura, S., Shimizu-Sasamata, M., Endres, M., Ma, J., Dalkara, T., Yuan, J., and Moskowitz, M.A. (1998). Prolonged therapeutic window for ischemic brain damage caused by delayed caspase activation. *J. Cereb. Blood Flow Metab.* 1810:1071–1076.

Fiskum, G. (2000). Mitochondrial participation in ischemic and traumatic neural cell death. *J. Neurotrauma* 1710:843–855.

Fong, D.K., Rao, A., Crump, F.T., and Craig, A. M. (2002). Rapid synaptic remodeling by protein kinase C: reciprocal translocation of NMDA receptors and calcium/calmodulin-dependent kinase II. *J. Neurosci.* 226:2153–2164.

Forster, C., Clark, H.B., Ross, M.E., and Iadecola, C. (1999). Inducible nitric oxide synthase expression in human cerebral infarcts. *Acta Neuropathol. (Berl.)* 973:215–220.

Frankel, A.D., and Pabo, C.O. (1988). Cellular uptake of the tat protein from human immunodeficiency virus. *Cell* 556:1189–1193.

Frerking, M., Malenka, R.C., and Nicoll, R.A. (1998). Synaptic activation of kainate receptors on hippocampal interneurons. *Nat. Neurosci.* 16:479–486.

Gadek, T.R., and Nicholas, J.B. (2003). Small molecule antagonists of proteins. *Biochem. Pharmacol.* 651:1–8.

Gerber, U., Sim, J.A., and Gahwiler, B.H. (1992). Reduction of potassium conductances mediated by metabotropic glutamate receptors in rat CA3 pyramidal cells does not require protein kinase C or protein kinase A. *Eur. J. Neurosci.* 49:792–797.

Ghosh, A., and Greenberg, M.E. (1995). Calcium signalling in neurons: molecular mechanisms and cellular consequences. *Science* 268:239–247.

Gilman, C.P., Chan, S.L., Guo, Z., Zhu, X., Greig, N., and Mattson, M.P. (2003). p53 is present in synapses where it mediates mitochondrial dysfunction and synaptic degeneration in response to DNA damage, and oxidative and excitotoxic insults. *Neuromolecular. Med.* 33:159–172.

Hallenbeck, J.M. (1997). Cytokines, macrophages, and leukocytes in brain ischemia. *Neurology* 495 (Suppl 4):S5–S9.

Hara, Y., Wakamori, M., Ishii, M., Maeno, E., Nishida, M., Yoshida, T., Yamada, H., Shimizu, S., Mori, E., Kudoh, J., Shimizu, N., Kurose, H., Okada, Y., Imoto, K., and Mori, Y. (2002). LTRPC2 Ca2+-permeable channel activated by changes in redox status confers susceptibility to cell death. *Mol. Cell* 91:163–173.

Hata, R., Maeda, K., Hermann, D., Mies, G., and Hossmann, K.A. (2000). Dynamics of regional brain metabolism and gene expression after middle cerebral artery occlusion in mice. *J. Cereb. Blood Flow Metab.* 202:306–315.

Hayashi, Y., Shi, S.H., Esteban, J.A., Piccini, A., Poncer, J.C., and Malinow, R. (2000). Driving AMPA receptors into synapses by LTP and CaMKII: requirement for GluR1 and PDZ domain interaction. *Science* 2875461:2262–2267.

Hengartner, M.O. (2000). The biochemistry of apoptosis. *Nature* 4076805:770–776.

Herb, A., Burnashev, N., Werner, P., Sakmann, B., Wisden, W., and Seeburg, P.H. (1992). The KA-2 subunit of excitatory amino acid receptors shows widespread expression in brain and forms ion channels with distantly related subunits. *Neuron* 84:775–785.

Hermann, D.M., Kilic, E., Kugler, S., Isenmann, S., and Bahr, M. (2001). Adenovirus-mediated GDNF and CNTF pretreatment protects against striatal injury following transient middle cerebral artery occlusion in mice. *Neurobiol. Dis.* 84:655–666.

Hill, I.E., MacManus, J.P., Rasquinha, I., and Tuor, U.I. (1995). DNA fragmentation indicative of apoptosis following unilateral cerebral hypoxia-ischemia in the neonatal rat. *Brain Res.* 6762:398–403.

Hoffpauir, B.K., and Gleason, E.L. (2002). Activation of mGluR5 modulates GABA(A) receptor function in retinal amacrine cells. *J. Neurophysiol.* 884:1766–1776.

Hollmann, M., and Heinemann, S. (1994). Cloned glutamate receptors. *Annu. Rev. Neurosci.* 17:31–108.

Hollmann, M., Hartley, M., and Heinemann, S. (1991). Ca2+ permeability of KA-AMPA–gated glutamate receptor channels depends on subunit composition. *Science* 2525007:851–853.

Hume, R.I., Dingledine, R., and Heinemann, S.F. (1991). Identification of a site in glutamate receptor subunits that controls calcium permeability. *Science* 2535023:1028–1031.

Hung, A.Y., and Sheng, M. (2002). PDZ domains: structural modules for protein complex assembly. *J. Biol. Chem.* 2778:5699–5702.

Iadecola, C. (1997). Bright and dark sides of nitric oxide in ischemic brain injury. *Trends Neurosci.* 203:132–139.

Iadecola, C., Forster, C., Nogawa, S., Clark, H.B., and Ross, M.E. (1999a). Cyclooxygenase-2 immunoreactivity in the human brain following cerebral ischemia. *Acta Neuropathol. (Berl.)* 981:9–14.

Iadecola, C., Salkowski, C.A., Zhang, F., Aber, T., Nagayama, M., Vogel, S.N., and Ross, M.E. (1999b). The transcription factor interferon regulatory factor 1 is expressed after cerebral ischemia and contributes to ischemic brain injury. *J. Exp. Med.* 1894:719–727.

Iihara, K., Joo, D.T., Henderson, J., Sattler, R., Taverna, F.A., Lourensen, S., Orser, B.A., Roder, J.C., and Tymianski, M. (2001). The influence of glutamate receptor 2 expression on excitotoxicity in Glur2 null mutant mice. *J. Neurosci.* 217:2224–2239.

Ikonomidou, C., and Turski, L. (2002). Why did NMDA receptor antagonists fail clinical trials for stroke and traumatic brain injury? *Lancet Neurol.* 16:383–386.

Ikonomidou, C., Bosch, F., Miksa, M., Bittigau, P., Vockler, J., Dikranian, K., Tenkova, T.I., Stefovska, V., Turski, L., and Olney, J.W. (1999). Blockade of NMDA receptors and apoptotic neurodegeneration in the developing brain. *Science* 2835398:70–74.

Ikonomidou, C., Stefovska, V., and Turski, L. (2000). Neuronal death enhanced by *N*-methyl-D-aspartate antagonists. *Proc. Natl. Acad. Sci. USA* 9723:12885–12890.

Ishimaru, M.J., Ikonomidou, C., Tenkova, T.I., Der, T.C., Dikranian, K., Sesma, M.A., and Olney, J.W. (1999). Distinguishing excitotoxic from apoptotic neurodegeneration in the developing rat brain. *J. Comp. Neurol.* 4084:461–476.

Johnson, J.W., and Ascher, P. (1987). Glycine potentiates the NMDA response in cultured mouse brain neurons. *Nature* 3256104:529–531.

Kadotani, H., Hirano, T., Masugi, M., Nakamura, K., Nakao, K., Katsuki, M., and Nakanishi, S. (1996). Motor discoordination results from combined gene disruption of the NMDA receptor NR2A and NR2C subunits, but not from single disruption of the NR2A or NR2C subunit. *J. Neurosci.* 1624:7859–7867.

Kass, I.S., and Lipton, P. (1982). Mechanisms involved in irreversible anoxic damage to the. *in vitro* rat hippocampal slice. *J. Physiol.* 332:459–472.

Kennedy, M.B. (1993). The postsynaptic density. *Curr. Opin. Neurobiol.* 35:732–737.

Kennedy, M.B. (1998). Signal transduction molecules at the glutamatergic postsynaptic membrane. *Brain Res. Brain Res. Rev.* 262-3:243–257.

Kieval, J.Z., Hubert, G.W., Charara, A., Pare, J.F., and Smith, Y. (2001). Subcellular and subsynaptic localization of presynaptic and postsynaptic kainate receptor subunits in the monkey striatum. *J. Neurosci.* 2122:8746–8757.

Kihara, S., Shiraishi, T., Nakagawa, S., Toda, K., and Tabuchi, K. (1994). Visualization of DNA double strand breaks in the gerbil hippocampal CA1 following transient ischemia. *Neurosci. Lett.* 1751-2:133–136.

Kilic, E., Dietz, G.P., Hermann, D.M., and Bahr, M. (2002). Intravenous TAT-Bcl-Xl is protective after middle cerebral artery occlusion in mice. *Ann. Neurol.* 525:617–622.

Kind, P.C., and Neumann, P.E. (2001). Plasticity: downstream of glutamate. *Trends Neurosci.* 2410:553–555.

Koppenol, W.H., Moreno, J.J., Pryor, W.A., Ischiropoulos, H., and Beckman, J.S. (1992). Peroxynitrite, a cloaked oxidant formed by nitric oxide and superoxide. *Chem. Res. Toxicol.* 56:834–842.

Kornau, H.C., Seeburg, P.H., and Kennedy, M.B. (1997). Interaction of ion channels and receptors with PDZ domain proteins. *Curr. Opin. Neurobiol.* 73:368–373.

Kraft, R., Grimm, C., Grosse, K., Hoffmann, A., Sauerbruch, S., Kettenmann, H., Schultz, G., and Harteneck, C. (2003). Hydrogen peroxide and ADP-ribose induce TRPM2-mediated calcium influx and cation currents in microglia. *Am. J. Physiol. Cell Physiol.*

Krupp, J.J., Vissel, B., Thomas, C.G., Heinemann, S.F., and Westbrook, G.L. (1999). Interactions of calmodulin and alpha-actinin with the NR1 subunit modulate Ca2+-dependent inactivation of NMDA receptors. *J. Neurosci.* 194:1165–1178.

Kumar, C., Okuda, M., Ikai, I., and Chance, B. (1990). Luminol enhanced chemiluminescence of the perfused rat heart during ischemia and reperfusion. *FEBS Lett.* 2721-2:121–124.

Landis, D.M., and Reese, T.S. (1974). Differences in membrane structure between excitatory and inhibitory synapses in the cerebellar cortex. *J. Comp. Neurol.* 1551:93–125.

Lauri, S.E., Bortolotto, Z.A., Bleakman, D., Ornstein, P.L., Lodge, D., Isaac, J.T., and Collingridge, G.L. (2001). A critical role of a facilitatory presynaptic kainate receptor in mossy fiber LTP. *Neuron* 324:697–709.

Le, D.A., Wu, Y., Huang, Z., Matsushita, K., Plesnila, N., Augustinack, J.C., Hyman, B.T., Yuan, J., Kuida, K., Flavell, R.A., and Moskowitz, M.A. (2002). Caspase activation and neuroprotection in caspase-3-deficient mice after *in vivo* cerebral ischemia and *in vitro* oxygen glucose deprivation. *Proc. Natl. Acad. Sci. USA* 9923:15188–15193.

Lee, H.J., and Pardridge, W.M. (2001). Pharmacokinetics and delivery of Tat and Tat-protein conjugates to tissues *in vivo*. *Bioconjugate Chem.* 12:995–999.

Lee, S.H., Liu, L., Wang, Y.T., and Sheng, M. (2002). Clathrin adaptor AP2 and NSF interact with overlapping sites of GluR2 and play distinct roles in AMPA receptor trafficking and hippocampal LTD. *Neuron* 364:661–674.

Lees, G.J. (1993). The possible contribution of microglia and macrophages to delayed neuronal death after ischemia. *J. Neurol. Sci.* 1142:119–122.

Lerma, J., Paternain, A.V., Rodriguez-Moreno, A., and Lopez-Garcia, J.C. (2001). Molecular physiology of kainate receptors. *Physiol. Rev.* 813:971–998.

Lewandowski, C., and Barsan, W. (2001). Treatment of acute ischemic stroke. *Ann. Emerg. Med.* 372:202–216.

Lieberman, D.N., and Mody, I. (1994). Regulation of NMDA channel function by endogenous Ca^{2+}-dependent phosphatase. *Nature* 3696477:235–239.

Liman, E.R., Corey, D.P., and Dulac, C. (1999). TRP2: a candidate transduction channel for mammalian pheromone sensory signaling. *Proc. Natl. Acad. Sci. USA* 9610:5791–5796.

Lindgren, M., Hallbrink, M., Prochiantz, A., and Langel, U. (2000). Cell-penetrating peptides. *Trends Pharmacol. Sci.* 213:99–103.

Liu, X., Kim, C.N., Yang, J., Jemmerson, R., and Wang, X. (1996). Induction of apoptotic program in cell-free extracts: requirement for dATP and cytochrome c. *Cell* 861:147–157.

Lu, J., Ashwell, K.W., and Waite, P. () Advances in secondary spinal cord injury: role of apoptosis.

Madden, D.R. (2002). The structure and function of glutamate receptor ion channels. *Nat. Rev. Neurosci.* 32:91–101.

Madden, K.P., Clark, W.M., Marcoux, F.W., Probert, A.W., Jr., Weber, M.L., Rivier, J., and Zivin, J.A. (1990). Treatment with conotoxin, an 'N-type' calcium channel blocker, in neuronal hypoxic-ischemic injury. *Brain Res.* 5371-2:256–262.

Madsen, U., Stensbol, T.B., and Krogsgaard-Larsen, P. (2001). Inhibitors of AMPA and kainate receptors. *Curr. Med. Chem.* 811:1291–1301.

Malinow, R., and Malenka, R.C. (2002). AMPA receptor trafficking and synaptic plasticity. *Annu. Rev. Neurosci.* 25:103–126.

Mandir, A.S., Poitras, M.F., Berliner, A.R., Herring, W.J., Guastella, D.B., Feldman, A., Poirier, G.G., Wang, Z.Q., Dawson, T.M., and Dawson, V.L. (2000). NMDA but not non-NMDA excitotoxicity is mediated by poly(ADP-ribose) polymerase. *J. Neurosci.* 2021:8005–8011.

Manev, H., Favaron, M., Guidotti, A., and Costa, E. (1989). Delayed increase of Ca^{2+} influx elicited by glutamate: role in neuronal death. *Mol. Pharmacol.* 36:106–112.

Manley, G.T., Fujimura, M., Ma, T., Noshita, N., Filiz, F., Bollen, A.W., Chan, P., and Verkman, A.S. (2000). Aquaporin-4 deletion in mice reduces brain edema after acute water intoxication and ischemic stroke. *Nat. Med.* 62:159–163.

Mann, D.A., and Frankel, A.D. (1991). Endocytosis and targeting of exogenous HIV-1 Tat protein. *EMBO J.* 107:1733–1739.

Marcoux, F.W., Probert, A.W., and Weber, M.L. (1989). Hypoxic neural injury in cell culture: Calcium accumulation blockade and neuroprotection by NMDA antagonists but not calcium channel antagonists In: Ginsberg, M.D. and Dietrich, W.D. (eds.), *Cerebrovascular Disease: Sixteenth Princeton Conference.* Raven Press, New York, pp.135–141.

Marcoux, F.W., Probert, A.W., Jr., and Weber, M.L. (1990). Hypoxic neuronal injury in tissue culture is associated with delayed calcium accumulation. *Stroke* 2111 (Suppl):III/1–III74.

Marcoux, F.W., Weber, M.L., Probert, A.W., Jr., and Dominick, M.A. (1992). Hypoxic neurodegeneration in culture: calcium influx, electron microscopy, and neuroprotection with excitatory amino acid antagonists. *Ann. NY Acad. Sci.* 648:303–305.

McDonald, B.J., Chung, H.J., and Huganir, R.L. (2001). Identification of protein kinase C phosphorylation sites within the AMPA receptor GluR2 subunit. *Neuropharmacology* 416:672–679.

McGovern R., and Rudd, A. (2003). Management of stroke. *Postgrad. Med. J.* 79928:87–92.

Migaud, M., Charlesworth, P., Dempster, M., Webster, L.C., Watabe, A.M., Makhinson, M., He, Y., Ramsay, M.F., Morris, R.G.M., Morrison, J.H., O'Dell, T.J., and Grandt, S.G.N. (1998). Enhanced long-term potentiation and impaired learning in mice with mutant postsynaptic density-95 protein. *Nature* 396:433–439.

Monteilh-Zoller, M.K., Hermosura, M.C., Nadler, M.J., Scharenberg, A.M., Penner, R., and Fleig, A. (2003). TRPM7 provides an ion channel mechanism for cellular entry of trace metal ions. *J. Gen. Physiol.* 1211:49–60.

Montell, C., Birnbaumer, L., and Flockerzi, V. (2002a). The TRP channels, a remarkably functional family. *Cell* 1085:595–598.

Montell, C., Birnbaumer, L., Flockerzi, V., Bindels, R.J., Bruford, E.A., Caterina, M.J., Clapham, D.E., Harteneck, C., Heller, S., Julius, D., Kojima, I., Mori, Y., Penner, R., Prawitt, D., Scharenberg, A.M., Schultz, G., Shimizu, N., and Zhu, M.X. (2002b). A unified nomenclature for the superfamily of TRP cation channels. *Mol. Cell* 92:229–231.

Moon, I.S., Apperson, M.L., and Kennedy, M.B. (1994). The major tyrosine-phosphorylated protein in the postsynaptic density fraction is *N*-methyl-D-aspartate receptor subunit 2B. *Proc. Natl. Acad. Sci. USA* 919:3954–3958.

Mori, H., and Mishina, M. (1995). Structure and function of the NMDA receptor channel. *Neuropharmacology* 3410:1219–1237.

Muizelaar, J.P. (1993). Cerebral ischemia-reperfusion injury after severe head injury and its possible treatment with polyethyleneglycol-superoxide dismutase. *Ann. Emerg. Med.* 22:1014–1021.

Nadler, M.J., Hermosura, M.C., Inabe, K., Perraud, A.L., Zhu, Q., Stokes, A.J., Kurosaki, T., Kinet, J.P., Penner, R., Scharenberg, A.M., and Fleig, A. (2001). LTRPC7 is a Mg.ATP-regulated divalent cation channel required for cell viability. *Nature* 4116837:590–595.

Naisbitt, S., Kim, E., Tu, J.C., Xiao, B., Sala, C., Valtschanoff, J., Weinberg, R.J., Worley, P.F., and Sheng, M. (1999). Shank, a novel family of postsynaptic density proteins that binds to the NMDA receptor/PSD-95/GKAP complex and cortactin. *Neuron* 233:569–582.

Nicholson, D.W., and Thornberry, N.A. (1997). Caspases: killer proteases. *Trends Biochem. Sci.* 228:299–306.

Nicoletti, F., Bruno, V., Catania, M.V., Battaglia, G., Copani, A., Barbagallo, G., Cena, V., Sanchez-Prieto, J., Spano, P.F., and Pizzi, M. (1999). Group-I metabotropic glutamate receptors: hypotheses to explain their dual role in neurotoxicity and neuroprotection. *Neuropharmacology* 3810:1477–1484.

Niethammer, M., and Sheng, M. (1998). Identification of ion channel-associated proteins using the yeast two-hybrid system. *Methods Enzymol.* 293:104–122.

Niethammer, M., Valtschanoff, J.G., Kapoor, T.M., Allison, D.W., Weinberg, T.M., Craig, A.M., and Sheng, M. (1998). CRIPT, a novel postsynaptic protein that binds to the third PDZ domain of PSD-95/SAP90. *Neuron* 204:693–707.

Nikam, S.S., and Kornberg, B.E. (2001). AMPA receptor antagonists. *Curr. Med. Chem.* 82:155–170.

Nogawa, S., Zhang, F., Ross, M.E., and Iadecola, C. (1997). Cyclo-oxygenase-2 gene expression in neurons contributes to ischemic brain damage. *J. Neurosci.* 178:2746–2755.

Olney, J.W. (1969). Brain lesions, obesity, and other disturbances in mice treated with monosodium glutamate. *Science* 164880:719–721.

O'Neill, L.A., and Kaltschmidt, C. (1997). NF-kappa B: a crucial transcription factor for glial and neuronal cell function. *Trends Neurosci.* 206:252–258.

Osten, P., Srivastava, S., Inman, G.J., Vilim, F.S., Khatri, L., Lee, L.M., States, B.A., Einheber, S., Milner, T.A., Hanson, P.I., and Ziff, E.B. (1998). The AMPA receptor GluR2 C terminus can mediate a reversible, ATP-dependent interaction with NSF and alpha- and beta-SNAPs. *Neuron* 211:99–110.

Osten, P., Khatri, L., Perez, J.L., Kohr, G., Giese, G., Daly, C., Schulz, T.W., Wensky, A., Lee, L.M., and Ziff, E.B. (2000). Mutagenesis reveals a role for ABP/GRIP binding to GluR2 in synaptic surface accumulation of the AMPA receptor. *Neuron* 272:313–325.

Pak, D.T., Yang, S., Rudolph-Correia, S., Kim, E., and Sheng, M. (2001). Regulation of dendritic spine morphology by SPAR, a PSD-95-associated RapGAP. *Neuron* 312:289–303.

Papadopoulos, M.C., Krishna, S., and Verkman, A.S. (2002). Aquaporin water channels and brain edema. *Mt. Sinai J. Med.* 694:242–248.

Passafaro, M., Sala, C., Niethammer, M., and Sheng, M. (1999). Microtubule binding by CRIPT and its potential role in the synaptic clustering of PSD-95. *Nat. Neurosci.* 212:1063–1069.

Paternain, A.V., Morales, M., and Lerma, J. (1995). Selective antagonism of AMPA receptors unmasks kainate receptor-mediated responses in hippocampal neurons *Neuron* 141:185–189.

Pawson, T. (1995). Protein modules and signalling networks. *Nature* 373:573–579.

Pellegrini-Giampietro, D.E., Gorter, J.A., Bennett, M.V., and Zukin, R.S. (1997). The GluR2 (GluR-B) hypothesis: Ca^{2+})-permeable AMPA receptors in neurological disorders. *Trends Neurosci.* 2010:464–470.

Pellegrini-Giampietro, D.E., Peruginelli, F., Meli, E., Cozzi, A., Albani-Torregrossa, S., Pellicciari, R., and Moroni, F. (1999). Protection with metabotropic glutamate 1 receptor antagonists in models of ischemic neuronal death: time-course and mechanisms. *Neuropharmacology* 3810:1607–1619.

Perez, J.L., Khatri, L., Chang, C., Srivastava, S., Osten, P., and Ziff, E.B. (2001). PICK1 targets activated protein kinase Calpha to AMPA receptor clusters in spines of hippocampal neurons and reduces surface levels of the AMPA-type glutamate receptor subunit 2. *J. Neurosci.* 2115:5417–5428.

Perraud, A.L., Fleig, A., Dunn, C.A., Bagley, L.A., Launay, P., Schmitz, C., Stokes, A.J., Zhu, Q., Bessman, M.J., Penner, R., Kinet, J.P., and Scharenberg, A.M. (2001). ADP-ribose gating of the calcium-permeable LTRPC2 channel revealed by Nudix motif homology. *Nature* 4116837:595–599.

Petralia, R.S., Wang, Y.X., and Wenthold, R.J. (1994). Histological and ultrastructural localization of the kainate receptor subunits, KA2 and GluR6/7, in the rat nervous system using selective antipeptide antibodies. *J. Comp. Neurol.* 3491:85–110.

Pin, J.P., and Duvoisin, R. (1995). The metabotropic glutamate receptors: structure and functions. *Neuropharmacology* 341:1–26.

Planas, A.M., Soriano, M.A., Berruezo, M., Justicia, C., Estrada, A., Pitarch, S., and Ferrer, I. (1996). Induction of Stat3, a signal transducer and transcription factor, in reactive microglia following transient focal cerebral ischaemia. *Eur. J. Neurosci.* 812:2612–2618.

Prochiantz, A. (2000). Messenger proteins: homeoproteins, TAT and others. *Curr. Opin. Cell Biol.* 124:400–406.

Radi, R., Beckman, J.S., Bush, K.M., and Freeman, B.A. (1991a). Peroxynitrite oxidation of sulfhydryls. The cytotoxic potential of superoxide and nitric oxide. *J. Biol. Chem.* 2667:4244–4250.

Radi, R., Beckman, J.S., Bush, K.M., and Freeman, B.A. (1991b). Peroxynitrite-induced membrane lipid peroxidation: the cytotoxic potential of superoxide and nitric oxide. *Arch. Biochem. Biophys.* 2882:481–487.

Rafiki, A., Bernard, A., Medina, I., Gozlan, H., and Khrestchatisky, M. (2000). Characterization in cultured cerebellar granule cells and in the developing rat brain of mRNA variants for the NMDA receptor 2C subunit. *J. Neurochem.* 745:1798–1808.

Raghupathi, R., Graham, D.I., and McIntosh, T.K. (2000). Apoptosis after traumatic brain injury. *J. Neurotrauma.* 1710:927–938.

Rahn, C.A., Bombick, D.W., and Doolittle, D.J. (1991). Assessment of mitochondrial membrane potential as an indicator of cytotoxicity. *Fundam. Appl. Toxicol.* 163:435–448.

Richter, C., and Kass, G.E. (1991). Oxidative stress in mitochondria: its relationship to cellular Ca^{2+} homeostasis, cell death, proliferation, and differentiation. *Chem. Biol. Interact.* 771:1–23.

Rodriguez, J., and Lazebnik, Y. (1999). Caspase-9 and APAF-1 form an active holoenzyme. *Genes Dev.* 1324:3179–3184.

Rodriguez-Moreno, A., and Lerma, J. (1998). Kainate receptor modulation of GABA release involves a metabotropic function. *Neuron* 206:1211–1218.

Rose, C.R., and Konnerth, A. (2000). Self-regulating synapses. *Nature* 4056785:413–415.

Rothman, S.M. (1983). Synaptic activity mediates death of hypoxic neurons. *Science* 2204596:536–537.

Rothwell, N.J., and Hopkins, S.J. (1995). Cytokines and the nervous system II: Actions and mechanisms of action. *Trends Neurosci.* 183:130–136.

Runnels, L.W., Yue, L., and Clapham, D.E. (2001). TRP-PLIK, a bifunctional protein with kinase and ion channel activities. *Science* 2915506:1043–1047.

Runnels, L.W., Yue, L., and Clapham, D.E. (2002). The TRPM7 channel is inactivated by PIP(2) hydrolysis. *Nat. Cell Biol.* 45:329–336.

Sakimura, K., Kutsuwada, T., Ito, I., Manabe, T., Takayama, C., Kushiya, E., Yagi, T., Aizawa, S., Inoue, Y., Sugiyama, H., and (1995). Reduced hippocampal LTP and spatial learning in mice lacking NMDA receptor epsilon 1 subunit. *Nature* 3736510:151–155.

Sakurada, K., Masu, M., and Nakanishi, S. (1993). Alteration of Ca2+ permeability and sensitivity to Mg2+ and channel blockers by a single amino acid substitution in the *N*-methyl-D-aspartate receptor. *J. Biol. Chem.* 2681:410–415.

Sattler, R., and Tymianski, M. (2001). Molecular mechanisms of glutamate receptor-mediated excitotoxic neuronal cell death. *Mol. Neurobiol.* 241-3:107–129.

Sattler, R., Charlton, M.P., Hafner, M., and Tymianski, M. (1998). Distinct influx pathways, not calcium load, determine neuronal vulnerability to calcium neurotoxicity. *J. Neurochem.* 716:2349–2364.

Sattler, R., Xiong, Z., Lu, W.Y., Hafner, M., MacDonald, J.F., and Tymianski, M. (1999). Specific coupling of NMDA receptor activation to nitric oxide neurotoxicity by PSD-95 protein. *Science* 2845421:1845–1848.

Sattler, R., Xiong, Z., Lu, W.Y., MacDonald, J.F., and Tymianski, M. (2000). Distinct roles of synaptic and extrasynaptic NMDA receptors in excitotoxicity. *J. Neurosci.* 201:22–33.

Saunders, D.E., Howe, F.A., van den, B.A., McLean, M.A., Griffiths, J.R., and Brown, M.M. (1995). Continuing ischemic damage after acute middle cerebral artery infarction in humans demonstrated by short-echo proton spectroscopy. *Stroke* 266:1007–1013.

Schilling, M., Besselmann, M., Leonhard, C., Mueller, M., Ringelstein, E.B., and Kiefer, R. (2003). Microglial activation precedes and predominates over macrophage infiltration in transient focal cerebral ischemia: a study in green fluorescent protein transgenic bone marrow chimeric mice. *Exp. Neurol.* 1831:25–33.

Schinder, A.F., Olson, E.C., Spitzer, N.C., and Montal, M. (1996). Mitochondrial dysfunction is a primary event in glutamate neurotoxicity. *J. Neurosci.* 1619:6125–6133.

Schmitz, D., Frerking, M., and Nicoll, R.A. (2000). Synaptic activation of presynaptic kainate receptors on hippocampal mossy fiber synapses. *Neuron* 272:327–338.

Schmitz, D., Mellor, J., Frerking, M., and Nicoll, R.A. (2001). Presynaptic kainate receptors at hippocampal mossy fiber synapses. *Proc. Natl. Acad. Sci. USA* 9820:11003–11008.

Schoepfer, R., Monyer, H., Sommer, B., Wisden, W., Sprengel, R., Kuner, T., Lomeli, H., Herb, A., Kohler, M., Burnashev, N., and (1994). Molecular biology of glutamate receptors. *Prog. Neurobiol.* 422:353–357.

Schwab, J.M., Nguyen, T.D., Postler, E., Meyermann, R., and Schluesener, H.J. (2000). Selective accumulation of cyclooxygenase-1-expressing microglial cells/macrophages in lesions of human focal cerebral ischemia. *Acta Neuropathol. (Berl.)* 996:609–614.

Schwarze, S.R., Ho, A., Vocero-Akbani, A., and Dowdy, S.F. (1999). In vivo protein transduction: delivery of a biologically active protein into the mouse [see comments]. *Science* 2855433:1569–1572.

Seabold, G.K., Burette, A., Lim, I.A., Weinberg, R.J., and Hell, J.W. (2003). Interaction of the tyrosine kinase Pyk2 with the *N*-methyl-D-aspartate receptor complex via the SH3 domains of PSD-95 and SAP102. *J. Biol. Chem.*

Shami, P.J., Sauls, D.L., and Weinberg, J.B. (1998). Schedule and concentration-dependent induction of apoptosis in leukemia cells by nitric oxide. *Leukemia* 129:1461–1466.

Shen, W.C., and Ryser, H.J. (1978). Conjugation of poly-L-lysine to albumin and horseradish peroxidase: a novel method of enhancing the cellular uptake of proteins. *Proc. Natl. Acad. Sci. USA* 754:1872 1876.

Sheng, M. (2001). Molecular organization of the postsynaptic specialization. *Proc. Natl. Acad. Sci. USA* 9813:7058–7061.

Sheng, M., and Kim, M.J. (2002). Postsynaptic signaling and plasticity mechanisms. *Science* 2985594: 776–780.

Shin, H., Hsueh, Y.P., Yang, F.C., Kim, E., and Sheng, M. (2000). An intramolecular interaction between Src homology 3 domain and guanylate kinase-like domain required for channel clustering by postsynaptic density-95/SAP90. *J. Neurosci.* 2010:3580–3587.

Smith, D.S., Rehncrona, S., and Siesjo, B.K. (1980). Barbiturates as protective agents in brain ischemia and as free radical scavengers *in vitro. Acta Physiol. Scand.* (Suppl.) 492:129–134.

Sommer, B., Keinanen, K., Verdoorn, T.A., Wisden, W., Burnashev, N., Herb, A., Kohler, M., Takagi, T., Sakmann, B., and Seeburg, P.H. (1990). Flip and flop: a cell-specific functional switch in glutamate-operated channels of the CNS. *Science* 2494976:1580–1585.

Sommer, B., Kohler, M., Sprengel, R., and Seeburg, P.H. (1991). RNA editing in brain controls a determinant of ion flow in glutamate-gated channels. *Cell* 671:11–19.

Sprengel, R., Suchanek, B., Amico, C., Brusa, R., Burnashev, N., Rozov, A., Hvalby, O., Jensen, V., Paulsen, O., Andersen, P., Kim, J.J., Thompson, R.F., Sun, W., Webster, L.C., Grant, S.G., Eilers, J., Konnerth, A., Li, J., McNamara, J.O., and Seeburg, P.H. (1998). Importance of the intracellular domain of NR2 subunits for NMDA receptor function *in vivo. Cell* 922:279–289.

Srivastava, S., Osten, P., Vilim, F.S., Khatri, L., Inman, G., States, B., Daly, C., DeSouza, S., Abagyan, R., Valtschanoff, J.G., Weinberg, R.J., and Ziff, E.B. (1998). Novel anchorage of GluR2/3 to the postsynaptic density by the AMPA receptor-binding protein ABP. *Neuron* 213:581–591.

Staudinger, J., Zhou, J., Burgess, R., Elledge, S.J., and Olson, E.N. (1995). PICK1: a perinuclear binding protein and substrate for protein kinase C isolated by the yeast two-hybrid system. *J. Cell Biol.* 1283:263–271.

Stoll, G., and Jander, S. (1999). The role of microglia and macrophages in the pathophysiology of the CNS. *Prog. Neurobiol.* 583:233–247.

Stricker, N.L., Christopherson, K.S., Yi, B.A., Schatz, P.J., Raab, R.W., Dawes, G., Bassett, D.E., Jr., Bredt, D.S., and Li, M. (1997). PDZ domain of neuronal nitric oxide synthase recognizes novel C-terminal peptide sequences. *Nat. Biotechnol.* 154:336–342.

Sucher, N.J., Akbarian, S., Chi, C.L., Leclerc, C.L., Awobuluyi, M., Deitcher, D.L., Wu, M.K., Yuan, J.P., Jones, E.G., and Lipton, S.A. (1995). Developmental and regional expression pattern of a novel NMDA receptor-like subunit (NMDAR-L) in the rodent brain. *J. Neurosci.* 1510: 6509–6520.

Sun, M.C., Honey, C.R., Berk, C., Wong, N.L., and Tsui, J.K. (2003). Regulation of aquaporin-4 in a traumatic brain injury model in rats. *J. Neurosurg.* 983:565–569.

Sussman, M.S., and Bulkley, G.B. (1990). Oxygen-derived free radicals in reperfusion injury. *Methods Enzymol.* 186:711–723.

Takuma, H., Kwak, S., Yoshizawa, T., and Kanazawa, I. (1999). Reduction of GluR2 RNA editing, a molecular change that increases calcium influx through AMPA receptors, selective in the spinal ventral gray of patients with amyotrophic lateral sclerosis. *Ann. Neurol.* 466:806–815.

Thornberry, N.A., and Lazebnik, Y. (1998). Caspases: enemies within. *Science* 2815381:1312–1316.

Thurman, D.J., Alverson, C., Dunn, K.A., Guerrero, J., and Sniezek, J.E. (1999). Traumatic brain injury in the United States: a public health perspective. *J. Head Trauma Rehabil.* 146:602–615.

Tominaga, T., Kure, S., Narisawa, K., and Yoshimoto, T. (1993). Endonuclease activation following focal ischemic injury in the rat brain. *Brain Res.* 6081:21–26.

Tomita, M., and Fukuuchi, Y. (1996). Leukocytes, macrophages and secondary brain damage following cerebral ischemia. *Acta Neurochir. Suppl. (Wien.)* 66:32–39.

Tu, J.C., Xiao, B., Naisbitt, S., Yuan, J.P., Petralia, R.S., Brakeman, P., Doan, A., Aakalu, V.K., Lanahan, A.A., Sheng, M., and Worley, P.F. (1999). Coupling of mGluR/Homer and PSD-95 complexes by the Shank family of postsynaptic density proteins. *Neuron* 233:583–592.

Tymianski, M. (1996). Cytosolic calcium concentrations and cell death *in vitro*. *Adv. Neurol.* 71:85–105.

Tymianski, M., Charlton, M.P., Carlen, P.L., and Tator, C.H. (1993). Source specificity of early calcium neurotoxicity in cultured embryonic spinal neurons. *J. Neurosci.* 135:2085–2104.

Venero, J.L., Vizuete, M.L., Ilundain, A.A., Machado, A., Echevarria, M., and Cano, J. (1999). Detailed localization of aquaporin-4 messenger RNA in the CNS: preferential expression in periventricular organs. *Neuroscience* 941:239–250.

Venero, J.L., Vizuete, M.L., Machado, A., and Cano, J. (2001). Aquaporins in the central nervous system. *Prog. Neurobiol.* 633:321–336.

Vignes, M., Bleakman, D., Lodge, D., and Collingridge, G.L. (1997). The synaptic activation of the GluR5 subtype of kainate receptor in area CA3 of the rat hippocampus. *Neuropharmacology* 3611-12:1477–1481.

Vizuete, M.L., Venero, J.L., Vargas, C., Ilundain, A.A., Echevarria, M., Machado, A., and Cano, J. (1999). Differential upregulation of aquaporin-4 mRNA expression in reactive astrocytes after brain injury: potential role in brain edema. *Neurobiol. Dis.* 64:245–258.

Wang, L.Y., Dudek, E.M., Browning, M.D., and MacDonald, J.F. (1994a). Modulation of AMPA/kainate receptors in cultured murine hippocampal neurones by protein kinase C. *J. Physiol* 4753:431–437.

Wang, L.Y., Orser, B.A., Brautigan, D.L., and MacDonald, J.F. (1994b). Regulation of NMDA receptors in cultured hippocampal neurons by protein phosphatases 1 and 2A. *Nature* 3696477:230–232.

Wang, X. (2001). The expanding role of mitochondria in apoptosis. *Genes Dev.* 1522:2922–2933.

Werner, P., Voigt, M., Keinanen, K., Wisden, W., and Seeburg, P.H. (1991). Cloning of a putative high-affinity kainate receptor expressed predominantly in hippocampal CA3 cells. *Nature* 3516329:742–744.

Wilding, T.J., and Huettner, J.E. (1995). Differential antagonism of alpha-amino-3-hydroxy-5-methyl-4-isoxazolepropionic acid-preferring and kainate-preferring receptors by 2,3-benzodiazepines. *Mol. Pharmacol.* 473:582–587.

Williams, K. (1993). Ifenprodil discriminates subtypes of the *N*-methyl-D-aspartate receptor: selectivity and mechanisms at recombinant heteromeric receptors. *Mol. Pharmacol.* 444:851–859.

Wyszynski, M., Lin, J., Rao, A., Nigh, E., Beggs, A.H., Craig, A.M., and Sheng, M. (1997). Competitive binding of alpha-actinin and calmodulin to the NMDA receptor. *Nature* 3856615:439–442.

Xia, J., Zhang, X., Staudinger, J., and Huganir, R.L. (1999). Clustering of AMPA receptors by the synaptic PDZ domain-containing protein PICK1. *Neuron* 221:179–187.

Xiang, H., Kinoshita, Y., Knudson, C.M., Korsmeyer, S.J., Schwartzkroin, P.A., and Morrison, R.S. (1998). Bax involvement in p53-mediated neuronal cell death. *J. Neurosci.* 184:1363–1373.

Yakovlev, A.G., Knoblach, S.M., Fan, L., Fox, G.B., Goodnight, R., and Faden, A.I. (1997). Activation of CPP32-like caspases contributes to neuronal apoptosis and neurological dysfunction after traumatic brain injury. *J. Neurosci.* 1719:7415–7424.

Yamasaki, Y., Matsuo, Y., Matsuura, N., Onodera, H., Itoyama, Y., and Kogure, K. (1995a). Transient increase of cytokine-induced neutrophil chemoattractant, a member of the interleukin-8 family, in ischemic brain areas after focal ischemia in rats. *Stroke* 262:318–322.

Yamasaki, Y., Matsuura, N., Shozuhara, H., Onodera, H., Itoyama, Y., and Kogure, K. (1995b). Interleukin-1 as a pathogenetic mediator of ischemic brain damage in rats. *Stroke* 264:676–680.

Yang, W., Guastella, J., Huang, J.C., Wang, Y., Zhang, L., Xue, D., Tran, M., Woodward, R., Kasibhatla, S., Tseng, B., Drewe, J., and Cai, S.X. (2003). MX1013, a dipeptide caspase inhibitor with potent *in vivo* antiapoptotic activity. *Br. J. Pharmacol.* 1402:402–412.

Ye, B., Liao, D., Zhang, X., Zhang, P., Dong, H., and Huganir, R.L. (2000). GRASP-1: a neuronal RasGEF associated with the AMPA receptor/GRIP complex. *Neuron* 263:603–617.

Ye, B., Sugo, N., Hurn, P.D., and Huganir, R.L. (2002). Physiological and pathological caspase cleavage of the neuronal RasGEF GRASP-1 as detected using a cleavage site-specific antibody. *Neuroscience* 1141:217–227.

Yu, X.M., Askalan, R., Keil, G.J., and Salter, M.W. (1997). NMDA channel regulation by channel-associated protein tyrosine kinase Src. *Science* 2755300:674–678.

Zhang, J., Dawson, V.L., Dawson, T.M., and Snyder, S.H. (1994). Nitric oxide activation of poly(ADP-Ribose) synthetase in neurotoxicity. *Science* 263:687 689.

Zhang, S., Ehlers, M.D., Bernhardt, J.P., Su, C.T., and Huganir, R.L. (1998). Calmodulin mediates calcium-dependent inactivation of N-methyl-D-aspartate receptors. *Neuron* 21:443–453.

Zhang, W., Vazquez, L., Apperson, M., and Kennedy, M.B. (1999). Citron binds to PSD-95 at glutamatergic synapses on inhibitory neurons in the hippocampus. *J. Neurosci.* 191:96–108.

Zhu, C., Qiu, L., Wang, X., Hallin, U., Cande, C., Kroemer, G., Hagberg, H., and Blomgren, K. (2003). Involvement of apoptosis-inducing factor in neuronal death after hypoxia-ischemia in the neonatal rat brain. *J. Neurochem.* 862:306–317.

Zukin, R.S., and Bennett, M.V. (1995). Alternatively spliced isoforms of the NMDARI receptor subunit. *Trends Neurosci.* 187:306–313.

13

A Thermodynamic Guide to Affinity Optimization of Drug Candidates

Ernesto Freire

ABSTRACT

A common starting point in drug development is the identification through screening or rational design of compounds that bind or exhibit some inhibitory activity against their intended targets. Often, those compounds bind to their targets with micromolar and sometimes weaker affinities. To become effective drugs, the binding affinities of those compounds need to be optimized by three or more orders of magnitude. This task is not a trivial one if one considers that it needs to be done while satisfying several stringent constraints, e.g., the molecular mass cannot substantially exceed 500 Da in order for the molecule to be orally bioavailable; the compound needs to exhibit appropriate target selectivity, appropriate membrane permeability and sufficient water solubility. Furthermore, the compound needs to exhibit an adequate pharmacokinetic profile, no toxicity, and so forth. These constraints considerably reduce the universe of chemical functionalities that can be utilized to achieve the optimization goals. In addition, at the thermodynamic level, chemical modifications that improve the binding enthalpy are usually accompanied by compensating entropy changes and vice versa, resulting in little or no gain in binding affinity. The identification of functionalities that carry the lowest enthalpy/entropy compensation is critical for affinity optimization. Since the binding affinity is the product of enthalpic and entropic contributions, it is possible for various ligands to have the same affinity but vastly different enthalpy/entropy profiles. While in theory, extremely high affinity can be achieved with arbitrary enthalpy/entropy combinations, the experience with HIV-1 protease inhibitors

ERNESTO FREIRE • Department of Biology, The Johns Hopkins University, Baltimore, MD, 21218, USA.

Proteomics and Protein–Protein Interactions: Biology, Chemistry, Bioinformatics, and Drug Design, edited by Waksman. Springer, New York, 2005.

indicates that a strong favorable binding enthalpy is necessary. Furthermore, enthalpically optimized inhibitors have been shown to respond better to target mutations associated with drug resistance or naturally occurring polymorphisms without losing selectivity towards unwanted targets. It is evident that high affinity inhibitors characterized by strong favorable binding enthalpies are highly desirable. Consequently, the development of accurate rules with the ability to guide the affinity and enthalpic optimization of drug candidates is extremely important. This is the subject of this chapter.

1. INTRODUCTION

The completion of the Human Genome Project as well as mapping the genomes of several pathogens has generated new challenges in ligand and drug design. The number of targets for drug development is expected to increase dramatically during the next few years. For each new target, lead compounds will need to be identified and optimized to achieve the required binding affinity, selectivity, bioavailability, and toxicological properties. This new reality accentuates the need for improved design paradigms and efficient ways of predicting not only binding affinities but also all relevant parameters that define a drug molecule.

Currently, the binding affinity is used as the main selection criteria in high-throughput screening and computational analysis. Nonetheless, the binding affinity is defined by the Gibbs energy of binding (ΔG), which, in turn, is determined by the enthalpy (ΔH) and the entropy (ΔS) changes ($\Delta G = \Delta H - T\Delta S$). In principle, many combinations of ΔH and ΔS can give rise to the same ΔG value and, therefore, to the same binding affinity to a given target. However, compounds with equal affinities but optimized enthalpically or entropically are not equivalent because the interactions that give rise to favorable binding enthalpies or entropies are different. Furthermore, the experience gained with the development of HIV-1 protease inhibitors strongly suggests that extremely high affinity can be achieved only when the binding enthalpy contributes favorably to binding. Since the enthalpy and entropy changes reflect different underlying interactions, the proportion in which these changes contribute to the binding affinity is also an important determinant of other characteristics such as the response to target or environmental variations, including those related to specificity and selectivity.

2. AFFINITY OPTIMIZATION

Compounds identified by screening or structure-based methods usually have binding affinities in the micromolar and sometimes weaker range. To become drug candidates, their binding affinities need to be optimized by several orders of magnitude. But, by how much does the affinity need to be improved? Or, more appropriately, is there an optimal affinity or is the highest affinity that can be achieved necessarily the better?

This question can be answered if we consider that the minimum amount of drug required to achieve complete target saturation must be at least equal to the amount of target molecules. Mathematically, the degree of target saturation, F_b, obeys the standard binding equation:

$$F_b = \frac{K_a[X]}{1 + K_a[X]} \tag{1}$$

where $[X]$ is the free concentration of ligand and K_a is the association constant of the ligand to its target. The standard equation, however, does not explicitly consider the protein target concentration, and therefore the total ligand concentration is not explicit. For the case in which binding to unwanted molecules is negligible, $[X]$ is given by the following expression:

$$[X] = \frac{-([P]K_a - K_a[X_{Total}] + 1) + \sqrt{([P]K_a - K_a[X_{Total}] + 1)^2 + 4K_a[X_{Total}]}}{2K_a} \tag{2}$$

where $[P]$ is the protein target concentration and $[X_{Total}]$ the total ligand concentration. Combination of Eqs. (1) and (2) allows calculation of the degree of saturation for any given target concentration.

In affinity optimization, the target concentration is a very important parameter, because it sets appropriate limits for the required binding affinity of the drug molecule. This is shown in Figure 13.1, where the inhibitor concentration required to achieve

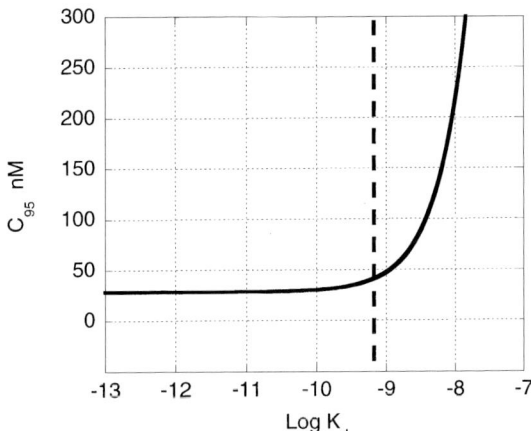

Figure 13.1. The inhibitor concentration necessary to achieve 95% saturation (C_{95}) as a function of its binding affinity expressed as the logarithm of the dissociation constant (log K_D). To the left of the *vertical dotted line*, essentially the same concentration of inhibitor is required to achieve 95% saturation regardless of binding affinity. To the right of the *vertical dotted line* the concentration required to achieve 95% saturation increases rapidly as the binding affinity decreases. The simulation was performed for a target concentration of 30 n*M*.

95% inhibition (C_{95}) has been plotted as a function of the inhibitor binding affinity assuming an effective target concentration of 30 nM. It is clear in the figure that independently of how tight the inhibitor binding affinity is, 30 nM is the minimal inhibitor concentration required for complete inhibition. Furthermore, the curve is essentially flat for all K_D's below approximately 0.5 nM (to the left of the vertical dotted line) indicating that, in this example, the same level of inhibition is achieved by a picomolar, femtomolar, or a 0.1 nM inhibitor. If this is the case, is there a preferred or optimal affinity?

Figure 13.1 demonstrates that the target concentration sets the range for the affinity required for effective inhibition. The optimal affinity does not depend on the target itself but on the selectivity that can be achieved against unwanted molecules or, conversely, the adaptability that might be required against different polymorphisms of the same target.

Selectivity is a critical parameter in drug development since it is related to the specificity with which a drug binds to its intended target. Binding selectivity can be quantitatively expressed in terms of the so-called K_D ratio, that is, the ratio of the binding affinity of the inhibitor toward the intended target, K_a, and the binding affinity of the inhibitor toward an unwanted target, K_{a_u}.

$$\text{Selectivity} = K_D \text{ ratio} = \frac{K_a}{K_{a_u}} = \frac{K_{D_u}}{K_D} \qquad (3)$$

The higher the K_D ratio the better the selectivity. Adaptability, on the other hand, is important when developing drugs against viral, bacterial, and other microbial targets, where naturally occurring polymorphisms or drug-resistant mutations are common and the objective is to achieve a broad spectrum of action (Ohtaka et al., 2003).

The optimal affinity is dictated by the desired balance between adaptability and selectivity. For example, if the target under consideration is a kinase or a blood co-agulation factor where selectivity rather than adaptability is critical, then the primary goal is to develop a molecule that effectively inhibits the intended target and no other related proteins. One of the main obstacles in the development of kinase inhibitors is that these enzymes are highly homologous and low selectivity factors (on the order of 10^2) are not uncommon. The same situation exists among coagulation factors, all of which are serine proteases. If this is the case, the optimal affinity against the target is not the highest possible affinity but one that is close to the vertical dotted line in Figure 13.1, such that it effectively inhibits the target but not unwanted molecules.

Equations (1) and (2) can be generalized to include binding of the inhibitor molecule to the intended target as well as other molecules in the organism.

$$[X]_{\text{Total}} = [X] + [P]\frac{K_a[X]}{1 + K_a[X]} + \sum_U [P_U]\frac{K_{a_U}[X]}{1 + K_{a_U}[X]} \qquad (4)$$

In the above equation, the summation on the right-hand side runs over all unwanted targets, $[P_U]$ is their concentration and K_{a_u} the binding affinity of the inhibitor to these molecules.

3. ENTHALPIC AND ENTROPIC LIGANDS

At the thermodynamic level, the main obstacle to affinity optimization is the ubiquitous phenomenon known as enthalpy/entropy compensation (Lumry and Rajender, 1970; Eftink et al., 1983). Briefly stated, it means that any gain in enthalpic contributions to binding is opposed by an accompanying loss in entropic contributions and vice versa. Enthalpy/entropy compensation is not perfect, however; otherwise affinity optimization would be impossible. From a purely thermodynamic standpoint, affinity optimization is accomplished by selecting chemical modifications that carry a low enthalpy/entropy compensation. For example, a hydrogen bond made with an unstructured or disordered region of the target carries a higher entropy penalty than the same hydrogen bond made with a structured region of the target. Rational affinity optimization requires identification of enthalpic or entropic interactions that carry the lowest enthalpy/entropy compensation.

The enthalpic or entropic character of a ligand is not determined by the target molecule. In fact, the experience accumulated with the development of HIV-1 protease inhibitors has demonstrated that both enthalpically and entropically driven ligands can be developed against the same binding site. If this is the case, it is important to identify the characteristics and behavior of enthalpically and entropically driven ligands so that a decision can be made about the preferred type for a given target. Figure 13.2 shows the experimental binding thermodynamics of HIV-1 protease inhibitors. Binding enthalpies range from $+8$ kcal/mol to -12 kcal/mol at 25°C. To appreciate the significance of the binding enthalpy spread, the difference of 20 kcal/mol between acetyl pepstatin and TMC-126 is equivalent to a 4×10^{14}-fold difference in binding affinity if there were not a compensating change in binding entropy. Most notably, all these inhibitors bind to exactly the same site and occupy essentially the same volume. A plot of the binding enthalpy versus molecular weight reveals no statistical correlation as shown in Figure 13.3A.

The magnitude and sign of the binding enthalpy do not correlate with global structural characteristics of the unbound inhibitors; e.g. there is no correlation between the binding enthalpy and the polarity of the inhibitors as reflected in their polar and nonpolar solvent accessible surface areas (Fig. 13.3B,C). Since the polarity is related to the solubility and bioavailability of the inhibitor, this observation indicates that the optimization of the binding enthalpy is not necessarily linked to a loss in solubility or bioavailability and that both parameters can be optimized independently. The binding enthalpy is also not correlated with LogP (Figure 13.3D), the octanol–water partition coefficient, calculated based on the MDL's QSAR program. Together, the results summarized in Figure 13.3 indicate that the thermodynamic parameters that optimize

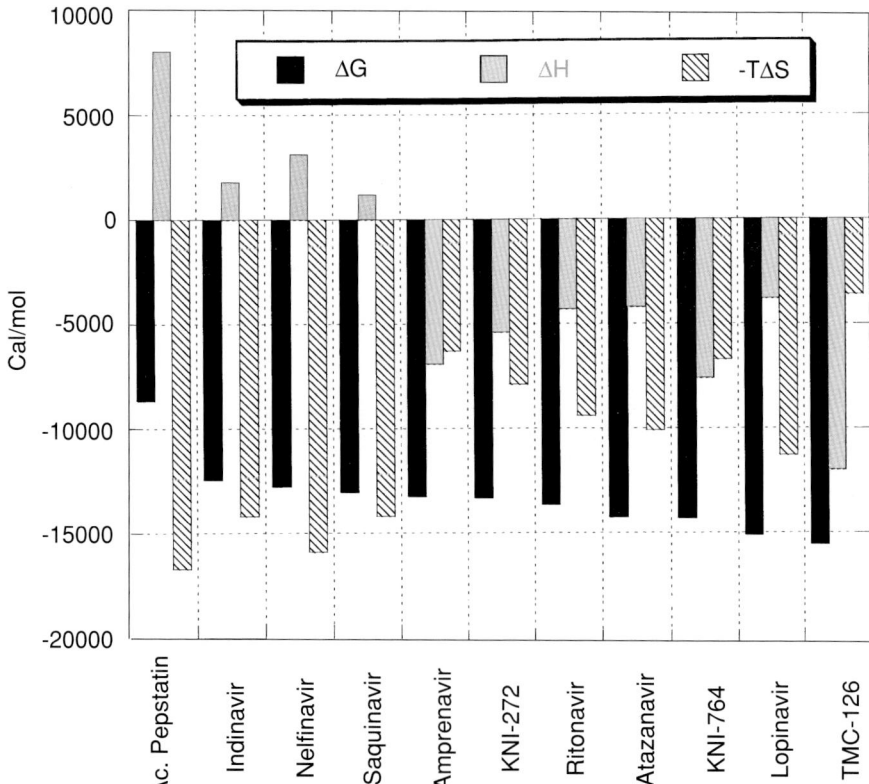

Figure 13.2. The binding thermodynamics of clinical and experimental HIV-1 protease inhibitors. The experimental isothermal titration calorimetry data clearly indicate that against the same target inhibitors can display strong favorable or unfavorable binding enthalpies. The development of new and more powerful inhibitors is reflected in the trend toward a predominance of enthalpic contributions to the binding affinity. All inhibitors with subnanomolar affinities are characterized by favorable enthalpic contributions to binding. All thermodynamic data are from this laboratory (Todd et al., 2000; Velazquez-Campoy et al., 2000a,b; 2001a,b,c,d; 2002; 2003a,b; Velazquez-Campoy and Freire, 2001; Ohtaka et al., 2002,2003). Atazanavir data (unpublished from Ohtaka).

binding enthalpy are not correlated to those parameters that have been shown to be important for solubility and permeability through the Lipinski's rules of five (Lipinski et al., 1997; Lipinski, 2000).

4. HIGH BINDING AFFINITY

In principle, many combinations of enthalpy and entropy changes can give rise to the same ΔG and therefore the same binding affinity. Experimentally, this is generally true for low-affinity binding but not so for extremely high affinity. The reason

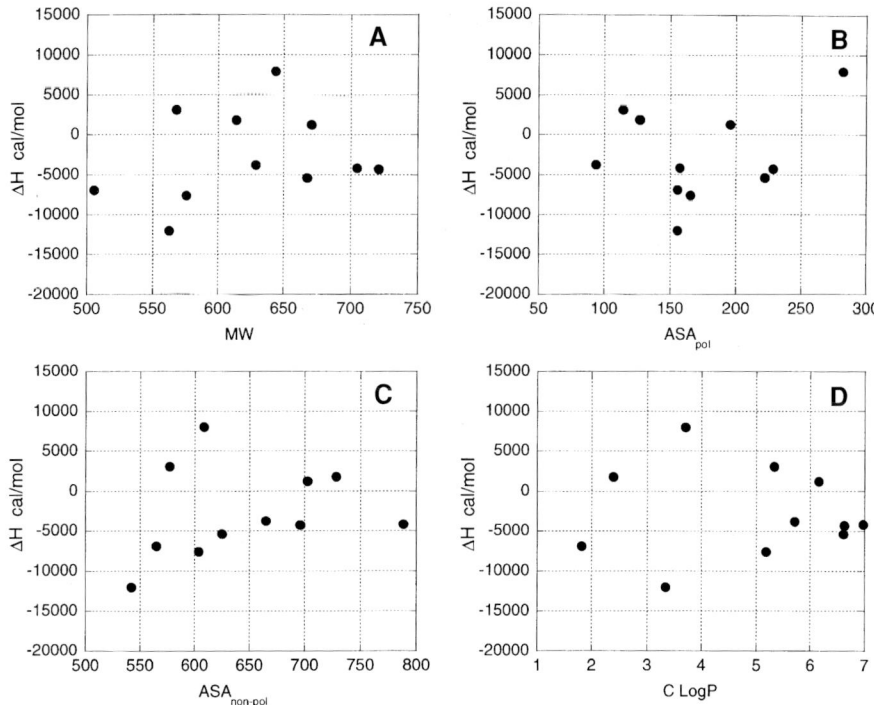

Figure 13.3. Correlation plots for the binding enthalpy of HIV-1 protease inhibitors and structural parameters of inhibitors. The experimental data indicate a lack of correlation between binding enthalpy and parameters traditionally associated with solubility and permeability. (**A**) Binding enthalpy and molecular weight. (**B**) Binding enthalpy and polar solvent accessible surface area of inhibitor. (**C**) Binding enthalpy and nonpolar solvent accessible surface area of inhibitor. (**D**) Binding enthalpy and $C \log P$, the calculated octanol–water partition coefficient.

for this phenomenon is that ΔH and ΔS reflect different types of interactions and, for small molecules binding to proteins the ensemble of interactions is finite and limited to the chemical functionalities involved in the binding reaction. Furthermore, solubility and permeability constraints impose additional limitations to the number of hydrophobic functionalities that can be added to a compound. Enthalpic contributions reflect the strength of the inhibitor interactions with the protein (H-bonds, van der Waals interactions) relative to those with the solvent (water). Entropic contributions to the binding affinity, on the other hand, are mainly due to a large increase in solvent entropy arising from the burial of hydrophobic groups on binding, and the loss of conformational degrees of freedom associated with binding.

The binding thermodynamics of the HIV-1 protease inhibitors in clinical use and some experimental inhibitors has been measured by isothermal titration microcalorimetry (Ohtaka et al., 2002, 2003; Todd et al., 2000; Velazquez-Campoy et al., 2000a,b; 2001a,c,d; 2002; 2003a,b; Velazquez-Campoy and Freire, 2001). If a

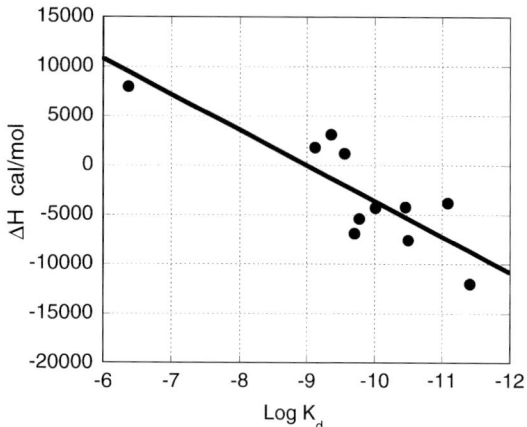

Figure 13.4. Correlation between binding affinity and binding enthalpy for HIV-1 protease inhibitors. While in theory high binding affinity can be achieved by many enthalpy/entropy combinations, in practice high binding affinity is correlated with a progressively more favorable binding enthalpy. Data from this laboratory (see Velazquez-Campoy et al. [2003] for a complete summary).

plot is made of the enthalpy and entropy contributions to the binding affinity to the wild type of all HIV-1 protease inhibitors for which thermodynamic data are available, the results shown in Figure 13.4 are obtained. It is clear that a correlation exists between high affinity and the proportion in which ΔH contributes to the binding affinity. The stronger the binding affinity, the more favorable the enthalpy contribution to the Gibbs energy. The experience derived from the development of HIV-1 protease inhibitors emphasizes the importance of optimizing the binding enthalpy to attain extremely high affinity.

5. ADAPTABILITY AND SELECTIVITY

By itself, the binding affinity of an inhibitor against its intended target does not provide an indication of the extent to which it will be affected by mutations associated with drug resistance or naturally occurring polymorphisms. These mutations or polymorphisms are common in viral, bacterial and other microbial targets (Freire, 2002; Nezami and Freire, 2002; Nezami et al., 2003; Ohtaka et al., 2003; Velazquez-Campoy et al., 2003a) and are generally conservative due to biological constraints for viability (e.g., the enzyme needs to maintain catalytic efficiency, binds the substrate, etc.) and usually result in geometric distortions of the binding cavity. Conformationally constrained inhibitors cannot adapt to those binding site deformations and lose significant affinity. Against these targets, adaptive inhibitors that are able to accommodate to target variations but still maintain selectivity against unwanted targets are highly desirable (Freire, 2002; Nezami et al., 2003). Against viral, bacterial and other microbial targets an ideal drug will be one that is extremely effective against a primary target (e.g., the wild-type enzyme) and that will exhibit a broad spectrum of action by maintaining its efficacy against the most important variants of the target molecule.

The adaptability of inhibitors to mutations associated with drug resistance or natural polymorphisms is not proportional to binding affinity, that is, a high binding affinity does not guarantee that an inhibitor will not lose significant affinity against mutations (Ohtaka et al., 2003; Velazquez-Campoy et al., 2003a). The situation appears to be different if the relative enthalpic contribution to the binding energy is taken into consideration. Figure 13.5 shows the susceptibility to mutations of different inhibitors, expressed as the logarithm of the K_D ratio $\equiv K_{D,\text{mutant}}/K_{D,\text{wt}}$, as a function of the relative enthalpic contribution to the Gibbs energy of binding to the wild-type protease ($\Delta H/\Delta G$). For all mutants for which data are available, the same trend is observed. As shown in Figure 13.5, enthalpically favorable and enthalpically unfavorable inhibitors appear to display opposite dependencies. The solid lines in the figure are the linear least squares fit to the data for entropic inhibitors ($\Delta H/\Delta G < 0$) and for enthalpically favorable inhibitors ($\Delta H/\Delta G > 0$). It is clear that the lowest susceptibilities to mutations are observed when the contributions to the Gibbs energy are either predominantly enthalpic or predominantly entropic. Inhibitors in which the enthalpy change contributes about 25% of the total Gibbs energy show the highest

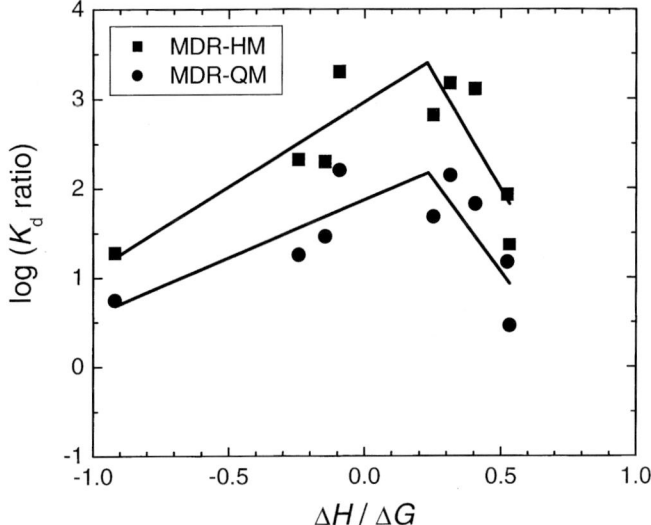

Figure 13.5. The susceptibility to mutations of different inhibitors, expressed as $\log(K_D$ ratio) $= \log(K_{D,\text{mutant}}/K_{D,\text{wild-type}})$, appears to be correlated with the proportion in which the enthalpy change contributes to the Gibbs energy ($\Delta H/\Delta G$) for the wild-type protease. Since ΔG is always negative, a positive value reflects a favorable enthalpy contribution and a negative value an unfavorable enthalpy contribution. The solid lines represent linear least-squares fits to the data for entropically driven ($\Delta H/\Delta G < 0$) and enthalpically favorable ($\Delta H/\Delta G > 0$) inhibitors. The data include the multidrug-resistant proteases MDR-HM (L10I/M46I/I54V/V82A/I84V/L90M) and MDR-QM (V82A/I84V/M46I/I54V) (Ohtaka et al., 2003).

susceptibility to mutations associated with drug resistance. The observed pattern is intriguing because it includes inhibitors with very different chemical scaffolds as well as inhibitors derived from the same scaffold. Nevertheless, if these correlations obtained for HIV-1 protease inhibitors could be extended to other systems, they may provide an experimental way of anticipating the response of an inhibitor to mutations associated with drug resistance using information derived from the main target alone.

Since high affinity is correlated with binding enthalpy (Figure 13.4) it seems that enthalpically favorable inhibitors must be preferred over entropic ones since the latter do not seem capable of eliciting extremely high affinity. For these inhibitors, the lowest susceptibility to mutations is predicted to occur when the enthalpy contribution to the Gibbs energy is maximal. This conclusion is supported by the published data for TMC-126, the most enthalpically favorable inhibitor for which data are available (Ohtaka et al., 2002). TMC-126 is characterized by a $\Delta H/\Delta G$ ratio of 0.53 and exhibits the lowest reported susceptibility to the active site mutation V82F/I84V (Ohtaka et al., 2002) and multidrug-resistant mutants (Ohtaka, unpublished data from this laboratory). Also, Yoshimura et al. (2002) studied TMC-126 against a panel of eight multidrug-resistant mutants isolated from inhibitor-experienced patients. These mutants included up to 14 different mutations. TMC-126 was the best performing inhibitor when compared to ritonavir, indinavir, saquinavir, nelfinavir, and amprenavir, and was shown to never lose more than eightfold potency against any of these multidrug mutations.

The high adaptability of HIV-1 protease inhibitors with a large $\Delta H/\Delta G$ ratio to mutations or polymorphisms associated with drug resistance apparently does not impair their selectivity with regard to unwanted targets, including related aspartic proteases such as cathepsin D or pepsin (Ohtaka, H., unpublished results from this laboratory).

6. ENTHALPIC OPTIMIZATION

The preceding discussion emphasizes the need to identify compounds that bind to their targets with strong binding enthalpies. Enthalpy optimization, however, is significantly more difficult to accomplish than entropy optimization. Entropy optimization essentially involves an increased hydrophobicity of the compound, which maximizes the favorable solvation entropy associated with binding, coupled with the introduction of conformational constraints, which minimizes the loss of conformational degrees of freedom upon binding (Luque and Freire, 1998). These entropic contributions are not stereospecific except from shape considerations and represent a common strategy (explicit or implicit) for affinity optimization as inferred, for example, from the thermodynamic signature of the first generation of HIV-1 protease inhibitors (Todd et al., 2000; Velazquez-Campoy et al., 2000b). Enthalpy optimization is inherently more difficult because it involves highly stereospecific interactions and because the strength of the resulting interaction needs to overcome the large positive enthalpy associated with the desolvation of polar groups. For example, while the enthalpy of desolvation of nonpolar groups is on the order of 0.7 kcal/mol, the desolvation of polar groups (e.g., NH_2, NH, OH) is on the order of 9 kcal/mol (Cabani et al., 1981). Accordingly, burying a polar group without establishing strong inter-actions may have a detrimental rather than a favorable impact on binding affinity. Because of this difficulty, it is preferable that the starting compound for optimization already binds to the target with a favorable binding enthalpy (throughout this chapter we refer to enthalpic contributions arising from ligand protein interactions and not from coupled reactions such as protonation/deprotonation processes).

A compound that binds with favorable enthalpy, even at the earliest stages of the optimization process, provides an indication that its binding is driven by favorable in-teractions with the target rather than by a nonspecific hydrophobic tendency to escape from the solvent. The identification and characterization of enthalpic compounds can be rigorously accomplished by isothermal titration calorimetry (ITC) (Wiseman et al., 1989). Once a primary screening has been performed and compounds that display activity have been identified, ITC can be used to measure their enthalpy/entropy pro-file. Those compounds that bind to the target with a favorable enthalpy have a higher probability of being successfully optimized to extremely high affinity. The thermody-namic profiling of a library of compounds also identifies functionalities that contribute favorably to affinity and enthalpy. This information can be incorporated into statis-tical protocols analogous to quantitative structure–activity relationships (QSAR) but that explicitly incorporate the energetics of binding (QESAR). The enthalpic opti-mization of a compound by QESAR does not require that the structure of the target molecule with the inhibitor is known as demonstrated by the successful optimization of plasmespsin II inhibitors (Nezami and Freire, 2002; Nezami et al., 2002,2003).

7. STRUCTURE-BASED OPTIMIZATION

Attempts to experimentally correlate binding enthalpies for small molecules with high resolution structures of inhibitor/protein complexes began as an extension of the parameterizations derived by several authors for the thermodynamics of stabilization of protein structures (Murphy and Gill,1991; Makhatadze and Privalov, 1995; Hilser et al., 1996; Robertson and Murphy, 1997; Luque and Freire, 1998; Henriques et al., 2000). In most cases, the enthalpy change is assumed to scale in terms of changes in solvent accessible surface areas (ASA) for different atom types associated with the transition between native and denatured states. According to this approach, the scaling coefficients for the changes in solvent accessibility are a function of the atomic packing density (δ) of the native structure (Hilser et al., 1996).

$$\Delta H(T, \delta) = \sum_i a_i(T, \delta) \times \Delta \, ASA_i \qquad (5)$$

where $\Delta \, ASA_i$ represents the changes in solvent accessible surface area for atoms of type i and $a_i(T, \delta)$ their corresponding scaling coefficients. The $a_i(T, \delta)$ coefficients are a function of temperature and the atomic packing density δ. There are two limits to these coefficients, the zero packing density limit in which the coefficients approximate the solvation enthalpy from the gas phase, and the high packing density limit in which the coefficients contain an additional contribution corresponding to the enthalpy of sublimation of the atom type under consideration (Hilser et al., 1996). Because the interior of proteins is highly homogeneous in terms of both atomic composition and packing density, it has been found that highly simplified forms of Eq. (5) are able to account for experimental data within error (Xie and Freire, 1994; Hilser et al., 1996; Robertson and Murphy, 1997). In particular, an equation that contains only polar and nonpolar atom types and assumes a constant packing density has been widely used (Xie and Freire, 1994; Hilser et al., 1996; Robertson and Murphy, 1997):

$$\Delta H(T) = a(T) \times \Delta ASA_{\text{ap}} + b(T) \times \Delta ASA_{\text{pol}} \tag{6}$$

where $a(T)$ and $b(T)$ are empirically determined coefficients and $\Delta ASA_{\text{apolar}}$ and $\Delta ASA_{\text{polar}}$ are the changes in solvent accessible surface area for nonpolar and polar atoms, respectively. Analysis of 60 proteins by Robertson and Murphy (1997) in terms of Eq. (6) yielded $a(60) = -1.9 \pm 2.6 \text{ cal·mol}^{-1}\cdot\text{Å}^{-2}$ and $b(60) = 20.6 \pm 4.1 \text{ cal·mol}^{-1}\cdot\text{Å}^{-2}$. The regression coefficient was 0.9 with a zero intercept and a slope of 0.96. The standard deviation between experimental and calculated values was close to 10 kcal/mol (about 10% of the mean denaturation enthalpy value).

When applied to the binding of small ligands, the structural parameterization of the enthalpy change derived from protein denaturation has yielded inconsistent results (Gomez and Freire, 1995; Luque et al., 1998; Edgcomb and Murphy, 2000; Henriques et al., 2000). Luque and Freire (Luque and Freire, 2002) concluded that the accuracy of the enthalpy parameterization could be improved by following a similar approach but utilizing a binding thermodynamics data set collected at 25°C (instead of the protein denaturation data set that has a median temperature of 60°C) and incorporating features that are critical to the binding of small ligands.

7.1. Protein Conformation

The binding of small ligands is usually associated with a change in protein conformation. This change in protein conformation is not necessarily a change between two well defined ordered conformations; in fact, most of the time it involves only a local rearrangement or stabilization of unstructured regions near the binding site (Luque and Freire, 2000). Small molecular weight ligands do not stick to the surface of proteins, they are buried in crevices or cavities that are often covered by loops or other structural components of the protein. This arrangement maximizes the number of contacts between ligand and protein and simultaneously buries a more significant surface area from the solvent. A published survey of 36 protein/ligand complexes indicated that small ligands (MW <800) bury $80 \pm 13\%$ of their surface area on binding (Luque and Freire, 2000). The regions of the protein that contribute to cover and shield the ligand from the solvent usually have low structural stability in the unbound protein. They are stabilized on binding and the energetics associated with this process needs to be taken into account.

In theory, the availability of the free and bound conformations provides the necessary information. In practice, however, the accumulation of small crystallographic differences (e.g., orientation of exposed side chains) is often larger than the effects due to the binding of low molecular weight ligands. In addition, in many cases the conformational change involves the stabilization of an unstructured region that cannot be represented as a single structure but as an ensemble of structures. Also, from a practical point of view, the crystallographic structure of the complex is often the only one available.

One way to address these issues is to use the structure of the complex alone and consider the conformational enthalpy as a fitting parameter in the parameterization equation. This approach, however, requires a minimum of two structure/thermodynamic datasets with different ligands for each protein, and the condition that the ligands induce the same bound conformation of the protein. The experimental enthalpy always corresponds to the one from the unbound ensemble to that of the complex observed in the crystallographic structure and the conformational component can be considered to be a constant:

$$\Delta H = \Delta H_{conformation} + \Delta H_{intrinsic} \tag{7}$$

where $\Delta H_{conformation}$ is the same for each ligand and $\Delta H_{intrinsic}$ depends on the specific interactions of each ligand with the target.

7.2. Linkage to Protonation/Deprotonation Reactions

Ligand binding is frequently coupled to the protonation or deprotonation of certain ionizable groups, either in the protein or the ligand itself (Gomez and Freire, 1995; Baker and Murphy, 1996; Velazquez-Campoy et al., 2000a). For small ligands, ionization enthalpies can be of the same order of magnitude as the intrinsic binding enthalpy itself and therefore need to be considered explicitly. At the present time, reliable ionization enthalpies for any given target can be obtained only experimentally. Explicit methods to dissect proton linkage contributions to the binding enthalpy were developed by Baker and Murphy (Baker and Murphy, 1996) and have been applied to various systems (Baker and Murphy, 1996; Velazquez-Campoy et al., 2000a). Once measured, these contributions can be subtracted from the experimental enthalpy in order to evaluate other effects.

7.3. Buried Water Molecules

The desolvation of the ligand and protein interfaces on binding is not always complete. Frequently, long-lived, buried water molecules are found at the binding interfaces of many complexes. Those buried water molecules may play a critical role in mediating protein–ligand interactions by serving as adapters that fill nonoccupied volumes, satisfying the hydrogen bonding potential of the ligand and the binding site, or assisting in the dissipation of charges. All these terms can be expected to contribute favorably to the binding enthalpy. Conversely, the incomplete desolvation of the ligand–protein interface will oppose this effect by a decrease in the solvation entropy (enthalpy/entropy compensation). Thus, the enthalpic effect of buried water can be expected to be significantly larger than the corresponding effect in the Gibbs energy.

8. STRUCTURAL PARAMETERIZATION OF BINDING ENTHALPY

According to the preceding discussion, as a first approximation Luque and Freire (2002) considered the experimental binding enthalpy of small ligands, ΔH_{exp} as the combination of three terms:

$$\Delta H_{exp} = \Delta H_{intrinsic} + \Delta H_{conformation} + \Delta H_{protonation} \tag{8}$$

The intrinsic enthalpy, $\Delta H_{intrinsic}$, is the most important term for ligand optimization since it reflects interactions between ligand and protein (H-bonds, van der Waals interactions, etc) and solvation changes upon binding. $\Delta H_{intrinsic}$ corresponds to the enthalpy that would be observed if protein and ligand had the same conformation in the free and bound states. These contributions are expected to scale with changes in accessible surface area (ΔASA) and the atomic packing density (δ) (Hilser et al., 1996).

The enthalpic contributions arising from conformational changes, $\Delta H_{conformation}$, in the protein cannot be easily parameterized in terms of changes in solvent accessibility on binding and therefore is left as an adjustable parameter to be determined from the thermodynamic data for a series of related compounds. This approach can be easily implemented during optimization since usually many derivatives of the same scaffold are tested. Also, since this term is essentially a constant for a family of compounds, it is not expected to play a major role in establishing an accurate ranking of drug candidates.

Contributions to the binding enthalpy arising from protonation/deprotonation processes, $\Delta H_{protonation}$ need to be evaluated experimentally by measuring the enthalpy of binding at different pH values and with buffers characterized by different ionization enthalpies (Baker and Murphy, 1996; Velazquez-Campoy et al., 2000a). Once corrected for protonation/deprotonation effects, the resulting protonation-independent binding enthalpy will be the combination of intrinsic contributions and the enthalpy associated with any possible conformational change in the protein and/or the ligand upon formation of the complex.

For those situations in which several ligands induce the same bound conformation in the protein, and assuming that any enthalpy associated with conformational changes in the ligand is small compared to that of the protein, the protonation-independent binding enthalpy at any given temperature, $\Delta H_{binding}(T)$, will be the sum of a constant term corresponding to the conformational enthalpy plus a ligand-specific term that accounts for the specific interactions of each ligand with the target:

$$\Delta H_{binding}(T) = \Delta H_{conformation}(T) + \Delta H_{intrinsic}(T) \tag{9a}$$
$$\Delta H_{binding}(T) = \Delta H_{conformation}(T) + a(T) \cdot \Delta ASA_{apolar} + b(T) \cdot \Delta ASA_{polar} \tag{9b}$$

where the intrinsic enthalpy is represented in terms of changes in solvent-accessible surface areas, and $a(T)$ and $b(T)$ are the scaling coefficients for nonpolar and polar groups, respectively.

Nonlinear least-squares analysis of the HIV-1 protease and other proteins (dihydrodipicolinate reductase, ribonuclease T1, streptavidin, pp60^{c-Src} SH2 domain, Hsp90, and β-Trypsin) were consistent with $a(25)$ and $b(25)$ coefficients of -7.35 ± 2.55 and 31.06 ± 6.32 cal $\cdot$ (mol $\cdot$ $\mathring{A}^2$)$^{-1}$, respectively and accounted for the experimental binding enthalpies of the systems considered with a standard deviation of ± 0.3 kcal/mol at 25°C. It must be noted that in the analysis, fully buried water molecules within 6 $\mathring{A}$ of the ligand molecules needed to be included in the analysis in order to obtain statistically satisfactory results. Two effects may be responsible for the buried water contribution to the binding enthalpy: first, interfacial water molecules improve the atomic packing density within the binding site, bringing it to values similar to those of the protein interior; second, water molecules that remain associated to the inhibitor diminish the unfavorable enthalpy associated with the complete desolvation of the ligand.

The studies discussed in this chapter and additional correlations involving other structural parameters and molecular descriptors (unpublished work from this laboratory) indicate that this type of statistical analysis, based on experimental thermodynamic data can be used to predict the binding enthalpy from structural parameters. These empirical correlations serve a practical purpose and are not assumed to provide a rigorous atomic level description of ligand–protein interactions. These correlations between binding thermodynamics and structural parameters are important in drug discovery and optimization because they provide a way to select chemical functionalities that will contribute favorably to affinity and binding enthalpy.

Acknowledgments

This work was supported by grants from the National Institutes of Health GM57144, GM56550, and the National Science Foundation MCB-0131241.

REFERENCES

Baker, B.M., and Murphy, K.P. (1996). Evaluation of linked protonation effects in protein binding using isothermal titration calorimetry. *Biophys. J.* 71:2049–2055.

Cabani, S., Gianni, P., Mollica, V., and Lepori, L. (1981). Group contributions to the thermodynamic properties of non-ionic organic solutes in dilute aqueous solution. *J. Solut. Chem.* 10:563–595.

Edgcomb, S.P., and Murphy, K.P. (2000). Structural energetics of protein folding and binding. *Curr. Opin. Biotechnol.* 11:62–66.

Eftink, M.R., Anusiem, A.C., and Biltonen, R.L. (1983). Enthalpy-entropy compensation and heat capacity changes for protein-ligand interactions: general thermodynamic models and data for the binding of nucleotides to ribonuclease A. *Biochemistry* 22:3884–3896.

Freire, E. (2002). Designing drugs against heterogeneous targets. *Nat. Biotechnol.* 20:15–16.

Gomez, J., and Freire, E. (1995). Thermodynamic mapping of the inhibitor site of the aspartic protease endothiapepsin. *J. Mol. Biol.* 252:337–350.

Henriques, D.A., Ladbury, J.E., and Jackson, R.M. (2000). Comparison of binding energetics of SrcSH2-phosphotyrosyl peptides with structure-based prediction using surface area based empirical parameterization. *Protein Sci.* 9:1975–1985.

Hilser, V.J., Gomez, J., and Freire, E. (1996). The enthalpy change in protein folding and binding. Refinement of parameters for structure based calculations. *Proteins* 26:123–133.

Lipinski, C.A. (2000). Drug-like properties and the causes of poor solubility and poor permeability. *J. Pharmacol. Toxicol. Methods* 44:235–249.

Lipinski, C.A., Lombardo, F., Dominy, B.W., and Feeney, P.J. (1997). Experimental and computational approaches to estimate solubility and permeability in drug discovery and development settings. *Adv. Drug Deliv. Rev.* 23:3–25.

Lumry, R., and Rajender, S. (1970). Enthalpy-entropy compensation phenomena in water solutions of proteins and small molecules: a ubiquitous property of water. *Biopolymers* 9:1125–1227.

Luque, I., and Freire, E. (1998). A system for the structure-based prediction of binding affinities and molecular design of peptide ligands. *Methods Enzymol.* 295:100–127.

Luque, I., and Freire, E. (2000). The structural stabilty of binding sites. Consequences for binding affinity and cooperativity. *Proteins* 4:63–71.

Luque, I., and Freire, E. (2002). Structural parameterization of the binding enthalpy of small ligands. *Proteins* 49:181–190.

Luque, I., Gomez, J., Semo, N., and Freire, E. (1998). Structure-based thermodynamic design of peptide ligands. Application to peptide inhibitors of the aspartic protease endothiapepsin. *Proteins* 30: 74–85.

Makhatadze, G.I., and Privalov, P.L. (1995). Energetics of protein structure. *Adv. Protein Chem.* 47:307–425.

Murphy, K.P., and Gill, S.J. (1991). Solid model compounds and the thermodynamics of protein unfolding. *J. Mol. Biol.* 222:699–709.

Nezami, A., and Freire, E. (2002). The integration of genomic and structural information in the development of high affinity plasmepsin inhibitors. *Int. J. Parasitol.* 32:1669–1676.

Nezami, A., Luque, I., Kimura, T., Kiso, Y., and Freire, E. (2002). Identification and characterization of allophenylnorstatine-based inhibitors of plasmepsin II, an anti-malarial target. *Biochemistry* 41:2273–2280.

Nezami, A., Kimura, T., Hidaka, K., Kiso, A., Liu, J., Kiso, Y., Goldberg, D.A., and Freire, E. (2003). High affinity inhibition of a family of *Plasmodium falciparum* proteases by a designed adaptive inhibitor. *Biochemistry* 42:8459–8464.

Ohtaka, H., Velazquez-Campoy, A., Xie, D., and Freire, E. (2002). Overcoming drug resistance in HIV-1 chemotherapy: the binding thermodynamics of amprenavir and TMC-126 to wild type and drug-resistant mutants of the HIV-1 protease. *Protein Sci.* 11:1908–1916.

Ohtaka, H., Schon, A., and Freire, E. (2003). Multi drug-resistance to HIV-1 protease inhibition requires cooperative coupling between distal mutations. *Biochemistry* 42:13659–13666.

Robertson, A.D., and Murphy, K.P. (1997). Protein structure and the energetics of protein stability. *Chem. Rev.* 97:1251–1267.

Todd, M.J., Luque, I., Velazquez-Campoy, A., and Freire, E. (2000). The thermodynamic basis of resistance to HIV-1 protease inhibition. Calorimetric analysis of the V82F/I84V active site resistant mutant. *Biochemistry* 39:11876–11883.

Velazquez-Campoy, A., and Freire, E. (2001). Incorporating target heterogeneity in drug design. *J. Cell. Biochem.* S37:82–88.

Velazquez-Campoy, A., Luque, I., Todd, M.J., Milutinovich, M., Kiso, Y., and Freire, E. (2000a). Thermodynamic dissection of the binding energetics of KNI-272, a powerful HIV-1 protease inhibitor. *Protein Sci.* 9:1801–1809.

Velazquez-Campoy, A., Todd, M.J., and Freire, E. (2000b). IIIV-1 protease inhibitors: enthalpic versus entropic optimization of the binding affinity. *Biochemistry* 39:2201–2207.

Velazquez-Campoy, A., Kiso, Y., and Freire, E. (2001a). The binding energetics of first and second generation HIV-1 protease inhibitors: implications for drug design. *Arch. Biochim. Biophys.* 390:169–175.

Velazquez-Campoy, A., Luque, I., and Freire, E. (2001b). The application of thermodynamic methods in drug design. *Thermochim. Acta* 380:217–227.

Velazquez-Campoy, A., Luque, I., and Freire, E. (2001c). The use of isothermal titration calorimetry in drug design: applications to high affinity binding and protonation/deprotonation coupling. *Netsu Sokutei* 28:68–73.

Velazquez-Campoy, A., Todd, M.J., Vega, S., and Freire, E. (2001d). Catalytic efficiency and vitality of HIV-1 proteases from African viral subtypes. *Proc. Natl. Acad. Sci. USA* 98:6062–6067.

Velazquez-Campoy, A., Vega, S., and Freire, E. (2002). Amplification of the effects of drug-resistance mutations by background polymorphisms in HIV-1 protease from African subtypes. *Biochemistry* 41:8613–8619.

Velazquez-Campoy, A., Muzammil, S., Ohtaka, H., Schon, A., Vega, S., and Freire, E. (2003a). Structural and thermodynamic basis of resistance to HIV-1 protease inhibition: implications for inhibitor design. *Curr. Drug Targets Infect. Disord.* 3:311–328.

Velazquez-Campoy, A., Vega, S., Fleming, E., Bacha, U., Sayed, Y., Dirr, H.W., and Freire, E. (2003b). Protease inhibition in African subtypes of HIV-1. *AIDS Rev.* 5:165–171.

Wiseman, T., Williston, S., Brandts, J.F., and Lin, L.N. (1989). Rapid measurement of binding constants and heats of binding using a new titration calorimeter. *Anal. Biochem.* 179:131–135.

Xie, D., and Freire, E. (1994). Molecular basis of cooperativity in protein folding. V. Thermodynamic and structural conditions for the stabilization of compact denatured states. *Proteins Struct. Funct. Genet.* 19:291–301.

Yoshimura, K., Kato, R., Kavlick, M.F., Nguyen, A., Maroun, V., Maeda, K., Hussain, K.A., Ghosh, A.K., Gulnik, S.V., Erickson, J.W., and Mitsuya, H. (2002). A potent human immunodeficiency virus type 1 protease inhibitor, UIC-94003 (TMC-126), and the selection of a novel (A28S) mutation in the protease active site. *J. Virology* 76:1349–1358.

Index